Topology Optimization via L-systems and Genetic Algorithms

Providing a succinct overview of Lindenmayer system (L-system) topology optimization, this book focuses on the methods and theory underlying this novel bioinspired approach. Starting from basic principles, the book outlines how topology optimization can be utilized at the conceptual design stage and shows how it offers straightforward applicability to multiobjective and/or multiphysical industrial problems.

Design strategies are clearly demonstrated using a host of case studies and real-world examples, and their potential challenges and solutions are discussed. Written from an optimization and design perspective, the authors both summarize the latest advances in this field and suggest potential avenues of research and development for future work.

This will be the ideal resource for engineering practitioners, researchers, and students wanting to gain a new perspective on using topology optimization to improve product design.

Brent R. Bielefeldt is a research engineer with the US Air Force Research Laboratory specializing in multidisciplinary design and optimization and adaptive structures. He received a PhD in Aerospace Engineering from Texas A&M University (2020) and is a recipient of the Texas A&M Distinguished Graduate Student Award for Excellence in Research.

Darren J. Hartl is an Associate Professor of Aerospace Engineering at Texas A&M University with a history of multifunctional structures research that includes government laboratory and industry projects. He leads both experimental and theoretical development efforts and was the 2016 winner of the ASME Gary Anderson Early Achievement Award.

Marcelo H. Kobayashi is a Professor at the University of Hawaii at Manoa. He received a PhD in Mechanical Engineering and PhD in Mathematics from IST/TU Lisbon. He was runner-up in the 2003 Portuguese Young Investigator Award in Applied and Computational Mechanics and Director of the Hawaii Open Supercomputing Center (2010–2012).

Topology Optimization via L-systems and Genetic Algorithms

Bioinspired Encoding for Generative Design

BRENT R. BIELEFELDT
US Air Force Research Laboratory

DARREN J. HARTL
Texas A&M University

MARCELO H. KOBAYASHI
University of Hawaii at Manoa

CAMBRIDGE
UNIVERSITY PRESS

CAMBRIDGE
UNIVERSITY PRESS

Shaftesbury Road, Cambridge CB2 8EA, United Kingdom

One Liberty Plaza, 20th Floor, New York, NY 10006, USA

477 Williamstown Road, Port Melbourne, VIC 3207, Australia

314–321, 3rd Floor, Plot 3, Splendor Forum, Jasola District Centre, New Delhi – 110025, India

103 Penang Road, #05–06/07, Visioncrest Commercial, Singapore 238467

Cambridge University Press is part of Cambridge University Press & Assessment, a department of the University of Cambridge.

We share the University's mission to contribute to society through the pursuit of education, learning and research at the highest international levels of excellence.

www.cambridge.org
Information on this title: www.cambridge.org/9781316515792

DOI: 10.1017/9781009026635

When citing this work, please include a reference to the DOI 10.1017/9781009026635

First published 2025

A catalogue record for this publication is available from the British Library

A Cataloging-in-Publication data record for this book is available from the Library of Congress

ISBN 978-1-316-51579-2 Hardback

Contents

1 Introduction

In solid mechanics, a *structure* is defined as a connected domain of material(s) (i.e., a body) capable of transferring loads from some point(s) of application to some other point(s) of fixity. These points of application and fixity, and indeed the limits of the structural domain itself, are defined by the scope and scale of the application being designed and its relationship to the engineer or designer. For example, during preliminary layout design an entire aircraft wing might be viewed as a structure, while later in the detailed design process a single rib might be carefully configured as a distinct structural component. Finally, those developing advanced materials might consider the distribution of voids and properties at the material level to represent a problem of "microstructure."

In engineering terms, all structures that can be designed in 2-D/3-D space have exterior boundaries, and, regardless of scale, the relative spatial placement of material(s) and voids within these boundaries defines a structure's **topology**. Mathematically, two objects are topologically equivalent if each can be continuously deformed into the other. Thus, a sphere in three dimensions is topologically equivalent to an ellipsoid but not to a doughnut, given that a hole would have to be cut in the sphere in the latter case. The determination of an appropriate, or even optimal, topology is typically one of the first steps toward developing a structural design. The goal of this book is to expose the reader to techniques that enable expedient design space exploration to identify novel, optimal topologies.

Once a topology has been determined, two secondary but important classes of structural design decision must be addressed, the most common motivations for which are simplification of engineering representation and definition, as for eventual fabrication. **Shape design** in structures refers to the refinement of topology to consider 2-D and 3-D shapes, usually of some canonical type, assigned to particular regions of a structure. Returning to the mathematical definition above, topological equivalence partitions a set of objects into disjoint sets of shape classes, with each member of a shape class being a continuous deformation of any object of the same shape class. Therefore, in shape design, the optimization is restricted to remain inside the given shape class and is thus greatly limited when compared with topology optimization. **Sizing design** is the subsequent process of assigning specific dimensions to these shapes (e.g., radii of holes and fillets, thickness of features). Therefore, in sizing, only a reduced set of deformations is allowed, further limiting the scope of the structural optimization.

Shape and sizing design are of course critical to the full design of a structure and will be addressed throughout this book beginning in Section 1.2.4 and much more rigorously in Chapters 2 and 4 (Sections 2.3 and 4.4.1, respectively). Further, the topology, shape, and sizing design phases are also addressed in other resources (Bendsøe & Sigmund 2003, Querin, Victoria, Alonso, Loyola & Montrull 2017). However, in this work we focus primarily on that first and most essential challenge: the selection and representation of topology.

1.1 Structural Topology and Representation

The aim of this overall work is to develop and demonstrate developmental/algorithmic approaches for structural representation and design that leverage biological analogies, allowing optimized structures and networks to be developed and evolved. Clearly, the problem of topological design is not limited to the engineering of solid bodies for the carrying of mechanical loads; all networks have as a fundamental property their topology.[1] Communication networks, electrical circuits, and fluidic and chemical processes all rely on selective connectivity between nodes to perform their intended functions. Topological design/optimization is the process of determining/optimizing these nodes and connectivities. Our goal is then to provide engineers and designers with a new set of tools for optimizing topologies across all topics of engineering (solid mechanics, electronics, communications, fluidics, etc.). Indeed, it will also be shown that extensions to the graphical and environmental design arts are also possible.

From design to physical realization, all bodies and networks subject to analysis and/or intended for precise replication must be deterministically represented. In illustrating such representation, we consider, in particular, the definition of solid bodies (e.g., those occupying nonzero two-dimensional area or three-dimensional volume) as both a general and a motivating case. Such a task is general in that it requires the consideration of faces or volumes, while the definition of networks and circuits requires the consideration of only nodes and edges. The task is motivating inasmuch as it will be seen to represent the majority of examples provided throughout this text. The topological and geometric representations that will be addressed throughout this work are illustrated for a two-dimensional domain in Figure 1.1.

A deterministic and quantitatively precise representation of a body allows a finite array of design variables to define the configuration of that body in such a way that (i) the effect of design changes on engineering performance can be assessed and (ii) the body may eventually be realized as a prototype or product. As shown in Figure 1.1,

[1] Throughout this work, a *structure* is defined as an entity composed of material and having nonzero mass and volume (though it may be represented as two-dimensional). Therefore, it is fundamentally characterized by its *geometric* features (e.g., lengths, angles, areas, volumes). A *network* (or *graph*) is purely *combinatorial*, consisting of nodes and their connections, where the concepts of geometry are fundamentally irrelevant to functional performance. Note that networks, if assigned geometric properties, can be interpreted as structures. The fact that topological representation is common to both structures and networks is central to this body of work, which addresses both.

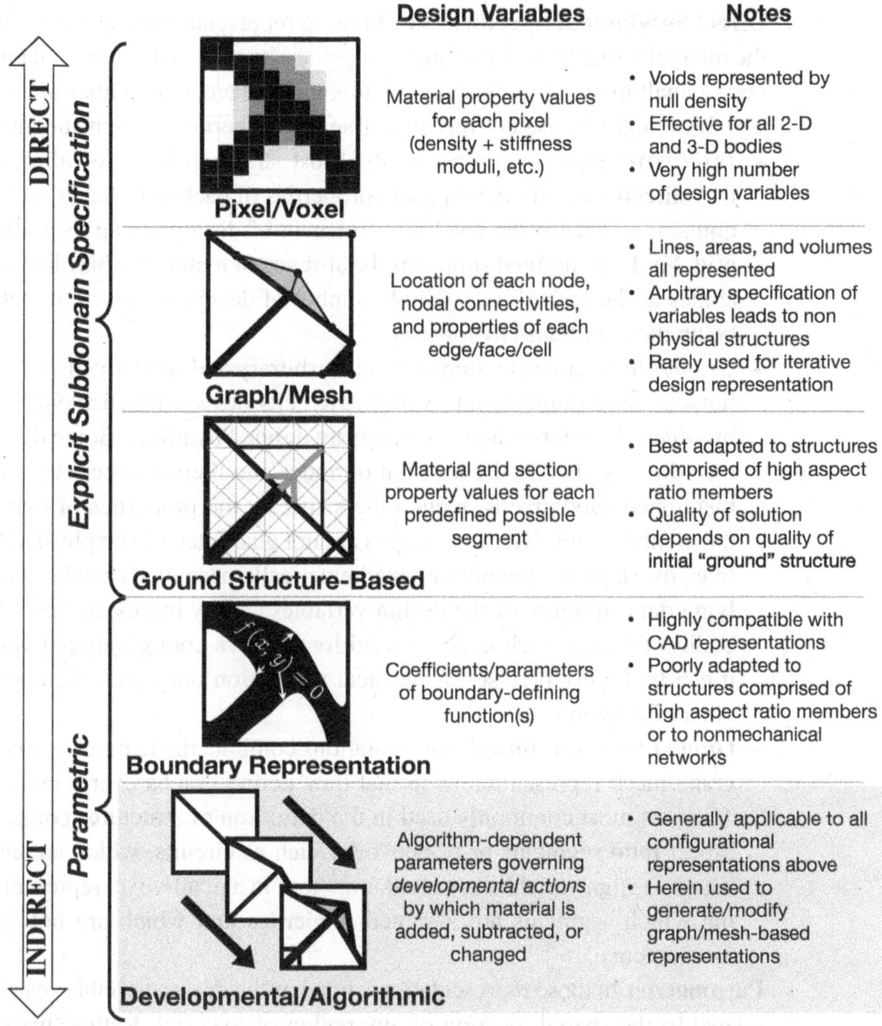

Figure 1.1 Approaches to structural and/or network representation used in engineering and design processes. Representations range from physical to abstract based on the extent to which design variables explicitly determine the spatial location of material and void regions. This work focuses on new *developmental* approaches using bioinspired algorithms.

such representations can be more *physical* in that design variables directly define the locations of material(s) and void(s), or they can be more *abstract*, in that no connection between the design variables and the resulting connected domain can be explicitly observed or intuited. Specifically, two classes of representation are proposed:[2]

[2] In the topological modeling community, only three representation options are often considered: *pixel/voxel*, *graph/mesh*, and *boundary representation*. While *ground structure-based* and *developmental/algorithmic* approaches might be considered as derivative of the other three, they are considered herein for the distinct and effective manner in which design variables are linked to geometric and topological configurations via these representations, especially for the purposes of iterative design.

- **Direct Subdomain Specification:** In these representations, at least some subset of the internal variables is 1-to-1 mapped to the actual spatial location of portions of the body (small material region, segment/ligament, face, etc.). Other associated design variables might be used to further define the properties of each subdomain.
  - *Pixel/Voxel* representations are the most straightforward of all options in that they directly specify the type and properties of each material at each point in the domain, subject to the level of refinement of discretization over a fixed regular grid. Voids are defined simply as the absence of material. This ultimate generality comes at the cost of a very high number of design variables, the physical limit being the continuum hypothesis.
  - *Graph/Mesh* representations consider arbitrary and spatially disparate discretizations of solid domains into volumetric cells or two-dimensional faces bounded by edges, the intersections of which are defined by nodes. Some design variables then directly specify the location of nodes and their connectivity into the aforementioned subdomains, while others specify the properties of materials within these subdomains. Because graphs are not guaranteed to be physically realizable (e.g., overlapping domains are mathematically permitted) and because the underlying data structure of the design variables can by highly complex for arbitrary bodies, this approach is rarely used for iterative configurational design, though it can be highly precise for physical realization once a configuration has been otherwise optimized.
  - *Ground Structure-Based* representations combine the features of pixel/voxel and graph/mesh representations in that they define graphs over fixed regular grids. They are most commonly used in the definition of structures comprised of high aspect ratio segments or of networks such as circuits, which are entirely edge-based. Design variables are similar to those in a pixel/voxel representation, defining which segments are assigned properties and which are not (and are thus nonexistent).
- **Parametric:** In these representations, none of the design variables can be explicitly related to the spatial location of any region of material. Rather, the variables are used as parameters in mathematical functions or algorithmic processes that then determine where material is, where it is not, and what material properties should be considered.
  - *Boundary Representations* define a body by the divisions between material and void, not by the spatial distribution of material explicitly. In general, this may be accomplished through any surface representation,[3] though in the context of shape modeling, computer-aided drafting (CAD), and the iterative design considered herein, this is usually accomplished through the use of mathematical functions capable of capturing linear/planar, cylindrical, and conic features, in addition to more generalized features through the use of such functions as nonuniform rational basis splines (NURBS). Because these approaches are specifically

[3] See, for example, the ubiquitous use of `.stl` files in additive manufacturing, which use triangular meshes to define closed surfaces, indicating where the material is and is not to be deposited.

formulated to define domains of nonzero area and volume, they suffer when local aspect ratios of regions become very high and are entirely inappropriate for the representation of edge-based networks.

- *Developmental/Algorithmic* representations are by far the most abstract in that the design variables, while fully sufficient to deterministically define a design configuration, are in no way linked directly to any observable geometric or topological feature of a body or network. Rather, they are explicitly used as *instructions* to an algorithm that creates or modifies any of the representations described above. The final configuration is then defined by the design variables as they inform cyclic developmental processes, each cycle usually increasing the complexity of the configuration.

1.2 Topology Optimization Toward Preliminary Structural Design

A well-chosen and well-developed structure or network representation allows for interdisciplinary technical communication, tractable quantitative design iteration, and ultimately the fabrication of systems and solutions. Of these, this work specifically focuses on the design implications of structural representation, especially as one seeks to determine optimal topologies in a rigorous, analysis-driven manner.

1.2.1 The Structural/Network Design Problem

The process of topological structure design is the means by which one answers the following motivating question:[4]

> How should one best configure material(s) into a geometric form to perform a needed *function* or *functions* subject to some *constraints* and given some metric for ranking design preferences, known as *objectives*?

These three design concepts are essential for rigorously posing and efficiently solving the structure/network design problem and thus should be clearly defined.

- **Functions** define the purpose of a structure or network, sometimes in the context of a larger system. In mechanical structures, this usually involves the transfer of loads from one or more points of application to one or more points of fixity. In compliant mechanisms, it may involve the transfer of displacements; in other networks, it usually involves the transfer of information between multiple points. Multifunctional structures might be tasked with one or more of the above.

[4] Alternatively, the process of *network* topological design is the process of answering the question "How should one best configure the connectivity between a (sometimes modifiable) set of nodes, and what should the properties of those connections be, to perform needed functions subject to some constraints and given some objectives?"

- **Constraints** are limits on global or local failure and/or gross dimensional restrictions and fixed geometric definitions (e.g., the particular locations of load application) that define the feasibility of a solution. They may also include limits on deformation in mechanical structures or on the loss of information or energy in other networks.
- **Objectives** are the goals that allow preferential ranking of designs, such as minimization of cost or of mass, minimization of manufacturing complexity, maximization of service life, and so on. Strict constraints can also be recast as less strict preferences in some cases, resulting in additional objectives, such as the minimization of deflection under load.

To answer our motivating question for a given functional need, an engineer or designer might take a number of approaches. Let us examine the following simple example.

Example 1.1 Consider the design of a simple structure intended to carry a shear load, as illustrated in Figure 1.2, where the function, constraints, and objectives associated with such a design can be clearly stated as follows:

Function: Transfer a total shear load F across some fixed distance L;

Constraint: Total structure should be at most $L \times L$ in size and no local material failure can occur under load;

Objective: Minimize total mass of material used (i.e., mass of the total structure).

Knowing the largest allowable size and the locations of load application and structural fixity, a designer might be guided by a number of influences as the preliminary ideation process begins (Figure 1.2).

- For many problems, some degree of **simplistic intuition** can provide guidance, and in this case the simple removal of material relative to the maximum allowed might allow one to arrive at a sufficiently lightweight design capable of carrying the required load without failure. Clearly, both the quantitative description of the design and subsequent fabrication via subtractive manufacturing can be very straightforward if simple geometric features are considered;
- It is very common for a designer of any engineering system or subsystem to leverage **historical successes** in configuring new solutions; the design of new structures or networks is often strongly influenced by such legacy design;
- Some degree of **rigorous quantitative analysis** can provide clear guidance toward optimal geometric configuration. For example, dominant principal stress paths computed via finite element analysis can provide clear guidance as to where material is needed and where it is not in a structure;
- Over the past few decades, rigorous methods of **topology optimization** have been proposed, demonstrated, and validated toward the automated preliminary design of structural layouts.

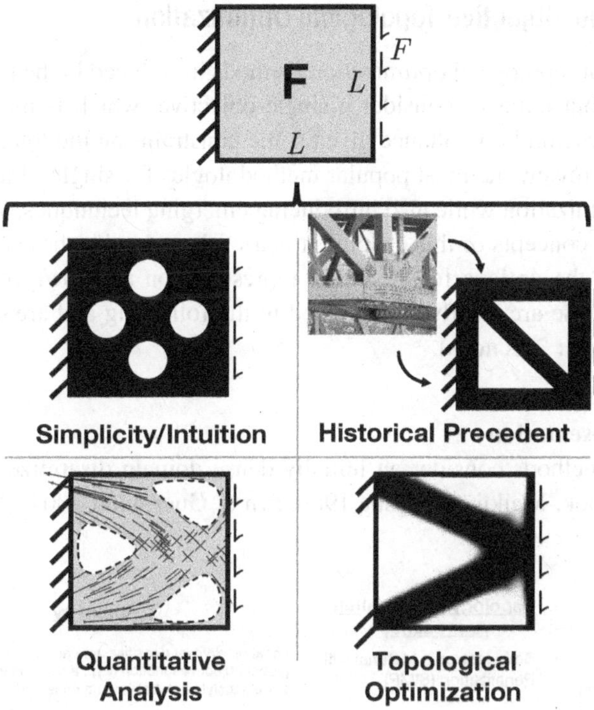

Figure 1.2 An engineer or designer may be guided by a number of different influences in configuring an optimal structure or network; this work addresses classical and novel methods of topological optimization especially for addressing design problems without historical precedent and for which intuition is lacking.

The relative strength of each influence during preliminary design will ultimately depend on such considerations as designer experience and design team capabilities, allowable design cycle cost and time, and problem complexity, among others.

As expected, the design of simple structures performing common functions has been considered across most engineering disciplines. Initial intuition and historical precedent, confirmed via quantitative analysis, are common throughout the engineering design landscape. The purpose of this book is to expand the understanding and application of the fourth option proposed: the more expansive use of topological optimization techniques throughout the engineering community. Toward that end, we next provide a brief review of the more well-known approaches, including their advantages and disadvantages. A more rigorous quantitative treatment is provided for each approach, as well as for others, in Chapter 2, but the primary intent overall is to introduce newer and sometimes much more capable bioinspired alternatives especially suited for multiobjective design problems. These problems are especially challenging as, being novel and usually highly nonlinear, they can be solved neither from intuition nor from precedent.

1.2.2 Methods of Single-Objective Topological Optimization

The vast majority of topological optimization methods introduced in the literature and employed throughout industry consider a single-objective, which is most often the minimization of structural compliance given some constraint on the total volume. In Chapter 2, we will review the most popular methodologies for single-objective structural topology optimization while also introducing emerging techniques. With respect to the introductory concepts of this chapter, it is useful to classify these optimization methods based on the options for structural representation as shown in Figure 1.3 (cf. Figure 1.1). These are briefly summarized in the following and are described in more detail in Chapter 2 as noted.

Pixel/Voxel Representation

In general, these methods consider an initially dense domain discretized into many finite elements (Cook, Malkus & Plesha 1989, Ern & Guermond 2004, Reddy 2019)

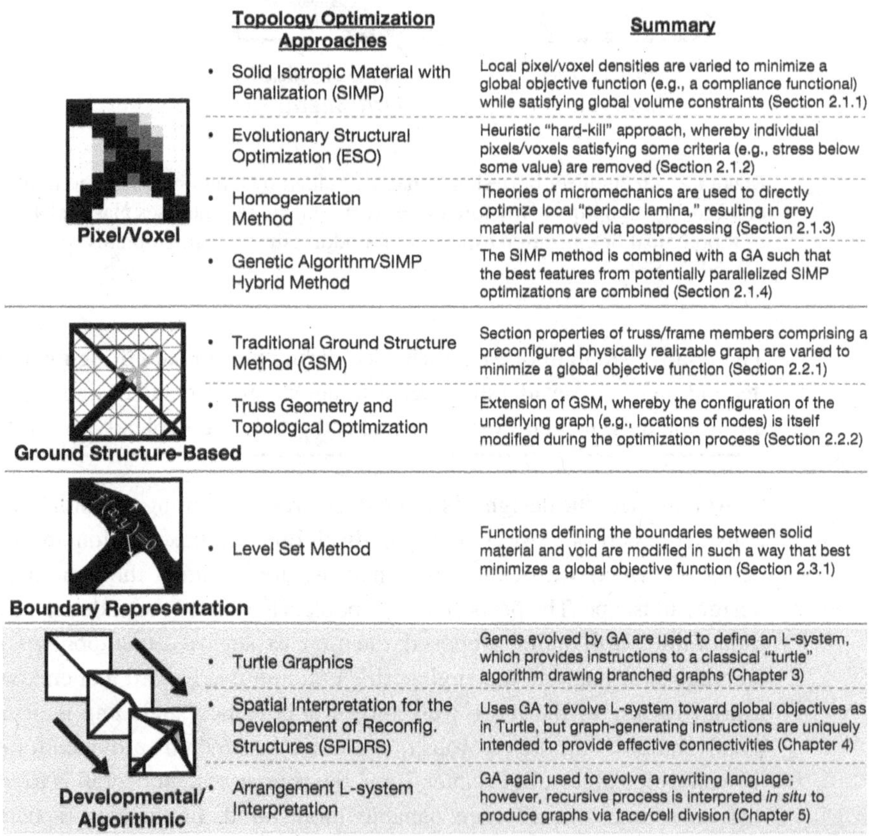

Figure 1.3 A range of single-objective topology optimization approaches has been proposed, each based on one of the structural representation options as introduced in Figure 1.1.

with the goal of minimizing an objective function by identifying whether a given element should consist of solid material or void. Design variables associated with each pixel/voxel/element may have discrete "0/1" or continuous values designating local density, depending on the method employed (Bendsøe 1989, Eschenauer & Olhoff 2001). By far the most popular approach is the solid isotropic material with penalization (SIMP) method (see Section 2.2.1), which often involves the mathematical minimization of a structural compliance metric via continuous variation of local densities, these being strongly penalized when taking on values other than 0 or 1. Two other approaches are heuristic in nature and thus can consider binary density values (i.e., full material or void). These include evolutionary structural optimization (ESO) and the Lamarckian genetic algorithm/SIMP hybrid method.

Evolutionary Structural Optimization (ESO) (see Section 2.2.2) is a "hard-kill" method based on the idealization that optimal structures are made up of material regions that are all fully stressed (Xie & Steven 1997*a*). ESO incrementally removes "inefficient" material from the design domain based on heuristic criteria such as elemental stress (Xie & Steven 1997*a*) or other sensitivity values (Liang, Xie & Steven 2000). A variant known as bidirectional ESO (BESO) also allows for material to be added in regions where elements have a high criterion value (Querin, Steven & Xie 1998, Querin, Young, Steven & Xie 2000). The Lamarckian genetic algorithm/SIMP hybrid method (see Section 2.2.4) uses a genetic algorithm (GA) to combine and evolve the results of possibly parallelized SIMP optimizations. Because GAs (introduced in Section 1.3.3) are heuristic and do not require gradient information to be calculated or approximated, a wide range of objectives and constraints can be considered.

Regardless of the method employed, however, all pixel/voxel approaches suffer from important challenges. Given that each 2-D pixel or 3-D voxel is associated with its own design variable, the dimensionality of the design space increases rapidly when considering that a higher resolution of the design space (which results in more accurate simulated results) requires an increased number of pixels/voxels (Deaton & Grandhi 2014, Kobayashi 2010). This problem is greatly compounded for three-dimensional structures, for which SIMP optimization requirements in terms of computational resources and time can explode (Aage, Andreassen, Lazarov & Sigmund 2017). At the same time, density-based methods alternatively suffer from a phenomenon known as "checkerboarding," resulting from high numerical stiffness of such patterns. This formation of adjacent solid/void elements arranged in a checkerboard pattern (Bendsøe 1989, Bourdin 2001, Deaton & Grandhi 2014), and indeed the generation of nonphysical or nonoptimal solutions, becomes prevalent at volume fractions that are prevalant in engineering practice (e.g., the $\sim 10\%$ volume fraction found in aerospace structures such as wings). These approaches are also notoriously mesh-dependent, meaning different optimal topologies can result from identical design domains of different discretization refinements. Finally, "gray" material resulting from local densities other than 0 or 1 can arise when the design conflict between mass and stiffness is weak and often requires postprocessing or "interpretation," which may greatly reduce the gains of the optimized, nonmanufacturable solution obtained with the density-based method.

The ESO/BESO implementations have also been shown to break down for simple structural problems (Zhou & Rozvany 2001) and utilize heuristic criteria that may not relate to the objective function (Rozvany 2009). Both ESO and the Lamarckian genetic algorithm/SIMP hybrid method approaches can result in singularities in associated finite element matrices (Deaton & Grandhi 2014), which can be avoided when discrete 1/0 design variables are replaced with continuous variables, but this results in the formation of gray "transition" material between solid and void regions (Guest, Prévost & Belytschko 2004, Sigmund 2007, Guest, Asadpoure & Ha 2011, Kawamoto et al. 2011). Various *regularization* techniques have been created to prevent numerical issues and control the quality of final results, such as filtering methods (Sigmund 1997, Sigmund & Petersson 1998, Bourdin 2001, Lazarov & Sigmund 2011), length-scale constraints (Guest et al. 2004, Guest 2009, Lazarov & Wang 2017), and projection schemes (Guest et al. 2004, Sigmund 2007, Guest et al. 2011, Kawamoto et al. 2011, Wang, Lazarov & Sigmund 2011).

Ground Structure-Based

The ground structure method (GSM; see Section 2.3.2) is based on the concept of incrementally updating a pre-existing graph by computing the sensitivities of a given structural performance metric to the section properties (e.g., thickness, area, etc.) of each line segment in that graph. As such, it only considers structures consisting of slender structural members (trusses or frames). The topology of the domain is then evolved by varying sectional properties of members to such an extent that they can essentially be removed from the structure (Ben-Tal & Bendsøe 1993, Bendsøe, Ben-Tal & Zowe 1994, Lee, Mueller & Fivet 2016). While GSMs are adept at finding solutions to truss and frame problems, the quality of optimized results strongly depends on the quality and refinement of the initial graph, where meaningful refinement can quickly increase the number of design variables required as well as the computational time needed to analyze initially dense configurations (Bendsøe et al. 1994, Hagishita & Ohsaki 2009). Combining the sizing optimization of the GSM with shape optimization in the form of node placement optimization can mitigate the need for initially dense ground structures (Achtziger 2007). Similar to BESO, a method known as the growing GSM (GGSM) allows for both the addition and removal of structural members, allowing for the initial ground structure to remain relatively sparse (Hagishita & Ohsaki 2009). Further, an extension to GSM whereby the initial graph can itself be incrementally modified through changes in nodal locations is discussed in Section 2.3.3.

Boundary Representation

The single approach to topology optimization based on this approach for quantifying structural configurations discussed herein is second in popularity only to SIMP. Known as the level set method (see Section 2.4.1), it optimizes over the space of parameters defining the scalar-valued level set function that is used to define domain boundaries (i.e., the interface of solid and void) (Sethian & Wiegmann 2000, Osher & Fedkiw 2003, Wang, Wang & Guo 2003, van Dijk, Maute, Langelaar & van Keulen 2013,

Deaton & Grandhi 2014). This allows for the convenient treatment of topological changes, as structural boundaries can be modified by using the physical problem and optimization conditions to control the output of the level set function. A level set function can be parameterized using finite element method (FEM) basis functions (Wang et al. 2003, Allaire, Jouve & Toader 2004, Amstutz & Andrä 2006, Xing, Wei & Wang 2010, van Dijk, Langelaar & Keulen 2012), radial basis functions (RBFs) (de Ruiter & van Keulen 2004, Wang & Wang 2006, Luo, Tong, Wang & Wang 2007, Kreissl, Pingen & Maute 2011), or Fourier series (Gomes & Suleman 2006), which determine the design freedom and precision of the material boundaries.

One challenge this approach faces is that performance analysis requires the mapping of a parameterized level set function to a structural model, which can impact the accuracy of the structural response and the quality of the final result of the optimization process. Currently utilized methods for this mapping procedure include conforming discretization (unique remeshing; most accurate but computationally expensive) (Ha & Cho 2008, Allaire, Dapogny & Frey 2011), immersed boundary techniques (extended FEA approaches; accurate but difficult to implement) (Kreissl et al. 2011, van Dijk et al. 2012), and density-based mapping (mapping of boundaries onto pixel/voxel representation; accurate with regularization techniques) (Wang et al. 2003, Allaire et al. 2004, de Ruiter & van Keulen 2004). Further, the level set method suffers from a poor rate of convergence, convergence to local minima, and difficulties in dealing with constraints (van Dijk et al. 2013), some of which can be prevented by using regularization techniques such as sensitivity and density filtering (Luo, Wang, Wang & Wei 2008, van Dijk et al. 2012, Zhu, Zhang & Fatikow 2015). Finally, as mentioned in Section 1.1, this approach can be ineffective as local aspect ratios become very high, as in structures composed of slender truss or frame components.

Developmental/Algorithmic

These approaches to topological design optimization are among the newest and comprise the primary topic of this work. They employ a number of different algorithms to develop a graph, and thus a structure, by executing generative actions based on instructions associated with a rewriting language. The inputs to the language then comprise the design variables of the optimization problem such that an algorithm seeks to find a set of instructions that generates the best-performing structure (or network) depending on functional goals. Turtle graphics, the simplest of these approaches, is used to introduce all critical algorithmic concepts in Chapter 3. The spatial interpretation for the development of reconfigurable structures (SPIDRS) algorithm generates much better-performing networks and is described in Chapter 4. Finally, arrangement L-system interpretations, which also generate high-performance solutions, are developed in Chapter 5.

All the various single-objective methods of topological optimization summarized above can be classified as *gradient-based* and *gradient-free* depending on whether they employ an explicit dependence on the derivatives of the objective function considered. SIMP, ground structure, and level set methods are all gradient-based methods, meaning that they can be numerically efficient but may have difficulties with problems containing multiple local optima, discontinuous design spaces, or discrete variables (Deb 1999). They also can explicitly consider a single-scalar-objective (and thus its derivatives) at a time. On the other hand, ESO, GA bitmap methods, and the three developmental algorithms are all gradient-free, specifically relying on metaheuristics to guide the search for better designs, while not being mathematically limited to a single scalar objective. This may require a large number of structural analyses compared to gradient-based methods, but it makes these methods robust against local optima or discontinuous design spaces. The choice of gradient-based or gradient-free algorithm will also have critical implications as one extends to consider multiple objectives simultaneously, as is the focus of this work and is described in the following sections.

1.2.3 Extension to Multiobjective Topological Optimization

While single-objective approaches have been and will continue to be used for a wide range of structural design problems, the optimization process is made considerably more difficult when the underlying problem has *multiple* performance objectives. These objectives are often conflicting, leading to a compromised topology if pursued simultaneously (Chen & Wu 1998). Thus, multiobjective topology optimization problems rarely possess a single "optimal" solution but instead exhibit a series of solutions that are classified as *Pareto-optimal.* To classify as such, a solution cannot be outperformed in all objectives by another design (Cohon 2003). This book, in particular, focuses on new bioinspired approaches for multiobjective topology optimization using algorithmic approaches, which are also briefly summarized in Figure 1.3 and comprise much of the remainder of this text. However, there are several existing multiobjective topology optimization approaches that should also be quickly reviewed prior to proceeding with the introduction of new algorithms.

Pareto-optimal solutions for topology optimization problems have been obtained by implementing the algorithms listed in Section 1.2.2 within an extended scheme. This applies to both gradient-based and gradient-free approaches. In the former, consideration of a single scalar quantity to be minimized/maximized is usually essential, and thus any desire for the maximization and/or minimization of *multiple* objectives must be accomplished by mapping these into a single scalar value. This is most commonly achieved using the method of weighted sums, where the single scalar quantity for which derivatives are computed is taken to be the sum of each desired performance objective multiplied by some predetermined weight based on relative importance or other factors (Turevsky & Suresh 2011, Zhu, Zhang & Fatikow 2014). Variants of the weighted sums method known as compromise programming (Chen & Wu 1998) and physical programming (Lin, Luo & Tong 2010) have also been used for multiobjective topology optimization. However, the determination of suitable weights is nontrivial (Das & Dennis 1997, Messac, Sundararaj, Tappeta & Renaud 2000, Messac & Ismail-Yahaya 2001) and often requires a general intuition for the problem beforehand

(Tamaki, Kita & Kobayashi 1996). The single scalar value goal of gradient methods also dictates that a single optimal topology can be determined per optimization, such that each design along a desired Pareto frontier must be generated by an individual optimization process using a unique combination of objective weights. Furthermore, gradient methods using weighted objective functions have been shown to fail in capturing Pareto-optimal solutions for problems with non-convex Pareto frontiers (Das & Dennis 1997, Chen & Wu 1998, Messac et al. 2000, Messac & Ismail-Yahaya 2001).

As the name suggests, gradient-free optimization approaches do not require gradient information and instead solely use function evaluations of the objective function(s) to converge to a solution. One of the most common non-gradient methods, and the one most commonly utilized in topology optimization, is the GA, which is described in more detail in Section 1.3.3. GAs are motivated by the principles of natural selection and represent an optimization procedure that only requires a means of mapping design variable values to performance metric(s). (Tamaki et al. 1996, Deb 1999). This mathematical and algorithmic concept and its essential role in the overall bioinspired design approach captured in this work is introduced in Section 1.3. In short, GAs begin with a randomly generated set (or "population") of design solutions that are evaluated, assigned a fitness value based on their performance, and then modified by three operators borrowed from the fundamental ideas of genetics (selection, crossover, and mutation) to create a new and hopefully better population (Hare, Nutini & Tesfamariam 2013). This process continues until some specified termination criterion is met.

GAs are attractive when considering multiobjective topology optimization because they work with a population of solutions rather than a single solution, meaning that the set of solutions comprising the Pareto frontier can be obtained simultaneously. Their stochastic behavior also generally allows for these algorithms to better search the global design space, thereby avoiding convergence to local minima. Additionally, the lack of reliance on gradients makes GAs an appealing option when dealing with discrete design variables and discontinuous design spaces (Deb 1999), and GAs are perfectly suited to parallel processing, especially advantageous when using massive parallel architectures.

GAs are clearly essential to such previously listed algorithms as the Lamarckian genetic algorithm/SIMP hybrid method but have also been successfully coupled with other topology optimization approaches, including the level set (Guirguis, Hamza, Aly, Hegazi & Saitou 2015, Yoshimura, Shimoyama, Misaka & Obayashi 2016), ESO/BESO (Liu, Yi, Li & Shen 2008, Zuo, Xie & Huang 2009), and ground structure methods (Prasad & Diaz 2005). Sigmund has argued against their use in problems where the topology is represented explicitly by the design variables (e.g., SIMP, ESO/BESO) (Sigmund 2011), citing their computational cost relative to gradient-based methods and the necessity of using coarse meshes that cannot correctly represent the underlying physical response of the structure. However, as will be introduced in Chapter 2 and then demonstrated throughout the remainder of this work, they are essential to the use of developmental algorithms in topology optimization (see also Section 1.3).

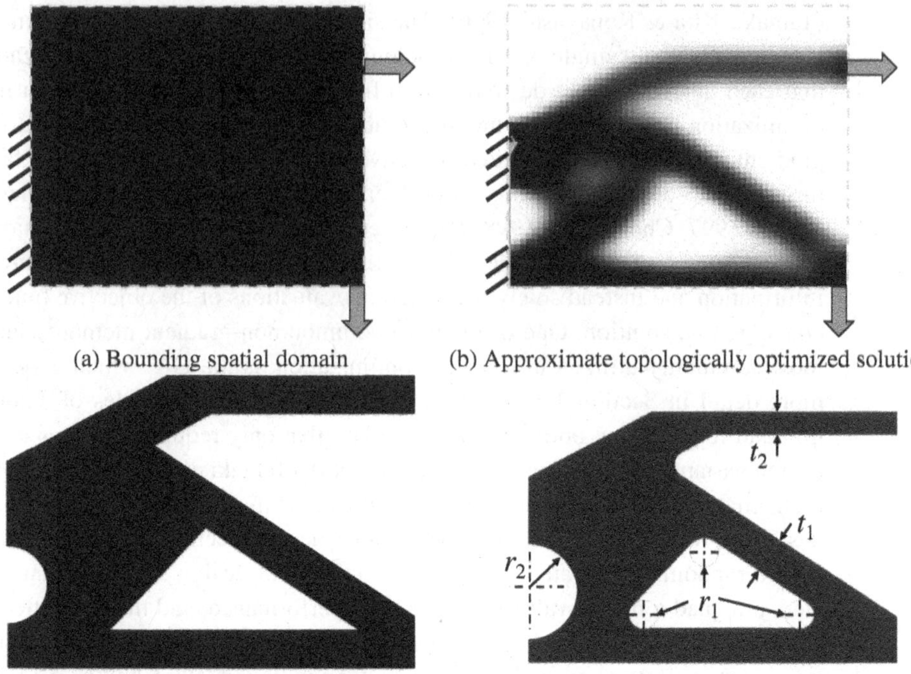

(a) Bounding spatial domain

(b) Approximate topologically optimized solution

(c) Preliminary geometrically parameterized design

(d) Dimensionally optimized final part

Figure 1.4 Stages of the topological design process applied to a generic structural part. Note that the details of (b) strongly depend on the topology optimization algorithm selected (cf. Figure 1.3).

1.2.4 From Preliminary Topology to Detailed Design

Regardless of the algorithm chosen or whether one or more objectives is considered, topology optimization is becoming more commonly implemented at the preliminary stage of structural design (see Section 1.2.1). In most cases, topology optimization can determine optimal geometric configurations without well-defined initial conditions and/or a historical intuition as to the best solution. This is exemplified in Figure 1.4, which illustrates how the design process of a generic structural part can include a topological optimization step but also shows how such a step is not sufficient for full component definition (i.e., as for traditional manufacturing). Therefore, it is important to comment briefly on the relationship between topological optimization and full design.

Given a spatial design domain, a material system or system(s) to be considered, and a series of predefined boundary conditions that define the function of the component (Figure 1.4(a)), topology optimization can be used to determine an optimal distribution of material within the design domain that maximizes the performance of the structure (e.g., here, its stiffness) while satisfying the specified boundary conditions and not exceeding preexisting dimensional limits. The resulting "optimal" topological configuration (Figure 1.4(b)) can then be interpreted by designers experienced in parameterized design optimization and/or manufacturing limitations to generate a parametric geometric representation (Figure 1.4(c)). Additional studies (experimental, numerical, etc.) considering additional objectives, constraints, safety factors, etc. can then be performed to arrive at a finalized design (Figure 1.4(d)) having optimized thicknesses, fillet radii, and other dimensional features. In this way, employing topology optimization early in the preliminary design phase can lead to well-defined structures performing functions unique to a given design team.

1.3 Bioinspiration in Topological Design: The "EvoDevo" Approach

To this point, we have introduced the concepts of structural (or network) representation and topological design. Existing approaches for accomplishing the latter have been introduced, and the ideas of gradient-based, gradient-free, single-objective, and multiobjective optimization have all been addressed. The novelty and uniqueness of the content provided in this book is to propose generative approaches that hold promise for negating many of the disadvantages of these other methods, especially in the context of highly nonlinear, nonintuitive, multifunctional systems.

1.3.1 Motivation for Bioinspiration

Regardless of the tools utilized, the preliminary stage of structural design (cf. Figures 1.2 and 1.4) is largely influenced by designers' previous experience with or intuition for the problem at hand (Kaldate, Thurston, Emamipour & Rood 2006, Fantini 2007). A closer inspection of several of the methods described in Sections 1.2.2 and 1.2.3 reveals that topology optimization is no different. SIMP, level set, and GSM implementations all require an initial volumetric constraint, meaning that designers must have knowledge of approximately how much material they want to use. Each of the three methods is also typically reliant upon gradient-based optimization approaches, which require that for multiobjective problems the objectives be combined into a single scalar score using the method of weighted sums or similar. However, as previously mentioned, the determination of these weights is nontrivial and still requires that designers have a level of intuition regarding the desired outcome of the solution (Tamaki, Kita & Kobayashi 1996). SIMP and GSM implementations also employ direct subdomain specification (cf. Figure 1.1), meaning that the design variables directly define the topology of the structure. As such, the accuracy of the analysis and thus of the resulting topology is a function of the domain discretization, and both methods are plagued by rapidly increasing design space dimensionality and computation time when accurate solutions are required. For simple structural problems with a limited design space such as the example shown in Figure 1.4, such topology optimization methodologies may be sufficient. However, the maturation of adaptive material technology and a growing interest in developing structures or other networks with increased multifunctionality may begin to render these methods ineffective.

The goal of designing structures with multiple functionalities is inherently multiobjective, and the vast design space associated with these problems renders intuition formation extremely difficult. The use of traditional approaches (SIMP, GSM, etc.) would require a single objective to be formulated from multiple objectives (e.g., through the method of weighted sums), practically limiting designers to a small subset of all potential solutions. Furthermore, multifunctional systems typically respond to various stimuli such as stress, heat, electrical current/voltage, magnetic field, moisture, or light, and the complex physics models required to accurately analyze these responses greatly complicate the derivative calculations necessary for gradient-based topology optimization approaches. Therefore, there is a growing need for an inherently multiobjective preliminary design approach capable of exploring a vast design space to identify well-performing solutions to problems for which designers have little/no intuition or experience.

As the title of this monograph indicates, we propose bioinspired solutions to the various shortcomings of the currently established topological optimization approaches. The tools introduced and demonstrated herein, while varying in their details and utilities, all fall under a common framework that abstracts the design process into a biological analogy that is described in the following section.

1.3.2 The Evolutionary-Developmental Approach to Systems Design

In biological systems, the problem to be solved is "sufficient procreation toward propagation of a species," which requires survival (and even good health) of population members despite the presence of resource limitations, adverse environmental conditions, predators, intraspecies competition, and other challenges. As a means of formally introducing such bioinspiration into the design processes of interest, we here introduce the evolutionary-developmental or *EvoDevo* approach. As will be described, this analogy goes beyond simple "survival of the fittest" in terms of the selection, crossover, and mutation that comprise the common GA approach to optimization (introduced in Section 1.3.3). Rather, it also considers how genes, environment, heredity, and the laws of nature combine to drive the unique development of each new individual seeking survival. The full EvoDevo approach thus combines the developmental processes that govern spatial configuration (topological layout) with the evolutionary processes that improve populations over time, permitting great diversity in the context of straightforward algorithms. It is the *development* in particular that defines the unique topology optimization schemes described in Chapters 3–6 of this book.

As is appropriate for biological analogies to design, the EvoDevo approach is best suited for multiobjective optimization. In nature, individuals must perform a diversity of tasks associated with survival and procreation (e.g., hunting, courting, etc.) as they avoid critical pitfalls (disease, death, etc.). Some engineering designs must likewise perform multiple tasks without failure. In this work we will often focus on such *multifunctionality*, especially considering the optimization of networks, structural and otherwise, that seek multiple objectives and satisfy multiple constraints across

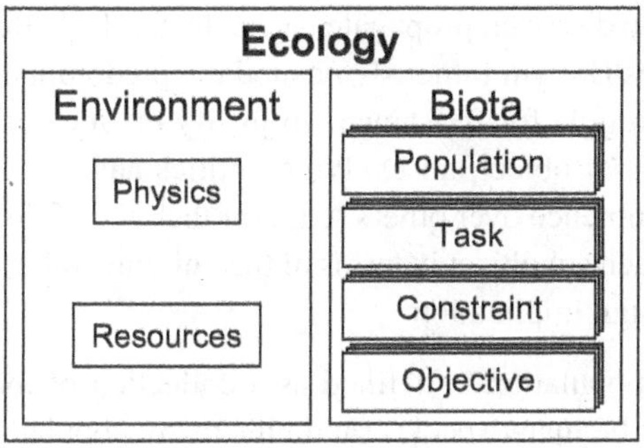

Figure 1.5 The EvoDevo abstraction considering both the living and nonliving aspects of the natural world, these concepts being interpreted for engineering design throughout this section.

multiple tasks.[5] This makes EvoDevo an effective alternative to traditional topological design approaches for multifunctional structures and systems.

The EvoDevo abstraction that captures the considerations of a multiobjective design problem at the highest level is depicted in Figure 1.5. Note that as we introduce these concepts and those that follow, we will use biological phrasing to represent engineering design concepts. At the top of the structure is the *Ecology* object, which is the cyber ecosystem within which design development and evolution occur. It represents the many choices of the designer in setting up a particular problem to be solved and, in this analogy, requires definition of two subobjects: the *Environment*, which represents the nonliving considerations within nature (physical laws and rigid constraints in engineering), and the *Biota*, which represents the relationships between living, and thus evolving, life-forms in nature (diversity of design options in engineering).

The *Environment* object abstracts the unchanging and impartial world that surrounds communities of living individuals. For us, it captures the designer specifications that apply to all processes and all individual designs. This object contains a *Physics* subobject, which includes the definition(s) of the physical phenomena to be considered, the mathematical models to be applied to these phenomena, the computational approaches employed in analyzing the models, and all physical constants. Also important are the *Resources* available to all designs and their assessment, including feasible ranges of all considered material properties, gross limitations on geometric parameters, and even computational resources (licenses, cores, time, etc.) available to the design problem.

On the other hand, the *Biota* object represents the relations of individuals to one another as populations of living entities evolve over time. While the *Environment* can be described as a negative directing force or constraint of evolutionary change, biotic

[5] To expand on the analogy, consider that a predatory bird's wing, for example, must enable both efficient soaring flight and high-speed controlled dives in adults, while permitting the compactness needed to fit inside an egg of minimum volume prior to hatching. Likewise, an airplane wing must generate lift at very low drag, carry these lifting loads without failing, and store fuel, among many other functions. In designing any engineering solution, multiple quantitative performance metrics can then be defined, many of them being contradictory with respect to designer's goals (e.g., a strong and robust wing that is also light).

competition helps populations to discover, propagate, and enhance the better individual designs over the poorer. Charles Darwin famously posited the predominance of biotic competition in organisms' "Struggle for Existence" to justify the idea of progress in nature (Gould 2002). In the *Biota* object, all evolving entities and the manner(s) in which some come to take preeminence over others (e.g., for the sake of final selection) are specified. Let us consider each subobject in terms of the guiding biological analogy and associated engineering design implication:

- *Population(s):* In nature, a population is defined as a collection of individuals of the same species; there may be many species (and thus many populations) in relationship with each other in a given ecology. In design, a population then represents a collection of individual designs associated with a given concept; there may be multiple design concepts (parameterized in a number of different ways) in the search for an optimal design.
- *Task(s):* Living organisms must complete a number of tasks to survive and procreate (e.g., eat, fight, court, travel, etc.), and their performance in these tasks relative to their peers and with respect to their environment determines their success. Engineered designs must likewise perform functions, where in this monograph we especially consider multifunctional designs as inspired by natural systems.
- *Constraint(s):* The performance of some tasks in nature must meet at least a minimum standard to allow survival and procreation. Animals must procure sufficient caloric intake and avoid contact with deadly predators, for example; these are limitations placed on them by their environment and biota. Most individual engineering designs are likewise developed to satisfy strict criteria, where failure to do so renders a design "infeasible."
- *Objective(s):* The success with which some tasks are completed by a living individual may increase or decrease the probability of procreation rather than allowing it or preventing it outright. Healthier (e.g., better nourished and less diseased) organisms produce more progeny, for example. In engineering design, an objective is a score by which individuals can be ranked and a "best" option is eventually selected for continued development. The consideration of multiple objectives, if applicable, explicitly precludes the down-selection of a single design but rather focuses the designer on a rational population of potential best choices (as during preliminary design).

Of course, the most complex entity in the entire ecology is the individual, a collection of which comprises a population. In nature, the unique definition of any individual is the result of both developmental and evolutionary processes; the information flow associated with these in the context of engineering topological design is illustrated in Figure 1.6. In traditional genetic design optimization of engineered systems, one will often formulate the modeled configuration of a given individual based *directly* on the design variables, neglecting the explicit consideration of developmental processes. As an engineering design text, this work likewise seeks to link design variables (genes) to expressed performance (fitness), but we intentionally propose a unique approach in which the *development* of the configurational layout as informed by genetic information is critically important. The inclusion of this developmental step is the

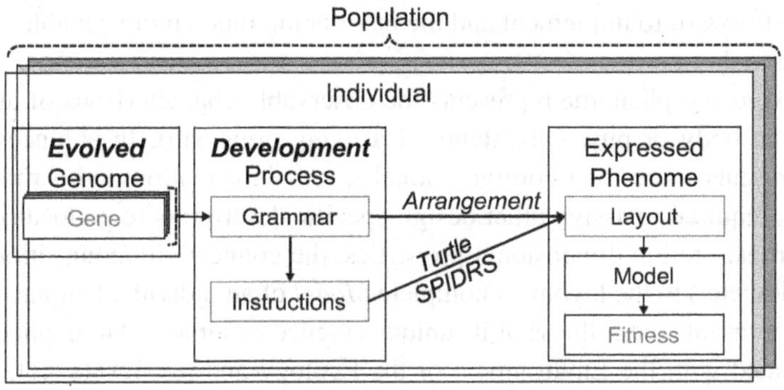

Figure 1.6 Population object and subobjects, specifically as interpreted in the context of engineering topology optimization.

essence of the *EvoDevo* approach and represents the primary bioinspiration driving this work.

As shown in Figure 1.6, the configurational definition of a design results from *Evolved* genetic information (*Evo*) as acted upon by a *Developmental* process (*Devo*). Together these result in the observable characteristics of the individual and, ultimately, the fitness thereof. Specifically, the *EvoDevo* approach taken throughout this book begins with the array of *Genes* (known as the *Genome*), which is determined through evolutionary processes and fully represents the uniqueness of a given individual. However, the genome alone is far from sufficient to fully define an individual and the full *EvoDevo* approach proceeds as follows:

1. Just as the position of every atom (or even every cell) in a natural organism is not defined by the genome but rather by the execution of chemical reactions and other physical processes based on genetic information, so too can information representing a unique engineering design be **translated** into a set of rules (i.e., a *grammar*) for developing a configurational layout. In the methods that follow, this will be achieved through the definition of the characters and rules of an *L-system* (see Chapter 3) or *arrangement L-system* (see Chapter 5).
2. Given these rules, the configurational (e.g., geometric) layout can then be accomplished in one of two **interpretation methods**. In nature, developmental processes (e.g., cellular division and differentiation) are directly informed by genetic information as they occur; in this way, the rules of development are iteratively applied directly to an iteratively changing configuration that, once complete, represents an individual. This direct grammar-to-layout approach in nature motivates the arrangement L-system topological representation described in Chapter 5. Alternatively, for reasons of algorithmic and conceptual simplicity, one can imagine an intermediate step during which gene-informed rules are fully interpreted into a long and highly complex set of layout-building instructions, which only upon completion are interpreted into a layout. This approach will be considered in Chapter 3 on Turtle graphics and Chapter 4 on the so-called SPIDRS algorithm, the former being quite

straightforward to implement and the latter being much more capable as a network design tool.

3. The expressed phenome represents the observable characteristics of an individual (i.e., the body or physical extents of a living organism). In engineering design, it represents more than configurational *layout* information and a **modeling** process is required to assign other design-specific descriptors (e.g., localized material properties, section dimensional properties, the connectivity/connection properties at nodes, etc.) to the layout. A complete *Model* of an individual engineering design then represents the fullness of its unique essence as subjected to applied conditions associated with the Environment of the Ecology and the Task(s) (cf. Figure 1.5) for which performance is to be assessed. In this work, for example, an FEA model might be used to fully represent a design (i.e., its geometry and material composition) assessed against a function of interest (i.e., via applied far field and boundary conditions).
4. The final expression of an individual is its performance when assessed against Task(s) of interest. This **analysis** of the model provides the overall *fitness* relative to the Objectives and Constraints of the Ecology (cf. Figure 1.5).

To summarize in the context of engineering topological design, an array of genes known as a genome (the design variables) is **translated** into a grammar (development rules) that are then **interpreted** either directly or via instructions into a layout (geometric or network configuration). The layout is then **modeled** via the consideration of additional information, and the resulting model is **analyzed** under a set of imposed conditions to provide a final fitness (functional performance).

Given that the fitness of each individual is compared against the objectives and constraints of the Ecology such that better-performing individuals can be separated from poorer-performing alternatives, the evolutionary programming then acts on the population to evolve it between generations, as described in more detail below. Here we use crossover, selection, and mutation to discover, enhance, and propagate good features in the population. This is the ultimate goal of the design process. Relative to other more traditional approaches, it is the development process explicitly considered herein that is key for the effective search using a GA, as it is responsible for both the improved quality of design solutions and the accelerated convergence of the overall methodology (Pedro, Kobayashi, Coimbra, & da Silva 2008, Kobayashi, Pedro, Coimbra & da Silva 2009, Kobayashi, Pedro, Kolonay & Reich 2009, Kolonay & Kobayashi 2010, Pedro and Kobayashi 2011, Stanford, Beran & Kobayashi 2012). The fact that this bioinspired process is indeed effective for the design of a wide range of engineering networks (structural, fluidic, electrical, etc.) is the foundational thesis of this book and will be repeatedly demonstrated.

1.3.3 Genetic Algorithm Overview

Evolutionary algorithms in general are applied to a wide range of application areas from engineering, to art, to biology, to physics, and others. In the field of structural topology optimization, these algorithms have been used to improve the exploration of complex search spaces (Chapman, Saitou & Jakiela 1994, Kane & Schoenauer 1996, Schoenauer 1996, Jakiela, Chapman, Duda, Adewuya & Saitou 2000, Tai & Chee 2000, Azid, Kwan & Seetharamu 2002, Fanjoy & Crossley 2002, Hamda, Jouve, Lutton, Schoenauer & Sebag 2002, Tai, Cui & Ray 2002, Cappello & Mancuso 2003, Wang & Tai 2004, Madeira, Rodrigues & Pina 2005, Tai & Akhtar 2005, Wang & Tai 2005*a*, Wang & Tai 2005*b*, Wang, Tai & Wang 2006, Tai & Prasad 2007, Balamurugan, Ramakrishnan & Singh 2008, Guan & Chun 2009, Zuo et al. 2009, Madeira, Pina & Rodrigues 2010).

Relative to the previous discussion, evolutionary optimization algorithms are biological metaphors that can produce high quality designs by identifying, recombining, and enhancing the best features present in a continuously adapting population of individual designs. Throughout this book, we will only make use of GAs , taken to be those heuristic approaches that specifically apply the processes of selection, crossover, and mutation to sequential populations of designs such that new, and hopefully improved, generations continue to be generated. The execution of such an algorithm encompasses all aspects of Figures 1.5 and 1.6. In the context of the GAs employed in this monograph, evolution starts with a population of individuals, often generated via a random or quasi-random manner (e.g., via Latin hypersquare sampling (Liefvendahl & Stocki 2006)), and then proceeds with the following steps:

1. Each member of a full population, being uniquely described by its individual genotype, is subject to the four-step translation–interpretation–modeling–analysis process described above such that all fitnesses can be quantified. In the context of engineering, this will involve the calculation of one or more performance metrics based on the physical response in question (e.g., structural, fluidic, electromagnetic, thermal).
2. The relative fitnesses of all members of a population are then compared, and a *selection* algorithm ranks the individuals, determining the best performers based on some predetermined criteria that may consider both response preferences and diversity preservation.
3. Selected designs are then assigned to pairs and their genotypic content combined and used to create one or more offspring based on a number of potential *crossover* algorithms.
4. Finally, the genotypes of the new offspring are most often subject to *mutation* to further increase design diversity. Sufficient crossover and mutation operations will eventually lead to a new population of desired size.

The cyclic evolutionary process continues when this new population is likewise evaluated to quantify its relative fitnesses. As this process emulates natural selection, populations tend to improve in the relative performance of their individuals until designs converge toward optimal configurations.

For the purpose of mathematically defining the kind of multiobjective optimization problem that will be solved throughout this text using the EvoDevo approach, we formally introduce the following four classes of variables:

- **Design Variables:** These are the components of $\boldsymbol{x}$ that represent all genotypic content subject to definition by a designer. In the methods on which this overall text is based, these will be shown in the following chapters to be either (i) aspects of the grammar (i.e., L-system) to be defined in the following chapters, (ii) graphing algorithm parameters, or (iii) physical parameters of the system to be modeled, especially as related to interpretation of genotype into phenotype (i.e., the modeling step of Figure 1.6).
- **Bounds:** These are the upper limits $\boldsymbol{x}^{ub}$ and lower limits $\boldsymbol{x}^{lb}$ on $\boldsymbol{x}$ such that $x_i^{lb} \leq x_i \leq x_i^{ub}\ \forall\ i$. In the current explorations, these will be defined in a straightforward manner relative to the L-system assumed or guided by *a priori* understandings regarding physical parameters (cf. *Resources* in Figure 1.5).
- **Objectives:** These components of the set $\boldsymbol{f}$ are the performance metrics with respect to which the designer has a preference (i.e., either minimize or maximize) but for which no strict requirements apply. Given an unchanging physical model considering an unchanging set of operational conditions, the objectives are fully defined by the selection of $\boldsymbol{x}$ such that one may express them as $\boldsymbol{f}(\boldsymbol{x})$ (cf. Figure 1.5).
- **Inequality Constraints:** These components of $\boldsymbol{g}$ quantify the satisfaction of strict design requirements by exploiting the fact that each requirement considered can be strictly expressed in the form $g_i \leq 0$. For example, if the maximum Mises stress $\bar{\sigma}$ calculated within a structure must be held to below a known failure stress σ_f, then the ith inequality constraint could be written as $g_i = \sigma_f - \bar{\sigma} \leq 0$. As with the objectives, these constraints (cf. Figure 1.5) may be expressed as $\boldsymbol{g}(\boldsymbol{x})$.

Note that a set of *equality constraints* (i.e., $\boldsymbol{h}_i = \boldsymbol{0}$) can also be considered but are rarely of utility in the design of engineered structures except as may be included implicitly in the calculation of $\boldsymbol{f}$ or $\boldsymbol{g}$.

With these variables defined, the multiobjective optimization problem can then be stated as follows:[6]

$$\begin{aligned} \boldsymbol{f}^* &= \min_{\boldsymbol{x}} \boldsymbol{f}(\boldsymbol{x}) = \boldsymbol{f}(\boldsymbol{x}^*), \\ \text{such that} \quad \boldsymbol{x}^{lb} &\leq \boldsymbol{x} \leq \boldsymbol{x}^{ub}, \\ \boldsymbol{g}(\boldsymbol{x}) &\leq \boldsymbol{0}. \end{aligned} \tag{1.1}$$

A design $\boldsymbol{f}^* = \boldsymbol{f}(\boldsymbol{x}^*)$ is said to satisfy Equation 1.1 if there is *no* $\boldsymbol{x} \neq \boldsymbol{x}^*$ such that

$$f_i(\boldsymbol{x}) \leq f_i(\boldsymbol{x}^*)\ \forall\ i.$$

Such a point cast into the space spanned by $\boldsymbol{f}$ is known as "nondominated." In the idealized case of full convergence, the set of all such nondominated points is known as the "Pareto frontier" (see, for instance, Coello 2002).

The evolution of sequential populations of 10 designs from an initial randomized guess toward a converged Pareto frontier is schematically shown in Figure 1.7. With

[6] Here we assume preference for the *minimization* of all objectives. This is without the loss of generality as any preference for maximization (e.g., maximum compliant displacement or maximum strength) can be mathematically converted to a preference for minimization via multiplication of such an objective by (-1).

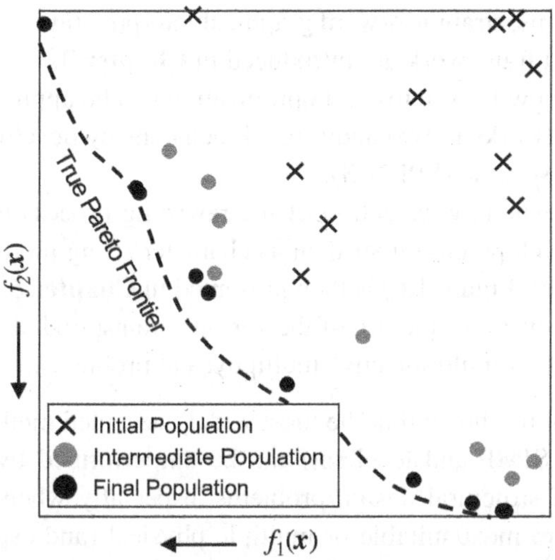

Figure 1.7 Schematic illustration of the evolution of a population of designs toward simultaneous minimization of two competing objectives f_1 and f_2. The idealized converged set of all such points accomplishing this goal (i.e., satisfying Equation 1.1) is known as a "Pareto frontier," the performance of which may not be reached by even the most evolved designs (black points).

reference to the general process for a GA previously described, designs that are on or near the current approximation of the Pareto frontier and that do not violate constraints (not shown) are retained for crossover, leading to an ever-improving population. Note that both simultaneous minimization of the components of $\boldsymbol{f}(\boldsymbol{x})$ and an even distribution of points along the resulting frontier (i.e., the maintenance of diversity) are both goals of many multiobjective GAs, including the widely used NSGA-II algorithm (Deb 2001). In general, as the true Pareto frontier is unknown, a genetic optimization process will terminate before its final population performs at this highest possible level, as shown in Figure 1.7.

1.4 Summary of This Book

The remainder of this book seeks to introduce and describe how various algorithmic schemes can be used to implement an EvoDevo approach for the design of multifunctional structures and other networks with the advantages of compact representation and complete general applicability across system physics and number of functions or objectives considered.

- To provide a comprehensive introduction to existing topological optimization approaches, **Chapter 2** first demonstrates the formulations and features of many of the options listed in Figure 1.3 in addition to others;

- The concepts of L-systems, straightforward graphical interpretation, and GA-based optimization over such a framework are introduced in **Chapter 3**;
- **Chapter 4** then shows how L-systems and optimization can be applied to produce much more effective networks in two and three dimensions by developing a novel graphical interpretation scheme (SPIDRS);
- A distinctive approach whereby execution of the rewriting process is manifested directly as graphical development rather than as character string modification (the arrangement L-system, cf. Figure 1.6) is then presented in **Chapter 5**;
- Finally, **Chapter 6** demonstrates the use of the various bioinspired algorithms presented in solving a range of multiobjective multiphysical problems.

Most importantly, it will be shown that the most widely accepted methods of topological optimization (i.e., SIMP and level set) can be outperformed by bioinspired approaches for a range of structural design problems, especially when low area or volume fraction designs are most suitable or multiple physical (and especially nonstructural) responses are of interest. It will also be shown that classical sensitivity approaches, such as the GSM can also leverage bioinspired approaches as the latter are capable of generating high-performing but nonintuitive ground structure configurations.

2 Review of Topology Optimization Methods

In Chapter 1 we introduced concepts of structural design, with emphasis on topological representations aligned with the approaches presented in this book. Historical approaches to the problem of topological optimization were discussed in Section 1.2.2. In this chapter we provide a more comprehensive quantitative review and demonstration of topological optimization approaches being used in contemporary design.

Rigorous quantitative structural optimization techniques date back to the early 1900s, when Michell (1904*a*) developed approaches to determine the mass-optimal layout of truss structures under stress constraints. Nearly 60 years later, Schmit (1960) posed this challenge as a nonlinear mathematical programming problem, applying the finite element method (FEM) to initially ill-defined domains. This was the genesis of modern structural optimization. As introduced in Chapter 1, topology optimization seeks the economical distribution of material (or, in a general network, connectivity) within some limiting domain such that some task(s) may be performed under physical and cost constraints. With the addition and subtraction of material connections, topology optimization alters structural/network boundaries, external and internal, giving it the freedom to eliminate or create voids within a domain.

Spurred by developments in additive manufacturing, there is currently intense research in topology optimization across a range of methods. Reviewing the full literature on the topic is well outside the scope of this book. Instead, in this chapter we review the three most popular topology optimization methods (i.e., solid isotropic material with penalization (SIMP), ground structure methods, and the level set method), in addition to other current and emerging alternative methods. To select among the representatives of these methods is itself a challenge. For didactic reasons, we chose educational versions of the selected methods that would come with a program readily available for the student to learn and test the method. We believe that quickly trying these methods in practical engineering problems is the most effective way to learn topology optimization and to motivate its study. Therefore, we have selected specific tools that have been clearly described by the educational journal articles devoted to topology optimization (Wang, Zhao, Zhou, Sigmund & Zhang 2021), with a particular interest in options for which functional codes have been provided.

For purposes of clarity and brevity, this chapter focuses on structural mechanics, and specifically on minimization of compliance under volume constraints. This is sufficient to provide an academic review of all approaches considered. Consideration of nonlinear effects and multiobjective structural and network optimization will comprise the remainder of this book. An excellent and more in-depth review of archival articles and educational programs implementing each method discussed in this chapter is available elsewhere in the literature (Wang et al. 2021).

To summarize, the overviews and demonstrations in this chapter are structured in a manner identical to that of Chapter 1 (cf. Figure 1.3) and thus will be organized as follows:

Pixel/voxel representation (Section 2.2)
- Established approach: solid isotropic material with penalization (SIMP) (Section 2.2.1)
- Newer approaches: evolutionary structural optimization (ESO) (Section 2.2.2); homogenization (Section 2.2.3); and Lamarckian genetic algorithm/SIMP hybrid method (Section 2.2.4)

Ground structure-based representation (Section 2.3)
- Established approach: traditional ground structure method (GSM) (Section 2.3.2)
- New approach: simultaneous optimization of truss/frame geometry and topology (Section 2.3.3)

Boundary representation (Section 2.4)
- Established approach: level set method (Section 2.4.1)

Emerging methods (Section 2.5)

For the purposes of brevity, essential mathematical preliminaries are collected in Appendix A. The reader is encouraged to review this content prior to proceeding to ensure efficient comprehension of the developments to follow.

2.1 The Topology Optimization Problem

To begin, let $\mathcal{B}$ denote a fixed domain in $\mathbb{R}^2$ or $\mathbb{R}^3$, here called the **design domain** and illustrated in Figure 2.1. This overall region is comprised of a continuously connected material domain $\mathcal{B}_m$ and potentially some number of void regions, which together comprise $\mathcal{B}_v$.[1] As $\mathcal{B}_m$ represents a solid body, it is capable of transferring loads from some point(s) of application to some point(s) of fixity (cf. Section 1.2.1). That portion of the boundary $\partial\mathcal{B}$ over which tractions $\mathbf{t}$ are known is denoted $\partial\mathcal{B}_N$ (i.e., the von Neumann boundary conditions), while that portion of the boundary over which displacements $\mathbf{u}$ are fixed is denoted $\partial\mathcal{B}_D$ (i.e., the Dirichlet boundary condition). Topological optimization is then the process of determining the configurations of $\mathcal{B}_m$ and its associated free boundaries that best satisfy some objective(s) (e.g., minimum compliance) under some constraint(s) (e.g., desired mass or volume), given some distribution of

[1] Due to the load-carrying requirement of the structural body as considered here, it is necessary that $\mathcal{B}_m$ be continuously connected; no such restriction exists for $\mathcal{B}_v$.

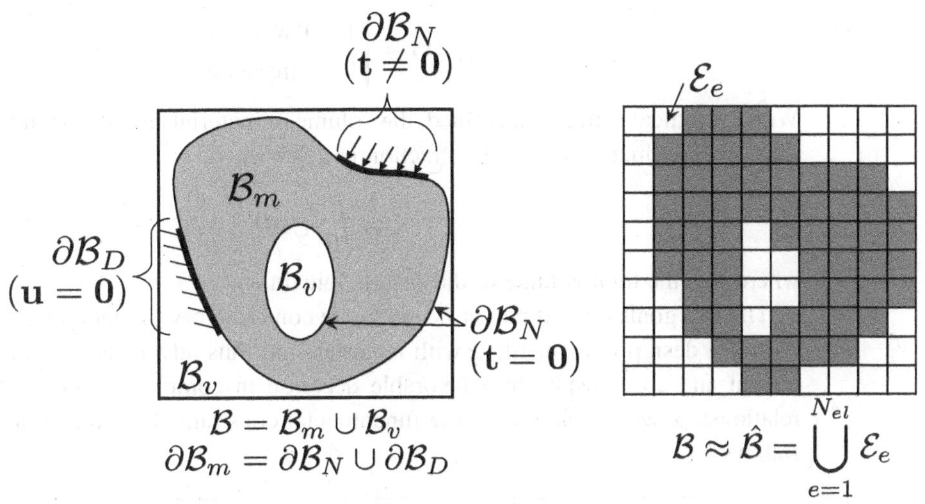

Figure 2.1 Abstract representation of the topological design domain $\mathcal{B}$, its decomposition into material regions $\mathcal{B}_m$ (gray) and void regions $\mathcal{B}_v$ (white), and its possible discretization into N_{el} element subdomains $\mathcal{E}_e$ for the purpose of optimization using pixel/voxel methods.

force-like boundary conditions $\mathbf{t}(\boldsymbol{x}) \neq \mathbf{0}$ for some $\boldsymbol{x} \in \partial\mathcal{B}_N$ and known points of fixity $\mathbf{u} = \mathbf{0}\ \forall\, \boldsymbol{x} \in \partial\mathcal{B}_D$.

While the goals of topological optimization may vary greatly throughout the practice of engineering design, the traditional approaches reviewed throughout this chapter have focused on the minimization of compliance. In this context, the compliance can be quantified by the action of all externally applied forces acting over the displacements $\mathbf{u}(\boldsymbol{x})$ that they cause. Assuming the possible existence of body forces $\mathbf{b}(\boldsymbol{x})$ throughout the body $\mathcal{B}$ acting in addition to the tractions $\mathbf{t}(\boldsymbol{x})$ distributed over portions of its surface, the structural compliance functional J_c is calculated using the approach of (Bendsøe and Sigmund 2003), and it monotonically increases as the deflection throughout a loaded structure increases such that

$$J_c\,(\mathbf{u}(\boldsymbol{x})) = \int_{\mathcal{B}} \mathbf{b}(\boldsymbol{x}) \cdot \mathbf{u}(\boldsymbol{x}) + \int_{\partial\mathcal{B}_N} \mathbf{t} \cdot \mathbf{u}. \tag{2.1}$$

Given the goals of topology optimization as previously stated, this functional provides a measure for quantified minimization. Obviously, the *least* compliant structures will result when the *most* material is placed within a given domain, but the goals of structural design (cf. Figure 1.2) most often include the search for an effective distribution of minimal material for the purposes of reduced cost and mass, among other advantages. Therefore, the search for minimum compliance must be set against the reduction of material usage relative to fully dense designs; in many algorithms including SIMP, this is accomplished via volume or mass constraints.

To measure the volume of material existing within a topological design domain $\mathcal{B}$, we introduce the indicator function $\mathbb{I}_{\mathcal{B}_m}$, denoting which regions of $\mathcal{B}$ are contained in $\mathcal{B}_m$ as per

$$\mathbb{I}_{\mathcal{B}_m}(\boldsymbol{x}) = \begin{cases} 1 & \text{if } \boldsymbol{x} \in \mathcal{B}_m, \\ 0 & \text{otherwise.} \end{cases} \tag{2.2}$$

With the indicator function defined, the volume of material can be cast in a normalized form by introducing the volume fraction v_f given as

$$v_f = \frac{1}{V} \int_{\mathcal{B}} \mathbb{I}_{\mathcal{B}_m} dV,$$

where V is the total volume of the design domain $\mathcal{B}$.

The two goals quantified via J_c and v_f are contradictory in that increasing the fraction of a design domain filled with material, and thus adversely affecting structural weight and cost, results in a favorable decrease in compliance. Given this inverse relationship, we can now pose the fundamental constrained optimization problem as follows:

Find the spatial distribution of material described by the spatial configuration of material region $\mathcal{B}_m$ and void region $\mathcal{B}_v$ that minimizes the compliance functional, J_c, subject to

- the satisfaction of equilibrium and
- $v_f = v_f^g$,

where v_f^g is a *goal* volume fraction, usually substantially less than one, that is to be maintained during the material location selection process.

As expected and observed over history (cf. Figure 1.2), solutions to the above problem will be associated with relatively complex geometric forms, where even the satisfaction of equilibrium under external loading generally cannot be computed using analytical methods applied to continuum structural representations. Rather, discrete numerical approaches such as the FEM are required to determine the meaningful distribution of deformations $\mathbf{u}(\boldsymbol{x})$ throughout the body. Discretization of the design domain also allows algorithmic decisions regarding material placement and removal to be made at a finite number of locations. This chapter will consider multiple FEM formulations as appropriate, from the continuum discretization used in the pixel/voxel or level set methods to the structural (truss/frame) FEM used in the ground structure approaches.

2.2 Pixel/Voxel Methods

The defining traits of the pixel/voxel methods are as follows: (i) they all approximate the design domain $\mathcal{B}$ (Figure 2.1) as $\hat{\mathcal{B}}$, which is a collection of regularly shaped continuum *elements* $\mathcal{E}_e$ for $e = 1 \ldots N_{el}$ and (ii) the collective assignment of material or void across all elements comprises the full description of the structural geometry (cf. also Figure 1.3). Effective analysis and design requires that $\mathcal{B} \approx \hat{\mathcal{B}}$ to within sufficient engineering precision such that

- satisfaction of equilibrium and the calculation of J_c can be accomplished via FEM to sufficient accuracy and
- geometric features associated with the distribution of material throughout the domain are accurately represented for the purposes of continued engineering development (e.g., fabrication). See also Figure 1.1.

A discretized domain approximation is illustrated in Figure 2.1, where we note that $\hat{\mathcal{B}} = \hat{\mathcal{B}}_m \cup \hat{\mathcal{B}}_v$. Clearly, the quality of the representation of the original $\mathcal{B}$, the accuracy of the equilibrium structural solution via FEM, and the effectiveness of any pixel/voxel topology optimization procedure all depend critically on the resolution of the discretization.

2.2.1 Solid Isotropic Material with Penalization (SIMP)

The SIMP (Bendsøe & Sigmund 2003) method is currently the most widely adopted topology optimization method in industry, with several commercially available software tools implementing some form of the algorithm (Schramm & Zhou 2006). Extensions of this method have been successfully applied to an array of problems in different disciplines and under different constraints. In this section, we briefly review the SIMP method as applied to the minimization of structural compliance under a volume constraint. For details on the method and other applications, the interested reader is directed to the seminal book on SIMP (Bendsøe & Sigmund 2003).

Given the general compliance minimization problem statement of Section 2.1 and the discretization shown in Figure 2.1, the remainder of SIMP development and implementation steps proceed by selecting or assuming discretized forms for compliance and volume fraction. Instead of the continuous field $\mathbf{u}(\boldsymbol{x})$, the fundamental solution entity becomes the vector of discrete displacements $\hat{\mathbf{u}}$ as calculated at each node (i.e., discrete points connecting elements, such as the corners connecting adjacent nodes in Figure 2.1). This vector is computed using common FEM approaches (Chandrupatla & Belegundu 2021). In the linear case considered here, a global matrix of constant stiffness terms $\hat{\mathbf{K}} = \sum_{e=1}^{N_{el}} \hat{\mathbf{K}}_e$ is formed based on the material and geometric properties of each element $\mathcal{E}_e$ and the distribution of elements throughout the void regions ($\mathcal{E}_e \in \hat{\mathcal{B}}_v$) and material regions ($\mathcal{E}_e \in \hat{\mathcal{B}}_m$). A vector of forces $\hat{\mathbf{f}}$ is likewise formed based on discretization of distributed body forces $\mathbf{b}(\boldsymbol{x})$ and surface tractions $\mathbf{t}(\boldsymbol{x})$. The discrete approximations for displacement throughout the entirety of $\hat{\mathcal{B}}$ are then found by solving the following system of linear equations:

$$\hat{\mathbf{K}}\hat{\mathbf{u}} = \hat{\mathbf{f}}. \tag{2.3}$$

The metric for compliance (Equation 2.1) is cast in a discrete form as

$$\hat{J}_c = \hat{\mathbf{f}}^T \hat{\mathbf{u}}. \tag{2.4}$$

Substituting Equation 2.3, which represents the satisfaction of equilibrium, into Equation 2.4 gives

$$\hat{J}_c = \hat{\mathbf{u}}^T \hat{\mathbf{K}} \hat{\mathbf{u}} = \sum_{e=1}^{N_{el}} \hat{\mathbf{u}}_e^T \hat{\mathbf{K}}_e \hat{\mathbf{u}}_e, \tag{2.5}$$

which computes the total compliance as the sum of the contributions from each element $\mathcal{E}_e$, where $\hat{\mathbf{u}}_e$ is the vector of deformations associated *only* with a given element.

To account for the fact that elements (pixels/voxels) comprise both the material and the void regions, we introduce a discrete form of the indicator function (cf. Equation 2.2), written as

$$\mathbb{I}^e_{\hat{\mathcal{B}}_m} = \begin{cases} 1 & \text{if } \mathcal{E}_e \in \hat{\mathcal{B}}_m, \\ 0 & \text{otherwise.} \end{cases}$$

This allows the local element stiffness matrix to be written as

$$\hat{\mathbf{K}}_e = \mathbb{I}^e_{\hat{\mathcal{B}}_m} \hat{\mathbf{K}}^0_e, \tag{2.6}$$

where $\hat{\mathbf{K}}^0_e$ is the FEM-derived stiffness of any $\mathcal{E}_e \in \hat{\mathcal{B}}_m$. The discretized approximation of the volume fraction can similarly be written as

$$v_f \approx \hat{v}_f = \frac{1}{V} \sum_{e=1}^{N_{el}} \mathbb{I}^e_{\hat{\mathcal{B}}_m} V_e,$$

where V_e represents the local volume of element $\mathcal{E}_e$.

With these approximations, the previous topology optimization problem can be rewritten in a more rigorous form.

Find the particular set of elements $\mathcal{E}_e \in \hat{\mathcal{B}}_m$ that minimizes the compliance functional, $\hat{J}_c$, subject to

- $\hat{\mathbf{K}}_e = \mathbb{I}^e_{\hat{\mathcal{B}}_m} \hat{\mathbf{K}}^0_e$,
- the satisfaction of $\left(\sum_{e=1}^{N_{el}} \hat{\mathbf{K}}_e \right) \hat{\mathbf{u}} = \hat{\mathbf{K}} \hat{\mathbf{u}} = \hat{\mathbf{f}}$, and
- $\hat{v}_f = 1/V \sum_{e=1}^{N_{el}} \mathbb{I}^e_{\hat{\mathcal{B}}_m} V_e = v_f^g$.

In this way the fundamental optimization problem then becomes an integer programming (IP) problem that, though posed in a straightforward manner, is notoriously more complicated and expensive to solve than a comparable continuous problem (Fletcher 1987). To eliminate this obstacle, the SIMP approach replaces the binary $\mathbb{I}^e_{\hat{\mathcal{B}}_m}$ with a *continuous* element-wise density variable $0 \leq \rho_e \leq 1$. This relaxes the binary nature of the topology optimization problem and the local element stiffness matrix is now written as

$$\hat{\mathbf{K}}_e = \rho_e \hat{\mathbf{K}}^0_e, \tag{2.7}$$

while the volume fraction becomes

$$\hat{v}_f = \frac{1}{V}\sum_{e=1}^{N_{el}} \rho_e V_e.$$

As conventionally implemented, pixel/voxel approaches employ elements of equal size and shape to prevent biased topological searches. This simplifies the volume fraction calculation to

$$\hat{v}_f = \frac{1}{N_{el}}\sum_{e=1}^{N_{el}} \rho_e. \tag{2.8}$$

However, except in very specialized cases, the existence of material with $0 < \rho_e < 1$ (i.e., "gray" material) is physically meaningless;[2] a local pixel/voxel (element $\mathcal{E}_e$) either *is* ($\rho_e = 1$) or *is not* ($\rho_e = 0$) composed of material. Posing the previous optimization problem while allowing a continuously variable ρ_e is easily shown to result in such nonphysical solutions as changes in structural compliance and total volume are linearly and oppositely dependent on this variable as per Equations 2.7 and 2.8, respectively. To mitigate this and drive the optimization process toward more physically representative (i.e., binary) solutions, SIMP introduces a *penalization* of the continuous density in the definition of the local elemental stiffness such that

$$\hat{\mathbf{K}}_e = (\rho_e)^p \hat{\mathbf{K}}_e^0, \tag{2.9}$$

where $p > 1$. This constant is chosen such that, as local material density decreases, ρ_e provides a nonlinearly diminishing contribution to the structural stiffness while its linear contribution to mass/volume is maintained. Indeed, for instance, for $p = 3$, an element with density $\rho_e = 0.1$ will have 10% of the mass of material added to the weight of the element, but only an increment of $0.001\hat{\mathbf{K}}_e^0$ to its element stiffness. Therefore, in design problems where there is a conflict between mass and stiffness, the optimizer will tend to avoid partial fractions for ρ_e since the ratio of increased stiffness per mass of the element will decrease with p when $\rho_e < 1$. This has the clearly observed effect of producing more distinct void/material regions as p is increased, even as numerical convergence difficulties eventually begin to preclude topological optimization altogether. In fact, in the limit as $p \to \infty$, the condition of Equation 2.6 is recovered, and the integer optimization problem again becomes unsolvable. Thus, in practice, a value such as $p = 3$ is often used as it is effective in eliminating gray material without preventing convergence to an optimal topology.[3]

It is critical to note that, in practice, the FEM becomes unstable if there exist any nodes exclusively associated with elements having $\hat{\mathbf{K}}_e = \mathbf{0}$ as when $\rho_e = 0$. In

[2] In (Bendsøe & Sigmund 2003) conditions under which gray material can be considered to represent a microstructure are investigated. When optimizing with preselected materials, as is common practice, the gray material is meaningless; however, if multiscale optimization is the intent there are more effective and efficient methodologies to use (Wu, Sigmund & Groen 2021).

[3] A value of $p = 3$ or similar will result in some limited gray materials being present in the domain, particularly at void boundaries. Such nonphysical material is usually removed from the final solution by a method of postprocessing logic (Bendsøe & Sigmund 2003) or a continuation of filtering, as addressed below.

isotropic materials, the components of $\hat{\mathbf{K}}_e$ are functions only of the Young's modulus E^0 and Poisson's ratio ν^0 of the dense material in addition to the local density. To prevent the numerical problems associated with empty elements while preserving the fidelity of the approach (i.e., $\rho_e \rightarrow 0$), a minimum Young's modulus $E_{min} > 0$ is usually introduced such that a lower limit for $\hat{\mathbf{K}}_e(E^0, \nu^0)$ is calculated, defined as $\hat{\mathbf{K}}_{min}(E_{min}, \nu^0)$. With this small stabilizing change, Equation 2.9 is rewritten as

$$\hat{\mathbf{K}}_e = \hat{\mathbf{K}}_{min} + (\rho_e)^p(\hat{\mathbf{K}}_e^0 - \hat{\mathbf{K}}_{min}), \tag{2.10}$$

and the final topology optimization problem solved by SIMP can then be stated as follows:

Find the vector of continuous local densities $\boldsymbol{\rho} = \{\rho_1 \ldots \rho_e \ldots \rho_{N_{el}}\}$ that minimizes the compliance functional

$$\hat{J}_c = \sum_{e=1}^{N_{el}} (\rho_e)^p \hat{\mathbf{u}}_e^T \hat{\mathbf{K}}_e^0 \hat{\mathbf{u}}_e,$$

subject to

- $0 \leq \rho_e \leq 1$,
- the satisfaction of $\sum_{e=1}^{N_{el}} \left(\hat{\mathbf{K}}_{min} + (\rho_e)^p(\hat{\mathbf{K}}_e^0 - \hat{\mathbf{K}}_{min}) \right) \hat{\mathbf{u}} = \hat{\mathbf{K}}\hat{\mathbf{u}} = \hat{\mathbf{f}}$,

  and
- $\hat{v}_f = \dfrac{1}{N_{el}} \sum_{e=1}^{N_{el}} \rho_e$.

This underlying optimization problem can be solved in a number of ways, where algorithms must be specifically chosen to address the very high number of components in the vector of design variables $\boldsymbol{\rho}$ (e.g., $10^3 < N_{el} < 10^5$ often). The development of the SIMP approach proceeds by converting this constrained optimization problem to an unconstrained one. Assuming (i) $\hat{\mathbf{K}}\hat{\mathbf{u}} = \hat{\mathbf{f}}$ can be identically satisfied for all $\boldsymbol{\rho}$, (ii) its solution is used to provide the displacement vector $\hat{\mathbf{u}}$, and (iii) only $0 \leq \rho_e \leq 1$ are considered, we can write the following Lagrangian function:

$$\mathcal{L}(\boldsymbol{\rho}) = \hat{J}_c(\boldsymbol{\rho}) + \lambda(\hat{v}_f - v_f^g), \tag{2.11}$$

which must be minimized, where λ is a Lagrange multiplier for maintaining the equality constraint on volume fraction.

Historically, the simplest way to minimize the total Lagrangian of Equation 2.11, especially given the substantial length of $\boldsymbol{\rho}$, is to introduce a heuristic approach based on its particular form (Bendsøe & Sigmund 2003). The goal is to iteratively reduce its value toward a minimum by cycling through all elements $\mathcal{E}_e$ ($e = 1 \ldots N_{el}$), updating each ρ_e per

$$\rho_e^{(k+1)} = \rho_e^{(k)} B_e^{\eta}, \tag{2.12}$$

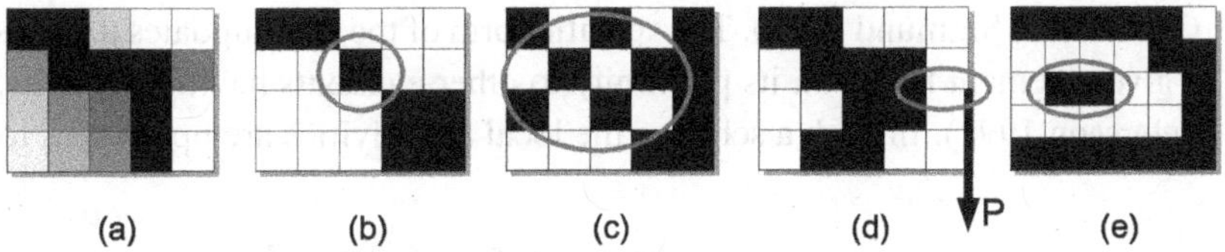

Figure 2.2 Issues with pixel/voxel representations and their use in methods such as SIMP: (a) "gray" material, (b) corner connection, (c) checkerboard pattern, (d) load disconnect, and (e) island sub-like structures. Reprinted from Pedro and Kobayashi (2011) with permission.

where k tracks the current iteration number, η is a numerical dampening exponent (often $\eta = 0.5$), and

$$B_e = -\frac{d\hat{J}_c}{d\rho_e}\left[\lambda\frac{d\hat{v}_f}{d\rho_e}\right]^{-1} = -\frac{d\hat{J}_c}{d\rho_e}\left[\lambda\frac{V_e}{V}\right]^{-1}. \tag{2.13}$$

According to Equations 2.5 and 2.10,

$$\frac{d\hat{J}_c}{d\rho_e} = -p(\rho_e)^{p-1}\hat{\mathbf{u}}_e^T\hat{\mathbf{K}}_e^0\hat{\mathbf{u}}_e. \tag{2.14}$$

Note that across all N_{el} evaluations of Equation 2.12, the value of a single common unknown λ must be provided. In the most common implementations of SIMP, $\lambda_{min} \leq \lambda \leq \lambda_{max}$ is found via a bisection method whereby various values are iteratively explored in an inner loop such that, given all N_{el} updates of Equation 2.12, the resulting volume fraction constraint $v_f = v_f^g$ is approximately satisfied. This will be clarified in the algorithm summary to follow.

Finally, one must consider the bounds on the heuristic update. Specifically, no update is permitted that

- violates $0 \leq \rho_e^{(k+1)} \leq 1$ or
- results in too large a step such that $\left|\rho_e^{(k+1)} - \rho_e^{(k)}\right| > m$,

where m is an algorithmic parameter.[4] These bounds can be combined with Equation 2.12 in a compact form to give

$$\rho_e^{(k+1)} = \begin{cases} \max(0, \rho_e^{(k)} - m) & \text{if } \rho_e^{(k)}B_e^\eta \leq \max(0, \rho_e^{(k)} - m), \\ \min(1, \rho_e^{(k)} + m) & \text{if } \rho_e^{(k)}B^\eta \geq \min(1, \rho_e^{(k)} + m), \\ \rho_e^{(k)}B_e^\eta & \text{otherwise.} \end{cases} \tag{2.15}$$

It is important to mention that pixel/voxel methods, and SIMP in particular, suffer from a number of disadvantages that must be algorithmically addressed. Some of these are illustrated in Figure 2.2. Gray material can be eliminated through proper use of the penalty factor p and additional postprocessing (Bendsøe & Sigmund 2003), but poor corner connections and "checkerboard patterns" can also result. The latter two results from artificially high numerical stiffness in the underlying optimization problem. Fortunately, a simple heuristic approach has been shown to mitigate this issue via a distance filter applied to the sensitivities used to update the design variables

[4] A value of $m = 0.2$ has shown to be effective in balancing algorithmic convergence and efficiency.

(Bendsøe & Sigmund 2003). The specific form of the filter updates the sensitivities for a given element based on its proximity to other elements having $\rho_e \approx 1$ (Sigmund & Petersson 1998). In such a scheme, the local sensitivities are updated as follows:

$$\frac{d\hat{J}_c}{d\rho_e} \leftarrow \sum_{i=1}^{N_{el}} \rho_i \omega_i \frac{d\hat{J}_c}{d\rho_i} \left[\rho_e \sum_{i=1}^{N_{el}} \omega_i \right]^{-1}, \tag{2.16}$$

where ω_i is a weighting factor for the ith element, which is given a simple form that linearly decreases as the distance from the element in process increases and falls to zero after a distance r_{min} is reached. Specifically,

$$\omega_i = \begin{cases} r_{min} - d(e,i) & \text{if } d(e,i) \leq r_{min}, \\ 0 & \text{otherwise,} \end{cases} \tag{2.17}$$

where $d(e,i)$ is the Euclidean distance between the centroids of elements e and i.

The basic SIMP process as introduced herein is summarized in Algorithm 1.

Algorithm 1 Basic algorithm for implementation of SIMP topology optimization

Require: A dense design domain $\hat{\mathcal{B}} = \bigcup_{e=1}^{N_{el}} \mathcal{E}_e$ and associated boundary conditions, the dense element stiffness matrices $\hat{\mathbf{K}}_e^0 \ \forall\ e = 1 \ldots N_{el}$ as computed from E^0 and ν^0, the minimum Young's modulus E_{min} to maintain numerical stability, the algorithmic parameters η, m, p, and v_f^g, the Lagrange multiplier bounds λ_{min} and λ_{max}, and algorithmic tolerances $\texttt{tol}_\rho$ and $\texttt{tol}_\lambda$

Initialize $k = 0$, $\rho^{(0)} = v_f^g$, $\Delta\rho_{max} = 10^8$
while $\Delta\rho_{max} > \texttt{tol}_\rho$ **do** ▷ Begin optimization loop
    $k \leftarrow k + 1$
    Compute $\hat{\mathbf{K}} = \sum_{e=1}^{N_{el}} \hat{\mathbf{K}}_{min} + (\rho_e)^p (\hat{\mathbf{K}}_e^0 - \hat{\mathbf{K}}_{min})$
    Solve $\hat{\mathbf{K}}\hat{\mathbf{u}} = \hat{\mathbf{f}}$
    Compute $d\hat{J}_c/d\rho_e$ per Equation 2.14
    (Optional: Apply sensitivity filtering to update $d\hat{J}_c/d\rho_e$ per Equations 2.16 and 2.17)
    Initialize $\lambda = (\lambda_{min} + \lambda_{max})/2$, $\epsilon_\lambda = 10^8$
    while $\epsilon_\lambda > \texttt{tol}_\lambda$ **do** ▷ Begin bisection to find λ
        for $e = 1 \ldots N_{el}$ **do**
            Update $\rho_e^{(k+1)}$ per Equations 2.13 and 2.15 using $d\hat{J}_c/d\rho_e$
        end for
        Update $\hat{v}_f$ per Equation 2.8
        Update λ per bisection method
        $\epsilon_\lambda = \left| \hat{v}_f - v_f^g \right|$
    end while
    $\Delta\rho_{max} = \max \left| \rho_e^{(k+1)} - \rho_e^{(k)} \right|$
end while

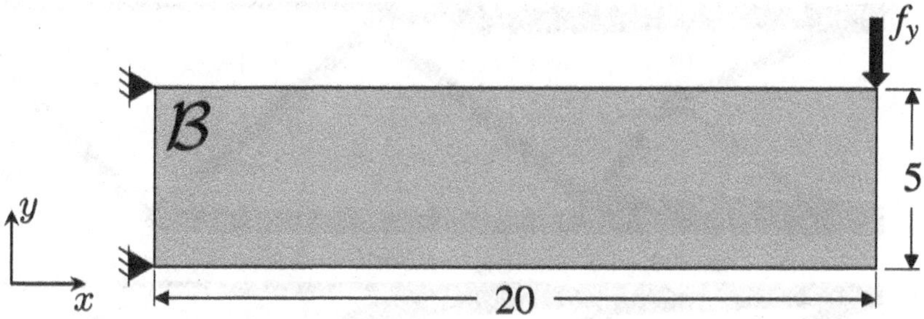

Figure 2.3 Design domain and boundary condition for the long cantilever benchmark problem.

Example 2.1 To illustrate SIMP in an application to structural mechanics, we consider the benchmark problem of optimizing a cantilever. The design domain for this problem is rectangular with width of 20 units and height of 5 units as illustrated in Figure 2.3. A vertical load of $f_y = 10$ is applied downward to the top right of the domain and zero displacement is imposed to the left side of the design domain. The Young's modulus of the material is $E^0 = 10^5$, $E_{min} = 10^{-9}$, and the Poisson ratio $\nu^0 = 0.3$. The computations shown below were carried out using the educational code described in Clausen et al. (2011) using a tolerance of 5×10^{-3} for the forward error (i.e., the change in the design variables) and a goal volume fraction of $v_f^g = 0.3$.

Figure 2.4 shows the resulting optimized layout for two mesh resolutions, a lower resolution of 800×200 cells and a higher resolution of 1660×400. As expected, the higher resolution mesh allows the development of slender bars and more complexity in the optimized solution. The resulting effect of the mesh refinement is a decrease of 3.8% in the compliance at the cost of an increase of 65.6% in the number of function evaluations needed for convergence. This corresponds to larger impact on the total optimization time since each function evaluation of the more refined mesh is more expensive than the coarser alternative.

2.2.2 Evolutionary Structural Optimization (ESO)

The SIMP algorithm is based on a rigorously formulated optimization problem where the set of design variables represents the local density of each element as a continuous number that tends toward 0 or 1 by virtue of the penalty method applied. However, the solution of the compliance minimization problem is, in the end, accomplished via a heuristic update over the very large set of continuous design variables. A popular alternative approach to pixel/voxel-based topology optimization algorithms is the set of "hard-kill" methods, which seek to solve the same fundamental compliance minimization problem addressed by SIMP, but instead address the underlying IP

(a) 800 × 200 mesh; $v_f = 0.30$; $\hat{J}_c = 0.58$; 598 iterations

(b) 1600 × 400 mesh; $v_f = 0.30$; $\hat{J}_c = 0.56$; 990 iterations

Figure 2.4 Optimized layouts for the long cantilever design problem of Example 2.1 using SIMP.

(i.e., binary elemental density) problem directly using a fully heuristic approach. The seminal work in this area (Xie & Steven 1997*b*) referred to this alternative approach as the ESO method, whereby elements are removed from $\hat{\mathcal{B}}_m$ until the satisfaction of $v_f = v_f^g$. A later variant of ESO, the so-called bi-directional ESO (BESO), (Querin et al. 2000) allowed new elements to be added as others were removed if the new elements were expected to decrease total structural compliance. These methods have become popular due to both their simplicity and intuitive appeal and, as well as the many successful examples for which the ESO-generated optimized solutions were shown to be comparable to known benchmarks solutions (Xie & Steven 1993, Chu, Xie, Hira & Steven 1996, Chu, Xie & Steven 1998). We will likewise show that for the particular example problem used in this chapter, a hard-kill approach generates a less compliant structure in less computational iterations when compared to SIMP.

While there exist many variations of hard-kill method in the literature, we briefly describe the open and educational version of the so-called sequential element rejection and admission (SERA) topology optimization approach as provided in the literature (Loyola, Querin, Jiménez & Gordoa 2018). Further details regarding this and other hard-kill approaches can be found in this previous guiding work and elsewhere (Rozvany 2009).

To begin this overview of SERA, we restate the IP problem of Section 2.2.1 with an alternative approach to ensuring numerical stability of the FEM problem: propose a minimum density $\rho_{min} > 0$ instead of a minimum Young's modulus. Noting that $\hat{J}_c$ is given in Equation 2.5 and again assuming all elements are of equal size, we have

---

Find the vector of continuous local densities

$$\rho = [\ \rho_1 \quad \ldots \quad \rho_e \quad \ldots \quad \rho_{N_{el}}\ ]$$

that minimizes the compliance functional

$$\hat{J}_c = \sum_{e=1}^{N_{el}} \rho_e \hat{\mathbf{u}}_e^T \hat{\mathbf{K}}_e^0 \hat{\mathbf{u}}_e$$

subject to:

- $0 < \rho_{min} \leq \rho_e \leq 1$,
- the satisfaction of $\left(\sum_{e=1}^{N_{el}} \rho_e \hat{\mathbf{K}}_e^0\right) \hat{\mathbf{u}} = \hat{\mathbf{K}}\hat{\mathbf{u}} = \hat{\mathbf{f}}$, and
- $\hat{v}_f = \dfrac{1}{N_{el}} \sum_{e=1}^{N_{el}} \rho_e$.

Note that $\rho_{min} \approx 0$ and is selected to be only large enough to allow the solution of $\hat{\mathbf{K}}\hat{\mathbf{u}} = \hat{\mathbf{f}}$ without numerical issue. The heuristic approach for solving this minimization problem then relies on the iterative execution of the following three-step process:

1. Determine the volume of material in $\hat{\mathcal{B}}_m$ needing to be removed toward satisfaction of the volume fraction criteria; moving material from one region to another is also permitted once $\hat{v}_f \approx v_f^g$ has been satisfied;
2. Determine the unique contribution to $\hat{J}_c$ associated with each element potentially switching assignment from material to void (i.e., $\rho = 1 \rightarrow \rho \approx 0$) or vice versa. Sort the elements according to their contribution;
3. Add, remove, or trade the number of elements needed to satisfy (1) based on the sorted list resulting from (2).

In other words, the particular elements selected to drive toward or maintain the goal volume fraction are chosen based on their expected contribution to the compliance per (cf. Equation 2.5)

$$\hat{J}_c = \sum_{e=1}^{N_{el}} \hat{\mathbf{u}}_e^T \hat{\mathbf{K}}_e \hat{\mathbf{u}}_e = \sum_{e=1}^{N_{el}} \hat{J}_c^{\,e}. \tag{2.18}$$

In the case of the SERA algorithm, the removal and addition of material occurs in two phases. Phase I begins with the fully dense design domain and proceeds until $\hat{v}_f \approx v_f^g$. During this phase, more material is removed in each iteration than added. When the volume constraint has been reached, Phase II then considers volume-preserving trades of elemental assignments between $\hat{\mathcal{B}}_v$ and $\hat{\mathcal{B}}_m$. Specifically, for all iterations k in Phase I, the volume fraction is updated through the addition

($\rho_e = \rho_{min} \rightarrow 1$) and, to a greater extent, removal ($\rho_e = 1 \rightarrow \rho_{min}$) of numerous elements as guided by the following global calculations:

$$\hat{v}_f^{(k)} = \max\left(\hat{v}_f^{(k-1)}(1 - C_1), v_f^g\right), \tag{2.19}$$

$$\Delta\hat{v}_f^{(k)} = \left|\hat{v}_f^{(k)} - \hat{v}_f^{(k-1)}\right|, \tag{2.20}$$

$$\Delta\hat{v}_{f,add}^{(k)} = \Delta\hat{v}_f^{(k)}(C_2 - 1), \tag{2.21}$$

$$\Delta\hat{v}_{f,rem}^{(k)} = \Delta\hat{v}_f^{(k)} C_2. \tag{2.22}$$

Once $\hat{v}_f \approx \hat{v}_f^g$ to within some tolerance, Phase II proceeds as an equal number of elements are converted from material to void and vice versa. Specifically,

$$\Delta\hat{v}_{f,add}^{(k)} = \Delta\hat{v}_{f,rem}^{(k)} = C_3\hat{v}_f^g. \tag{2.23}$$

With regard to the algorithmic constants, past foundational works such as Loyola et al. (2018) have found success with values in the following ranges: $0.01 \leq C_1 \leq 0.05$, $1.2 \leq C_2 \leq 1.4$, and $0.001 \leq C_3 \leq 0.005$. Note that any value $C_2 > 1$ results in a net removal of material from the domain as desired during Phase I. Choosing values for these constants requires consideration of the trade-off between accelerating the convergence and obtaining a satisfactory optimized solution for a given problem. Larger values not only allow for larger incremental changes and may reduce the number of iterations for convergence, but may also affect negatively the performance of the optimized solution.

As previously mentioned, development of the SERA methodology proceeds by assuming a regular underlying mesh such that all elements have the same volume. Given this, the desired increases and decreases in volume for a given increment can be directly converted to a *number of elements* added (N_{add}) or removed (N_{rem}) from the material subdomain (cf. Equation 2.8). It is left only to determine which elements are converted from material to void, which from void to material, and which are left unchanged. These decisions are cleverly based on simply determining the following:

- Among all elements having $\rho_e = 1$, which would most reduce $\hat{J}_c$ if they were instead assigned $\rho_e \approx 0$?
- Among all elements having $\rho_e = \rho_{min} \approx 0$, which would most reduce $\hat{J}_c$ if they were instead assigned $\rho_e = 1$?

To approximate the sensitivity of $\hat{J}_c$ to a binary change in the density of a given element ϵ_e, the enabling assumption is made that the nodal displacements $\hat{\mathbf{u}}_e$ associated with ϵ_e being calculated through the solution of the global problem $\hat{\mathbf{K}}\hat{\mathbf{u}} = \hat{\mathbf{f}}$ (with all N_{el} elements being considered thereby) are *not* greatly altered by the removal or addition of element e alone. This allows the sensitivity of the local elemental contribution to the structural compliance with respect to the local density change $d\hat{J}_c^e/d\rho_e$ to be written (after Equation 2.18) as

$$\frac{d\hat{J}_c^e}{d\rho_e} \approx \hat{\mathbf{u}}_e^T \Delta\hat{\mathbf{K}}_e \hat{\mathbf{u}}_e.$$

Noting that the SERA approach only allows for fully dense or (nearly) fully empty elements and thus that $\Delta\hat{\mathbf{K}}_e = \hat{\mathbf{K}}_e^0$ in elements being converted from void to material and $\Delta\hat{\mathbf{K}}_e = -\hat{\mathbf{K}}_e^0$ in the opposite case, we have

$$\left.\frac{d\hat{J}_c^e}{d\rho_e}\right|_{\hat{\mathcal{B}}_v} \approx \hat{\mathbf{u}}_e^T \hat{\mathbf{K}}_e^0 \hat{\mathbf{u}}_e, \tag{2.24}$$

for elements e in the void region, and

$$\left.\frac{d\hat{J}_c^e}{d\rho_e}\right|_{\hat{\mathcal{B}}_m} \approx -\hat{\mathbf{u}}_e^T \hat{\mathbf{K}}_e^0 \hat{\mathbf{u}}_e, \tag{2.25}$$

for elements in the material region.

Once these sensitivities have been computed for each element based on its current void/material assignment, the elements in each region are sorted according to sensitivity magnitude. In this way the N_{rem} specific elements to be removed and the N_{add} specific elements to be added can be selected from the tails of these sorted lists.

An appropriate criteria for declaring convergence of this iterative approach would be the stagnation of $\hat{J}_c$ as it comes to a minimum (given satisfaction of $\hat{v}_f \approx v_f^g$). However, by virtue of its heuristic choices of Equations 2.19–2.23, the SERA algorithm absolutely dictates a change in global compliance for every iterate (k). Therefore, convergence is tested using a backward error estimate based on *accumulated* changes in the compliance over some (even) number of iterations N_{accum}. Specifically, at the end of any iteration $(k > N_{accum})$, the accumulated error $\epsilon^{(k)}$ is calculated per

$$\epsilon^{(k)} = \frac{\displaystyle\sum_{i=k-N_{accum}-1}^{k-N_{accum}/2} \hat{J}_c^{\,i} - \sum_{i=k-N_{accum}/2-1}^{k} \hat{J}_c^{\,i}}{\displaystyle\sum_{i=k-N_{accum}/2-1}^{k} \hat{J}_c^{\,k}}, \tag{2.26}$$

where we test convergence against a prescribed tolerance $\epsilon^{(k)} < \mathtt{tol}_\epsilon$. Often $N_{accum} = 20$ is used.

As with the SIMP method, this pixel/voxel-based approach is known to produce corner connections and checkerboard patterns. The same filtering captured by Equations 2.16 and 2.17 can again be used to update the computed $\hat{J}_c^{\,e}$ values such that the tendency for an element to be added/removed is tempered according to the density state of neighboring elements.

To summarize, the SERA method proceeds according to Algorithm 2.

Example 2.2 The benchmark problem for a long cantilever in Example 2.1 is now addressed using a MATLAB implementation of the SERA method (Loyola et al. 2018). The resulting optimized layout is shown in Figure 2.5 and features a lower compliance value than the SIMP solution in Figure 2.4b ($\hat{J}_c = 0.54$ versus $\hat{J}_c = 0.56$) obtained in fewer iterations (230 for SERA, 990 for SIMP).

Algorithm 2 Basic algorithm for implementation of SERA topology optimization

Require: A dense design domain $\hat{\mathcal{B}} = \bigcup_{e=1}^{N_{el}} \mathcal{E}_e$ and associated boundary conditions, the dense element stiffness matrices $\hat{\mathbf{K}}_e^0 \; \forall \; e = 1 \ldots N_{el}$, algorithmic parameters C_1, C_2, C_3, and v_f^g, the accumulated convergence range N_{accum}, and algorithmic tolerance $\mathtt{tol}_\epsilon$

1: Initialize $k = 0$, $\boldsymbol{\rho}^{(0)} = \mathbf{1}$, $\epsilon^{(0)} = 10^8$, $\hat{v}_f = 1.0$
2: **while** $\epsilon^{(k)} > \mathtt{tol}_\epsilon$ **do** ▷ Begin optimization loop
3: $\quad k \leftarrow k + 1$
4: $\quad$ Compute $\hat{\mathbf{K}} = \sum_{e=1}^{N_{el}} \rho_e \hat{\mathbf{K}}_e^0$
5: $\quad$ Solve $\hat{\mathbf{K}}\hat{\mathbf{u}} = \hat{\mathbf{f}}$
6: $\quad$ **for** $e = 1 \ldots N_{el}$ **do**
7: $\quad\quad$ Compute $d\hat{J}_c^e / d\rho_e$ per Equations 2.24 and 2.25; store in two lists based on if element e is in $\hat{\mathcal{B}}_m$ ($\rho_e^{(k-1)} = 1$) or in $\hat{\mathcal{B}}_v$ ($\rho_e^{(k-1)} = \rho_{min}$)
8: $\quad$ **end for**
9: $\quad$ (Optional: Apply sensitivity filtering to update $d\hat{J}_c^e / d\rho_e$ per Equations 2.16 and 2.17)
10: $\quad$ **if** $\hat{v}_f > v_f^g$ **then** ▷ Phase I volume update
11: $\quad\quad$ Compute volume update requirements per Equations 2.19–2.22
12: $\quad$ **else** ▷ Phase II volume update
13: $\quad\quad$ Compute volume update requirements per Equation 2.23
14: $\quad$ **end if**
15: $\quad$ Convert update requirements from volume to number of elements
16: $\quad$ Sort each of the two lists of element $\hat{J}_c^e$ values computed in `Line 7`
17: $\quad$ Based on sorted lists and number of elements to be added/removed, select distinct element indices for which ρ_e should be updated (i.e., assigned $\rho_e = \rho_{min} \rightarrow 1$ or $\rho_e = 1 \rightarrow \rho_{min}$)
18: $\quad$ **if** $k > N_{accum}$ **then**
19: $\quad\quad$ Compute $\epsilon^{(k)}$ via Equation 2.26
20: $\quad$ **else**
21: $\quad\quad$ $\epsilon^{(k)} = 10^8$
22: $\quad$ **end if**
23: **end while**

2.2.3 Homogenization Method

To continue with the pixel/voxel approaches, it is worth considering a common observation when a SIMP or ESO/BESO method is applied to a mesh of increasing refinement. The topology does not converge toward a single configuration, but rather, an ever-increasing number and complexity of structural features and voids is observed, these voids having a diversity of orientations and aspect ratios. An example of this in the context of SIMP is shown in Figure 2.6. In the limit, such topologies could be thought of as having changed scales from *macro*structural to *micro*structural. The associated design problem transitions from one of determining the geometric nature of

Figure 2.5 Optimized layout for the long cantilever design problem of Example 2.2 (cf. Example 2.1) using the ESO/BESO (SERA) Method (1600 × 400 mesh; $v_f = 0.30$; $\hat{J}_c = 0.54$; 230 iterations).

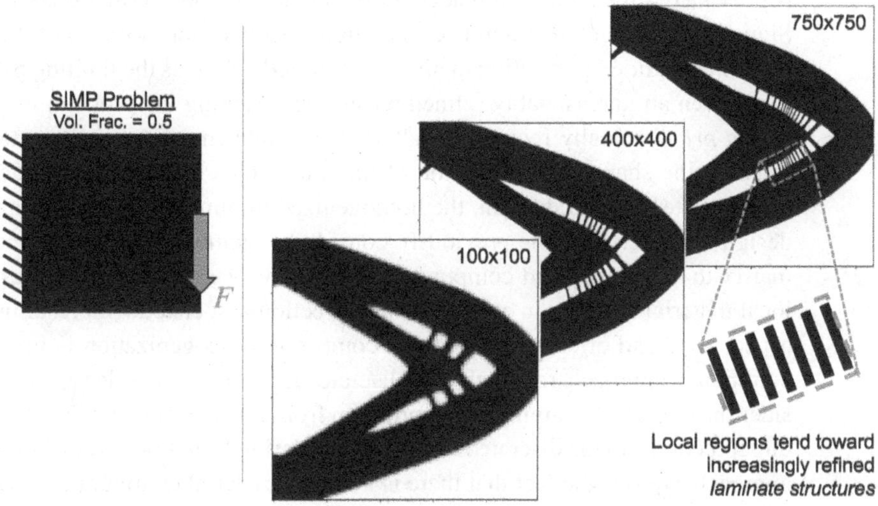

Figure 2.6 The homogenization approach to optimization is compatible with the observation that pixel/voxel-based approaches do not topologically converge with increasing mesh refinement but rather tend toward increasingly refined composite lamina, which can be locally assigned homogenized anisotropic properties.

solid/void boundaries to one of determining a spatially heterogeneous distribution of *composite* material properties, this composite consisting of microvoids in an otherwise dense matrix.

In solid mechanics, homogenization is the procedure used to compute macroscopic properties of composite materials from their microscopic constituents, considering both the local constitutive properties and geometric configurations of small domains (e.g., particles or fibers) distributed throughout a matrix (Berdichevsky 2009). Specifically, in homogenization, we consider averaged local stress tensors σ and strain tensors[5] ε. Then $\sigma = C\varepsilon$, and the primary goal is to determine C locally given the relative distribution of microfeatures distributed throughout a matrix material. This approach can also be applied to microvoids of various sizes distributed throughout a matrix,

[5] In this review, we consider only infinitesimal strains such that $\varepsilon = \frac{1}{2}[\nabla \mathbf{u} + (\nabla \mathbf{u}^T)]$.

and thus could be used to represent the kind of spatially variable mechanical response that SIMP solutions tend toward as meshes are increasingly refined (see Figure 2.6). In topology optimization, the homogenization approach is then a method for directly seeking the solution that SIMP might lead one toward given sufficient refinement. However, as an engineering solution, a load-bearing body consisting of essentially porous material generally has little merit, and heuristic postprocessing steps have been introduced to convert the matrix-void composite material back into a more meaningful structure (i.e., to avoid the gray material problem of Figure 2.2).

There are several variants of the homogenization method reported in the literature. Below, a brief description of the educational version reported in Allaire & Pantz (2006) is presented – the interested reader is directed to the books (Allaire 2002, Bendsøe & Sigmund 2003) and the references therein for details about variants of the method.

While the homogenization method may directly address the limiting SIMP solution (i.e., given an unreasonably refined mesh), approaching the problem in such a direct way *a priori* actually requires an alternative statement of the optimization problem. Because the shape and orientation of the microvoids can and do change throughout a composite material domain, the homogenized minimization problem must admit as design variables both the previously considered element densities ρ_e (now the local matrix-to-void ratio) and components of a vector or matrix that captures the relative local material stiffness in each principle direction associated with resulting from void aspect ratio and orientation. Because composite homogenization is fundamentally a continuum approach rather than a discrete one, we must at least temporarily consider the underlying minimization problem from this perspective (i.e., we do not make direct recourse to the discrete $\hat{\mathbf{K}}\hat{\mathbf{u}} = \hat{\mathbf{f}}$). Considering Equation 2.1, the homogenization approach exploits the fact that there is a known material configuration that minimizes this compliance: the *sequential laminate*.

The sequential laminate is a periodic microstructure that is composed of successive layers (laminae) of void and matrix material along orthogonal directions spanning the space of interest (i.e., 2-space or 3-space) at appropriate proportions. To minimize compliance, at all points $\boldsymbol{x}$ the laminae are oriented along the orthogonal directions of the eigenvectors of $\boldsymbol{\sigma}(\boldsymbol{x})$, denoted $\mathbf{e}_i(\boldsymbol{x})$, while the relative proportions of the two lamina are calculated from the local eigenvalues of $\boldsymbol{\sigma}(\boldsymbol{x})$, denoted $\sigma_i(\boldsymbol{x})$. Considering only two-dimensional problems in this brief review, the local laminate proportions known to minimize compliance are

$$m_1(\boldsymbol{x}) = \frac{|\sigma_2(\boldsymbol{x})|}{|\sigma_1(\boldsymbol{x})| + |\sigma_2(\boldsymbol{x})|}, \; m_2(\boldsymbol{x}) = \frac{|\sigma_1(\boldsymbol{x})|}{|\sigma_1(\boldsymbol{x})| + |\sigma_2(\boldsymbol{x})|}. \tag{2.27}$$

It has been further shown that the optimal solution will have a particular distribution of density $\rho^{opt}(\boldsymbol{x})$ that can be locally calculated from the eigenvalues of $\boldsymbol{\sigma}(\boldsymbol{x})$ in each element via

$$\rho^{opt}(\boldsymbol{x}) = \min\left[1, \sqrt{\frac{\lambda_0 + 2\mu_0}{4\mu_0(\lambda_0 + \mu_0)\Lambda}}\,(|\sigma_1(\boldsymbol{x})| + |\sigma_2(\boldsymbol{x})|)\right], \tag{2.28}$$

where λ_0 and μ_0 are the Lamé constants of the fully dense isotropic material with stiffness tensor[6] C_0, and Λ is the single global Lagrange multiplier that governs the total volume fraction as ρ^{opt} is integrated over the volume.

To sharpen the boundaries of the members in the optimized solution, a heuristic penalization is applied to the homogenization during the process of convergence, driving it back toward a spatial configuration having only $\rho(\boldsymbol{x}) = 0$ and $\rho(\boldsymbol{x}) = 1$ (e.g., dense material regions and void regions are reasonably sized at the macroscale). The instability of structural solvers considering zero-density (i.e., zero-stiffness) material is addressed by constraining $\rho_{min} \leq \rho(x) \leq 1$, as in the ESO method (Section 2.2.2). The final penalized density is then calculated via

$$\rho^{pen}(\boldsymbol{x}) = \max\left[\rho_{min}, \frac{1 - \cos\left(\pi \rho^{opt}(\boldsymbol{x})\right)}{2}\right], \tag{2.29}$$

where $\rho^{pen}(\boldsymbol{x})$ is the updated penalized density. The choice of the cosine is arbitrary, and alternatives exist (Allaire 2002).

Unfortunately, due to the mathematical complexity and rigor of the underlying derivations, it is not possible in the scope of this brief review to directly link the remaining key relations back to previously introduced governing equations (e.g., Equations 2.3 and 2.5). Rather, the reader interested in this approach is referred to the works of Allaire and Pantz (2006) and the codes they freely share. For the sake of conceptual completeness, however, we note that there is also a known solution for the inverse of the homogenized elasticity tensor $C(\boldsymbol{x})$. It can be calculated from the known orientation of the optimal laminate microstructure and the average local density of the homogenized materials (Equations 2.27 and 2.28) per[7]

$$C_{opt}^{-1} = C_0^{-1} + \frac{1 - \rho^{pen}}{\rho^{pen}} \left[m_1 \mathbf{f}_0^c(\mathbf{e}_1) + m_2 \mathbf{f}_0^c(\mathbf{e}_2)\right]^{-1}, \tag{2.30}$$

where C_0 is the elastic stiffness of a fully dense material mu_0 and $\mathbf{f}_0^c(e_i)$ is defined only in the context of its final implementation (Allaire & Pantz 2006) as

$$\mathbf{f}_0^c(\mathbf{e}_i)\xi : \xi = C_0\xi : \xi - \mu_0^{-1} C_0\xi : C_0\xi + \frac{\mu_0 + \lambda_0}{\mu_0(2\mu_0 + \lambda_0)} [(C_0\xi)\mathbf{e}_i \cdot \mathbf{e}_i]^2,$$

for any symmetric matrix ξ.

Up to this point, the introduction of the homogenization approach to topology optimization has been presented in the continuum sense, all critical equations being written for any local point $\boldsymbol{x}$. In practice, this approach is implemented as a pixel/voxel scheme, the local optimization (cf. Equations 2.28 and 2.30) being applied on an element-by-element basis in a manner compatible with Sections 2.2.1 and 2.2.2. Any local quantity $f(\boldsymbol{x})$ is then represented by its element-wise approximation via $f(\boldsymbol{x}) \approx f_e$ for

6 As per common isotropic material relations, the Lamé constants can be calculated from the more commonly used engineering constants Young's modulus E and Poisson's ratio ν via the following: $\lambda = \nu E/[(1 + \nu)(1 - 2\nu)]$ and $\mu = G = E/(2(1 + \nu))$, where G is also known as the shear modulus.

7 Note that the location argument $\boldsymbol{x}$ has been omitted for brevity and clarity of notation.

any element $\mathcal{E}_e$ centered at $\boldsymbol{x}$. The relations described above calculate the microstructural configuration and homogenized properties of the compliance-minimizing laminate for each element. They are implemented within a simple and direct iterative solution approach toward finding an optimal topography for $\mathcal{B}$. An initially fully dense material domain is assumed, and the distributed $\boldsymbol{\sigma}_e$ is computed (e.g., via a variational method or FEA). The solution is then used in Equations 2.28–2.30 where the global multiplier Λ applied to ρ_e^{opt} must be found to ensure satisfaction of $\hat{v}_f = v_f^g$, where

$$\hat{v}_f = \frac{1}{N_{el}} \sum_{e=1}^{N_{el}} \rho_e^{pen}, \tag{2.31}$$

when all elements are of the same size. Convergence is declared when the estimate of the forward error

$$\epsilon_{fwd} = \max_e \left| \rho_e^{pen(k+1)} - \rho_e^{pen(k)} \right| \tag{2.32}$$

and the estimate of the backward error

$$\epsilon_{back} = \frac{\int_{\mathcal{B}} (C_{opt}^{(k+1)})^{-1} \sigma^{(k+1)} : \sigma^{(k+1)} + \Lambda \hat{v}_f^{(k+1)}}{\int_{\mathcal{B}} (C_{opt}^{(k)})^{-1} \sigma^{(k)} : \sigma^{(k)} + \Lambda \hat{v}_f^{(k)}} \tag{2.33}$$

are both smaller than prescribed tolerances, subject also to $\hat{v}_f \approx v_f^g$. The full iterative approach is summarized in Algorithm 3, where we again represent the vector of all (penalized) element densities as $\boldsymbol{\rho}^{pen}$.

Example 2.3 For comparison with the methods previously discussed, the benchmark problem in Example 2.1 is now solved using an implementation of the homogenization method using FreeFem++ (Allaire & Pantz 2006). The material is the same as in Example 2.1 with $E = 10^5$, $\nu = 0.3$, and the parameters for the simulation are as follows: ersatz material with $E_{ersatz} = E/1000$, Lagrange $P2$ FE for the displacement field and Lagrange $P1$ FE for the stress field, a triangular mesh built from a Cartesian mesh comprising 600 × 150 elements. The results shown below were obtained after performing 400 iterations without penalization and 50 iterations with penalization. With these parameters and an initial uniform density of 0.3, an optimized solution with compliance $J_c = 0.494$ and volume fraction $v_f = 0.303$ is obtained as shown in Figure 2.7. Thus, the homogenization obtains a solution with a smaller compliance, though the final layout has a slightly higher volume fraction and requires interpretation.

2.2.4 Lamarckian Genetic Algorithm/SIMP Hybrid Method

As a final example of a pixel/voxel method, we expand upon a straightforward approach preexisting in the literature, whereby the vector of design variables representing the density of each element in $\hat{\mathcal{B}}$ is directly evolved by a genetic algorithm

Algorithm 3 Basic algorithm for implementation of topology optimization via homogenization

Require: A dense design domain $\hat{\mathcal{B}} = \bigcup_{e=1}^{N_{el}} \mathcal{E}_e$ and associated boundary conditions, the dense element elastic constants λ_0, μ_0, the minimum allowable local density ρ_{min}, the goal volume fraction v_f^g, the initial Lagrange multiplier $\Lambda = 1.0$, and algorithmic tolerances $\texttt{tol}_{fwd}$, $\texttt{tol}_{back}$, and $\texttt{tol}_{\Lambda}$

```
Initialize k = 0, ρ^{pen(0)} = 1, ε_fwd = ε_back = 10^8, C_opt = C_0(μ_0, λ_0), v̂_f = 1.0
while ε_fwd > tol_fwd AND ε_back > tol_back do        ▷ Begin optimization loop
    k ← k + 1
    Using C_opt, compute u_e and σ_e for all elements
    Compute laminate proportions m_1 and m_2 for each element per Equation 2.27
    while |v̂_f − v_f^g| > tol_Λ do                      ▷ Begin bisection to find Λ
        Compute optimal density ρ^opt per Equation 2.28
        Estimate current volume fraction v̂_f by using ρ^opt in Equation 2.31
        Update Λ per a bisection scheme
    end while
    Apply heuristic penalization to ρ^opt per Equation 2.29 to find ρ^pen
    Update optimal periodic laminate stiffness C_opt per Equation 2.30
    Compute errors for convergence ε_fwd and ε_back per Equations 2.32 and 2.33,
  respectively
end while
```

(a) Without penalization ($v_f = 0.23$; $\hat{J}_c = 0.46$)

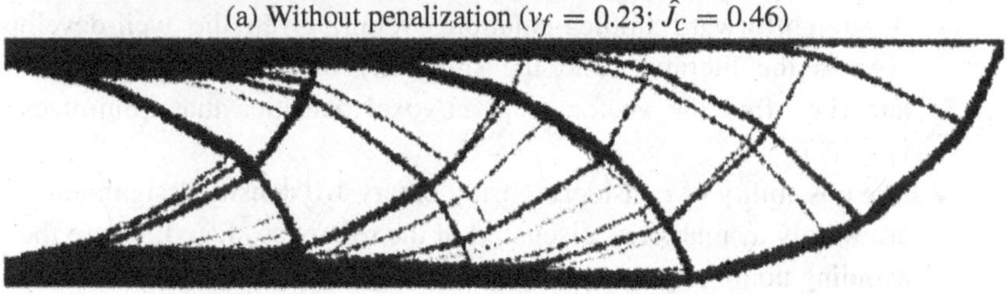

(b) With application of penalization ($v_f = 0.30$; $\hat{J}_c = 0.49$)

Figure 2.7 Nonpenalized and fully optimized layouts for the long cantilever design problem of Example 2.3 (cf. Example 2.1) using the homogenization method (600 × 150 mesh).

(GA)[8] (Chapman et al. 1994, Jakiela et al. 2000, Fanjoy & Crossley 2002, Cappello & Mancuso 2003, Madeira et al. 2005, Wang & Tai 2005*b*, Wang et al. 2006,

[8] See Section 1.3.3 for a review of GAs.

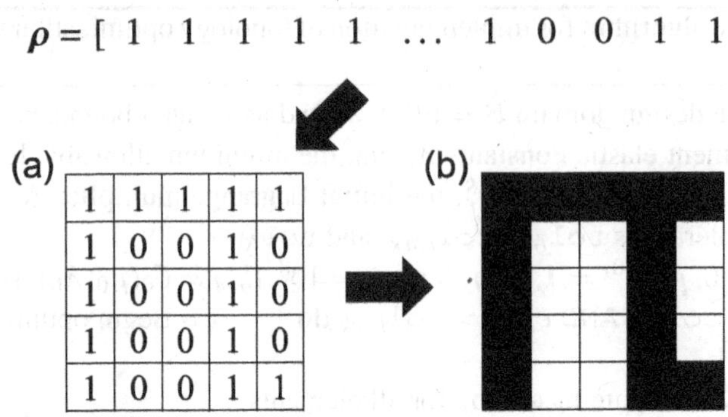

Figure 2.8 As demonstrated in the literature, GAs can directly consider one-dimensional arrays of 0/1 bits that are mapped into two dimensions (a), each entry corresponding to material and void assignments across pixels/voxels (b). However, connectivity toward structural feasibility is not generally guaranteed (cf. Figure 2.2).

Balamurugan et al. 2008, Madeira et al. 2010). With the pixel densities being directly evolved by a heuristic algorithm, these schemes need not consider intermediate densities, and a truly "bitmap" representation of topology is possible as shown in Figure 2.8. A "GA bitmap" approach can be applied to such structural representations to optimize performance while satisfying volume fraction constraints, which provides a number of advantages:

- A great diversity of designs can be considered, especially relative to the deterministic rule-based update of topology from an initial (e.g., fully dense) configuration to a final configuration that is characteristic of SIMP, SERA, and other approaches. GAs are widely known to be highly effective in avoiding local minima and finding globally optimized solutions.
- A straightforward implementation overall, given the well-developed nature of GAs in the literature and the simplicity of the underlying optimization problem (i.e., find the vector of pixel/voxel densities that minimizes $\hat{J}_c$ subject to $\hat{v}_f = v_f^g$).
- The possibility of considering truly binary 1/0 density assignments, as binary GAs are widely available, as discussed in the references listed, where the advantages of avoiding nonbinary "gray" assignments are clear (cf. 2.2(a)).

However, there is a substantial disadvantage to the direct implementation of a GA to the pixel/voxel representation. Structural feasibility is not guaranteed at any iteration of the optimization process. In particular, though gray material is disallowed, corner connections, checkerboarding, and island substructures (see Figure 2.2) are highly likely. This issue is exacerbated by the highly random nature of early generations processed by a GA, and one can imagine more elements being isolated than connected when $v_f^g < 0.5$ (Wang & Tai 2005*b*). The existence of material in regions of load application or deformation is also not guaranteed (Figure 2.2(d)).

To address these problems with connectivity, we relax the stipulation that density assignments be binary and take inspiration from past methods that have looked to the morphology of vertebrates (Tai & Chee 2000, Tai et al. 2002, Wang & Tai 2004, Tai & Akhtar 2005). They start by creating smooth paths that connect all critical points on the structure (support regions and loaded regions), which form a kind of "skeleton" for the topology. These paths are then surrounded by additional stiffening material just as flesh surrounds a skeleton in a vertebrate. In a similar way, SIMP can be used to add connectivity to any topology initially lacking this critical property. Thus, the strengths of SIMP, previously introduced in Section 2.2.1, can be used to mitigate the weaknesses of a direct GA bitmap approach and vice versa. By combining the two approaches, one can develop an algorithm that both:

- exploits the diversity naturally provided by a GA to establish a wide range of initial structures for SIMP sub-optimizations and
- leverages the effective compliance minimization abilities of SIMP, especially with filtering options activated, to iteratively reconfigure infeasible GA-selected designs toward promising topologies.

Such a hybridization of a global genetic evolution process with a local (individual-specific) sub-optimization produces what is known as a "Lamarckian GA" (Katoch, Chauhan & Kumar 2021), since it was Lamarck who proposed that the qualities of an individual were both inherited at birth and uniquely developed throughout life, and further that any new traits acquired could be passed on to offspring. As a first application of a GA to topology optimization, we present a hierarchical Lamarckian GA/SIMP hybrid method as shown in Figure 2.9. In an outer loop, the GA drives the exploration of the design space through selection, crossover, and mutation, while the iterative SIMP

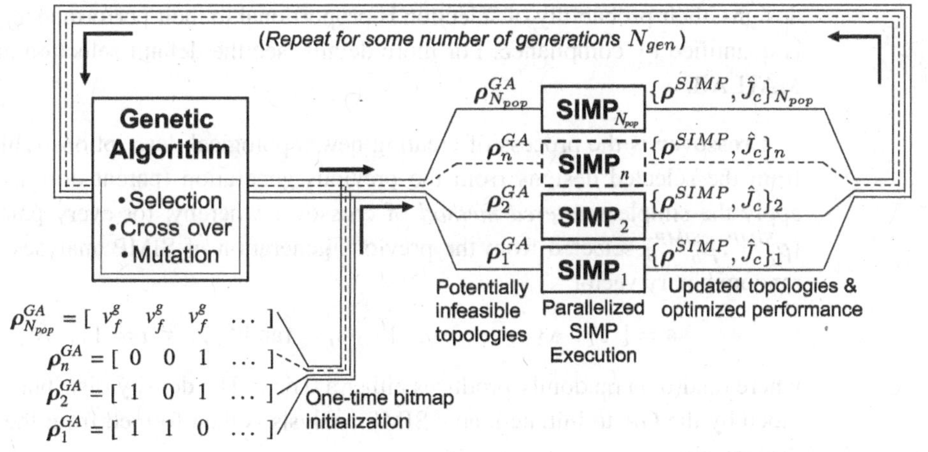

Figure 2.9 An appropriate "Lamarckian" combination of GAs with SIMP topological optimization is straightforward to implement and can leverage the strengths and mitigate the disadvantages of each while exploiting parallelization over N_{pop} designs simultaneously to reduce computational time.

method is used to greatly improve the performance of each member of a population, just as each individual member of a natural population seeks maximum benefit and minimum harm. Because SIMP and the GA use the same set of design variables (i.e., the set of all pixel/voxel densities), the results of any SIMP process can be used to update the genetic information defining each design. In this way, individualized SIMP iterations produce viable individuals (topological designs) and greatly speed up the exploitation capabilities of the overall method by searching for local optima in the region of the design space near individuals generated by the GA. The outer GA loop is repeated for some predetermined number of generations and each internal SIMP analysis is iterated until convergence (cf. Algorithm 1) or until a maximum number of iterations N_{SIMP} is reached. This approach represents the first example of an evolution-development scheme as introduced in Section 1.3.2.

Note that the process allows perfect parallelization of the SIMP analyses. Therefore, while this approach requires substantially more CPUs than previous methods to fully exercise its advantages, the actual run time is limited only by the number of cores available and the numbers of generations considered. The latter need not be high (e.g., 10 in the examples provided in this section).

There are many options for specifying the subprocesses of the global GA as employed in this hybrid scheme and illustrated in Figure 2.9 (i.e., selection, crossover, and mutation). The specific selections demonstrated here are briefly described in the following, though the reader is encouraged to explore modifications.

Selection is the process of choosing which designs from the previous generation provide the best performance and thus should be denoted as parents, their density distributions being combined to form designs for evaluation in the next generation. Here we use an elitist approach whereby some number of designs, N_{elite}, are carried forward unchanged into the next generation.[9] For the remainder of the high-performing designs, their probability of selection is proportional to their performance, which here is quantified by compliance. For more details, see the default selection algorithm in MATLAB.

Crossover is the process of creating new topological descriptions (child designs) from the selected designs from the previous generation (parent designs). Here we apply the simple *scattered method* of crossover whereby, for every pair of parents $\{\rho_a^{SIMP}, \rho_b^{SIMP}\}$ selected from the previous generation of SIMP analyses, we form a random binary vector

$$\mathbf{s} = [\ s_1 \quad s_2 \quad \dots \quad s_{N_{el}}\ ]^T, \quad s_i = \text{rand}(0,1)\ \forall\ i = 1 \dots N_{el},$$

where rand(0, 1) randomly produces either a 0 or 1. The density distribution ρ^{GA} provided by the GA to initiate a new SIMP analysis is then formed from the two parent designs via

[9] In the context of this hybrid scheme, this dictates that some high-performing designs are subjected to more total SIMP iterations as they are passed forward unaltered by the GA and thus subjected to SIMP multiple times.

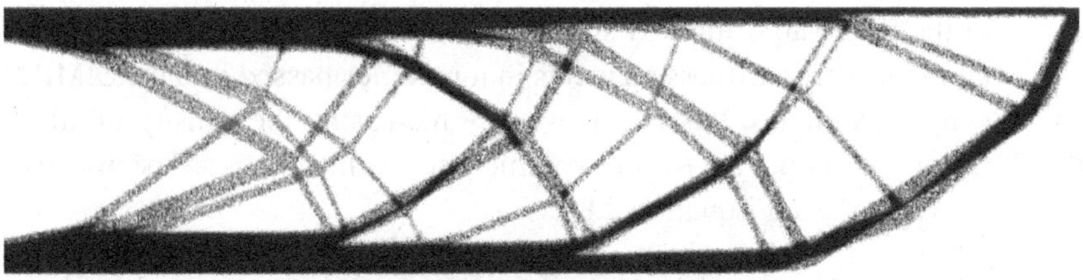

(a) Result of direct crossover

(b) Direct crossover result subjected to 25 SIMP iterations (algorithm failed)

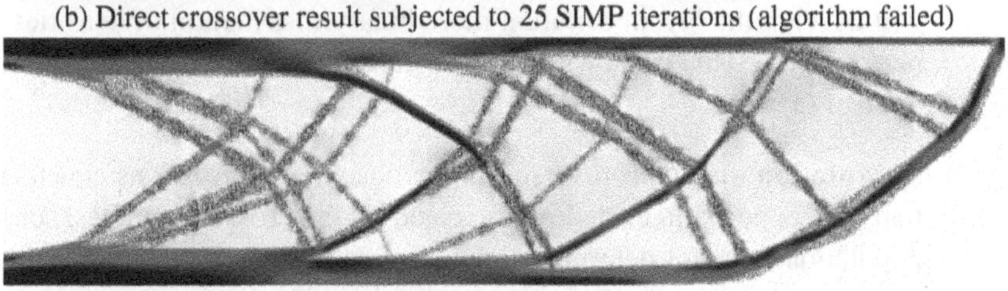

(c) Result of crossover augmented with the application of Equation 2.35 ($\Delta\rho_{stab} = 0.1$)

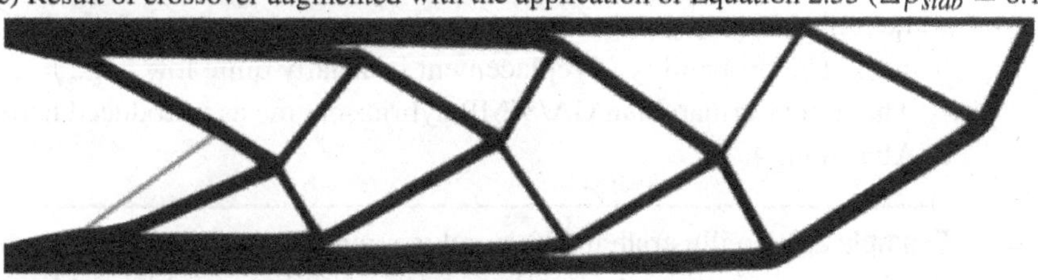

(d) Final successful result after SIMP applied to augmented crossover ($N_{SIMP} = 40$)

Figure 2.10 Structural topologies associated with possible crossover strategies within the Lamarckian GA/SIMP hybrid method. Note the importance of crossover augmentation.

$$\rho^{GA} = \mathbf{s} \odot \rho_a^{SIMP} + (\mathbf{1} - \mathbf{s}) \odot \rho_b^{SIMP}, \tag{2.34}$$

where $\mathbf{1}$ is the 1-vector and $\odot$ is the Hadamard (element-wise) product operator.[10] In this way, a gene is selected from ρ_a^{SIMP} or from ρ_b^{SIMP} if the component of $\mathbf{s}$ associated with that gene is a 1 or 0, respectively.

One complication arising from direct crossover when used to initiate subsequent SIMP processes is that the random combination of fully dense and fully empty pixels from two topologically distinct parents results in configurations with sparse and/or checkerboard-like density distributions. This is seen in Figure 2.10(a). When such a child configuration is subjected to the SIMP process, FEA calculations often fail. Figure 2.10(b) shows the topology developed from Figure 2.10(a) just before failure

[10] For any two vectors $\boldsymbol{x}$ and $\boldsymbol{y}$ of length N, $\boldsymbol{x} \odot \boldsymbol{y} = [\ x_1y_1 \quad x_2y_2 \quad \dots \quad x_Ny_N\ ]^T$.

of the SIMP algorithm. A straightforward approach for mitigating this problem is to eliminate zero-stiffness elements in topologies passed into the SIMP algorithm. In the current examples, this is achieved by increasing the density of all pixels by a small amount such that the overall volume fraction matches v_f^g. Specifically, this update can be written using Equation 2.8 as

$$\rho_e \leftarrow \min\left[(\rho_e + \Delta\rho_{stab})v_f^g\left(\frac{1}{N_{el}}\sum_{e=1}^{N_{el}}(\rho_e + \Delta\rho_{stab})\right)^{-1}, 1\right], \tag{2.35}$$

where $\Delta\rho_{stab}$ is a small stabilizing (temporary) increase in all densities and local densities greater than 1.0 are disallowed. In this section, $\Delta\rho_{stab} = 0.1$ is employed. An example of such augmentation is shown in Figure 2.10(c) (cf. Figure 2.10(a)). When such an augmented crossover topology is then used to initialize SIMP, the algorithm will develop the layout toward a more binary distribution of densities as per its usual performance. This is shown in Figure 2.10(d), which has clearly inherited some features of (a).

Mutation maintains diversity in the design configurations considered as it potentially alters genes in each design created by the crossover process, ensuring that later populations are not deterministically defined by the genes of the initial population. In this section, simple uniform mutation is used whereby any given gene in ρ^{GA} has some probability p_{mut} of being replaced by a random number flatly distributed between 0 and 1. The probability of replacement is usually quite low (e.g., $p_{mut} = 0.01$).

The basic Larmarckian GA/SIMP hybrid scheme as introduced here is summarized in Algorithm 4.

Example 2.4 To illustrate the Lamarckian genetic algorithm/SIMP hybrid method for topology optimization, the benchmark long cantilever problem of Example 2.1 is revisited. MATLAB's GA toolbox with default parameters was used together with a population of 20 individuals evolved over 10 generations. The GA options for the study are: stochastic selection algorithm, crossover fraction of 0.8, and a mutation rate $p_{mut} = 0.01$. An elite count of $N_{elite} = 2$ is also used. The initial population is random and binary, except for the last individual, which encodes the conventional starting condition for SIMP, namely, $\rho_e = v_f \; \forall \; e = 1 \ldots N_{el}$ (see Figure 2.9). The crossover stabilization parameter is set to $\Delta\rho_{stab} = 0.1$. The material properties and SIMP parameters are the same as those used in Example 2.1 and a maximum of $N_{SIMP} = 40$ iterations are performed for each individual.

Figure 2.11(a) shows the optimized solution obtained using this Lamarckian GA/SIMP hybrid method. The final compliance of $\hat{J}_c = 0.55$ is 2.7% lower (i.e., stiffer) than that obtained with the conventional SIMP (see Figure 2.4). The two optimized layouts are clearly different, with the GA/SIMP hybrid solution exhibiting more features than the comparable SIMP solution. Another layout also obtained in the last generation of the evolutionary process is shown in Figure 2.11(b). This second topology is distinct from that of Figure 2.11(a) even though the quantified

Algorithm 4 Basic algorithm for implementation of Lamarckian GA/SIMP hybrid topology optimization (cf. Figure 2.9); given the great variety of subprocess options in heuristic GAs, a range of other algorithmic choices may be made

Require: A design domain $\hat{\mathcal{B}} = \bigcup_{e=1}^{N_{el}} \mathcal{E}_e$ and associated boundary conditions, the dense element stiffness matrices $\hat{\mathbf{K}}_e^0\ \forall\ e = 1 \ldots N_{el}$, the set of GA parameters $\{N_{elite}, p_{mut}, N_{pop}, N_{gen}\}$, the set of SIMP parameters, bounds, and tolerances (cf. Algorithm 1) $\{N_{SIMP}, \eta, m, p, v_f^g, \lambda_{min}, \lambda_{max}, \texttt{tol}_\rho, \texttt{tol}_\lambda\}$, and the crossover stabilizing density augmentation $\Delta\rho_{stab}$

Initialize $k = 0$, $\Delta\rho_{max} = 10^8$
Initialize general FEA equations for each element
Initialize $N_{pop} - 1$ random bitmap (0/1) topologies $[\ \rho_1^{GA}\ \ \rho_2^{GA}\ \ \ldots\ \ \rho_{N_{pop}-1}^{GA}\ ]^{(0)}$; initialize all densities in the last entry $\rho_{N_{pop}}^{GA(0)}$ to v_f^g
while $k < N_{gen}$ **do** ▷ Begin GA loop
    $k \leftarrow k + 1$
    while $j < N_{pop}$ **do** ▷ Parallel SIMP execution if possible
        Perform SIMP on the jth design per Algorithm 1, where the initial vector of element densities is taken from $\rho_j^{GA(k-1)}$ and the number of iterations is limited to a maximum of N_{SIMP}. Obtain updated topology and compliance $\{\rho^{SIMP}, \hat{J}_c\}_j^{(k)}$
    end while
    Perform selection over all $[\ \rho_1^{SIMP}\ \ \rho_2^{SIMP}\ \ \ldots\ \ \rho_{N_{pop}}^{SIMP}\ ]^{(k)}$ to form parental pairs, where probability of selection is proportional to compliance score. Using elite approach, select N_{elite} best performing topologies to pass directly to next generation such that $\rho_i^{SIMP(k)} \to \rho_i^{GA(k)}$ identically
    Using the scattered method per Equation 2.34, use all selected parents to generate remaining required designs $\rho_i^{GA(k)}$, $i = 1 \ldots (N_{pop} - N_{elite})$ for use in the next evolutionary cycle
    Augment all densities $\rho_i^{GA(k)}$ per Equation 2.35 to ensure SIMP stability
    Consider mutation across all components (i.e., every element density) in all topological designs $[\ \rho_1^{GA}\ \ \rho_2^{GA}\ \ \ldots\ \ \rho_{N_{pop}}^{GA}\ ]^{(k)}$, where the probability of mutation is p_{mut}.
end while

compliance is nearly equivalent, illustrating that a variety of designs can and do give similar performance results. Such a phenomenon is regularly observed in nature.

Finally, it is noteworthy that using $N_{pop} = 20$ processors, the clock time to obtain the solution with the GA/SIMP hybrid approach would be smaller than that needed to compute the solution using the conventional SIMP because (i) the GA is perfectly parallel, (ii) N_{SIMP} is limited to 40, and (iii) N_{gen} is limited to 10. This highlights the efficiency of this hybrid scheme as implemented; better designs can be generated in less time.

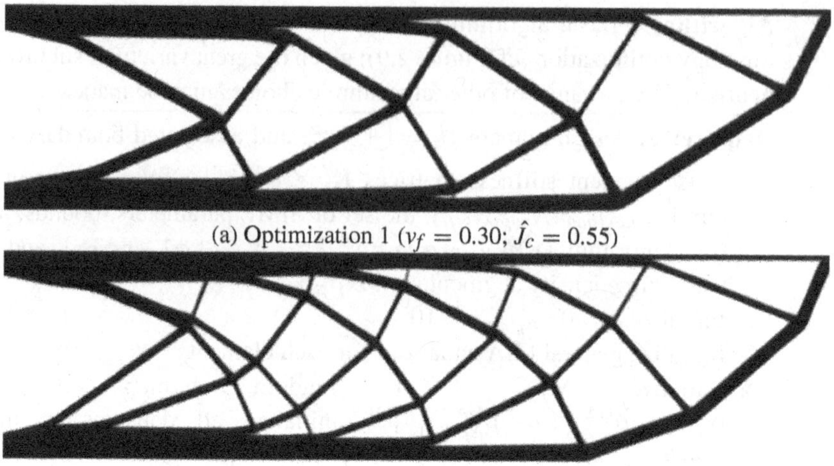

(a) Optimization 1 ($v_f = 0.30$; $\hat{J}_c = 0.55$)

(b) Optimization 2 ($v_f = 0.30$; $\hat{J}_c = 0.55$)

Figure 2.11 Optimized layouts for the long cantilever design problem of Example 2.4 (cf. Example 2.1) using the Lamarckian GA/SIMP hybrid method as obtained from two independent runs (1600 × 400 mesh; 20 individuals for 10 generations). Note how the random nature of the GA leads to different solutions.

While the GA/SIMP hybrid method has clear advantages in its own right, its introduction here is also enlightening relative to the multiple bioinspired approaches that represent the primary theme of this book. Chapters 3–6 all consider topology optimization approaches, whereby GAs of one type or another are used to direct configurational exploration. There is a key difference, however. In the current algorithm, the GA selects topologies that may be infeasible with the expectation that a subsequent algorithm will make necessary adjustments (and perhaps even find an optimized design). The chapters to follow will instead develop and demonstrate multiple processes by which GAs are used to direct structure/network generation algorithms that are preconfigured to provide feasible designs that improve in performance as generations pass.

2.3 Ground Structure-Based Methods

With reference to Figure 1.3 (cf. also Figure 1.1), we now progress to ground structure-based topology optimization, often referred to as the *ground structure method* (GSM). Recall that the structural representation underlying the pixel/voxel methods is some collection of very small continuum subdomains (elements). We will now consider an alternative representational basis: a collection of structural members possibly placed in one of many predetermined positions that span a spatial design domain in a patterned and relatively dense manner (as shown in Figure 2.12). The general approach to optimization is similar to that of the pixel/voxel-based approaches in that some of the (original) fully populated set is eliminated and some is retained such that meaningful and hopefully efficient load paths are determined. Ground structure methods are fundamen-

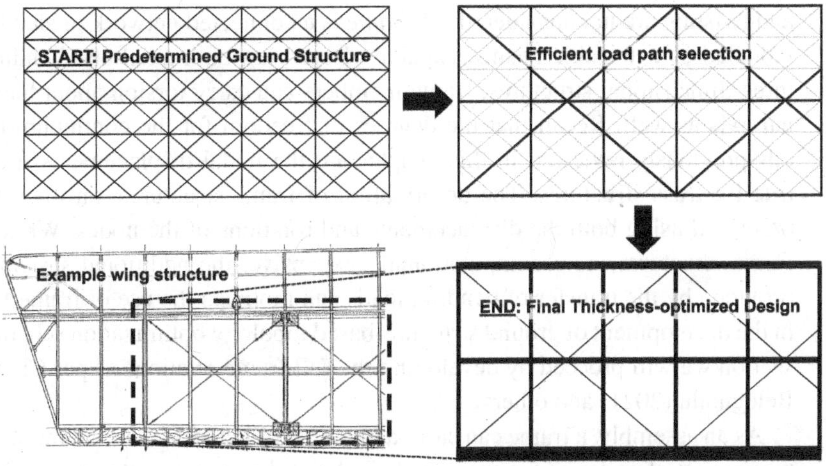

Figure 2.12 The process of topological optimization from ground structures involves assuming some initially fully populated truss or frame structure and then optimizing thicknesses while maintaining volume or mass constraints. Very thin members are removed altogether, and results are compatible with the kinds of variable thickness truss and frame structures observed across engineering sectors.

tally different, however, in that the underlying set of design variables contains section information for all local structural members (e.g., areas, radii, thicknesses) rather than their densities. This suggests a physical meaningfulness and even engineering utility to each individual design variable taking *intermediate* values; it is evident from structures across the application spectrum that a diversity of local component thicknesses can be highly advantageous in design (see Figure 2.12; example wing structure). In other words, the "gray material" problem is no longer a factor and no manner of "penalizing" methods will be needed to enforce that design variables take values of ≈ 0.0 or 1.0.

It is important to consider that in most compliance minimization problems, as the goal volume fraction is driven to values well below 0.5, the structural configurations obtained begin to resemble that of trusses or frames (i.e., long, slender internal members of constant cross-section, sometimes carrying pure tensile or compressive loads). This can be seen in the numerous pixel/voxel-based solutions provided in previous sections and in the literature (Bendsøe & Sigmund 2003). However, for such low-volume solutions, the efficiency of all such methods quickly deteriorates for two reasons: (i) the pixel/voxel resolution required to accurately represent slender rods increases the dimension of the problem and (ii) most of the computational effort is expended solving the equilibrium equations for the *void* regions of the design domain (i.e., all $\mathcal{E}_e \in \hat{\mathcal{B}}_v$, where a *near* zero stiffness has been assigned for numerical stability). For low-volume fractions, it then becomes advantageous to directly consider the topological optimization of assemblages of long slender members already known to be structurally efficient rather than the more generalized pixel/voxel-based approaches.

To be clear as we proceed, a *truss* is defined as a pin joint-connected network of slender members modeled as capable of only axial deformation (i.e., local bending and twisting are not considered). A *frame* is a connected network of slender members modeled as extensible beams, capable of carrying both axial and bending loads. In three-dimensional structures, frame members can carry torsion about their longitudinal axis as well. Recall that the degrees of freedom for the continuum-based methods previously reviewed in this chapter are the nodal displacements $\hat{\mathbf{u}}$; the same is true for truss structures. The deformation of frame structures, on the other hand, is described using both the displacements and rotations of the nodes. While this makes frame structures more computationally expensive, the additional structural stability provided by the transfer of bending loads can motivate the use of frame formulations in the development of ground structure-based topology optimization schemes.[11] In this section we will proceed by developing the full frame relations as per Chandrupatla & Belegundu (2021) and others.

As an assembly, a frame can carry a given set of loads (forces and moments) applied to some set of nodes and is supported via defined displacements and rotations (usually fixed to be zero) over a different set of nodes (Hemp 1973). As per the previous discussion, topological optimization of frames begins by considering all members of a *ground structure* to be active and to be of some homogeneous initial thickness, as seen in the "START" image of Figure 2.12. The procedure continues by redistributing thicknesses such that structural performance is maximized while volume fraction preferences are maintained. Consistent with the other approaches reviewed in this chapter, we continue to seek the minimization of compliance in this section.

2.3.1 Structural Governing Equations

In continuing this review chapter, it would be possible to introduce GSM in terms of general stiffness matrices as with the pixel/voxel methods previously described. However, it is most common for the bioinspired topological optimization methods that comprise the remainder of this book to also make recourse to truss- or frame-based structural assemblies. Further, aspects of the GSM approach will later be used to improve the more novel algorithms. For these reasons, we will provide a more complete mathematical introduction to GSM and will refer back to it as appropriate.

As we proceed with this review it is important to note that many truss and frame analysis and optimization algorithms and codes found in the open literature have been optimized for numerical efficiency and decreased run time. Here, we will present the essential relations in a clear and expanded form toward concept mastery; efficient implementation is left to the reader, where the provided codes may also be of assistance.

To begin, the fundamental system of equations for a finite element representation of a full frame assembly is given as

[11] The bending stiffness provided by frame elements also allows for the potential subdivision of long, slender, straight structural members into smaller elements, which can be useful for FEA accuracy. Subdivided truss members cannot be subjected to compression lest they immediately buckle for lack of bending stiffness.

$$\hat{\mathbf{K}}\hat{\mathbf{q}} = \hat{\mathbf{Q}}. \tag{2.36}$$

Note the distinction between Equation 2.3 and this new form. The former considers the nodal displacements $\hat{\mathbf{u}}$ as degrees of freedom in a discretized domain such that

$$\hat{\mathbf{u}} = [\ u^1 \quad v^1 \quad \dots \quad u^n \quad v^n \quad \dots \quad u^{N_{nodes}} \quad v^{N_{nodes}}\ ]^T,$$

for an assembly of N_{nodes} nodes in two-dimensional x–y space, where u^i is the translation of node i in the x-direction and v^i is the translation of node i in the y-direction. In the current system, $\hat{\mathbf{q}}$ contains both the translational and rotational degrees of freedom, such that

$$\hat{\mathbf{q}} = [\ u^1 \quad v^1 \quad \theta^1 \quad \dots \quad u^n \quad v^n \quad \theta^n \quad \dots \quad u^{N_{nodes}} \quad v^{N_{nodes}} \quad \theta^{N_{nodes}}\ ]^T,$$

where θ^i is the rotation of node i about the z-direction in the same two-dimensional space. Likewise, $\hat{\mathbf{Q}}$ contains both the forces and moments energetically conjugate to the deformations captured in $\hat{\mathbf{q}}$. The metric for compliance, which is to be minimized, is updated from Equation 2.4 to consider also rotations of nodes under applied moments such that

$$\hat{J}_c = \hat{\mathbf{Q}}^T\hat{\mathbf{q}}. \tag{2.37}$$

Finally, we must consider the volume fraction constraint for ground structure-based optimization of trusses and frames. In general, we have

$$\hat{v}_f = \frac{1}{V}\sum_{e=1}^{N_{el}} L_e A_e, \tag{2.38}$$

where L_e and A_e are the length and cross-sectional area, respectively, of the eth truss or frame element and V is the total volume of the design domain.[12]

For the sake of detailed development, we consider the specific case that member cross-sections are rectangular, and the cross-sectional area is then given as $A_e = wt_e$. For straightforward development of the equations to follow, we will further assume that the out-of-plane width w is constant while the in-plane thickness t_e is an optimization variable (i.e., as in a two-dimensional problem considering a design domain with area A). The volume fraction can then be written as

$$\hat{v}_f = \frac{1}{A}\sum_{e=1}^{N_{el}} L_e t_e. \tag{2.39}$$

Recall that the GSM is most appropriate and becomes most effective for relatively low volume fractions as compared to other approaches (e.g., $\hat{v}_f \leq 0.3$).

[12] Note that, given the representation of long slender structural components in terms of their neutral axis and section properties, relations such as Equation 2.38 are strictly speaking only approximations of the volume fraction. The overlap of material at nodes is not rigorously considered and material may be double-counted in these regions. The accuracy of these approximations increases as the aspect ratio of structural members increases.

Given these definitions and continuing to assume a rectangular cross-section in a two-dimensional design domain, we can state the overall ground structure optimization problem as follows:

Find the vector of local member thicknesses $\mathbf{t} = [t_1 \ldots t_e \ldots t_{N_{el}}]$ that minimizes the truss or frame compliance functional

$$\hat{J}_c = \hat{\mathbf{Q}}^T \hat{\mathbf{q}}$$

subject to:

- $t_e \geq 0$,
- the satisfaction of $\hat{\mathbf{K}}\hat{\mathbf{q}} = \hat{\mathbf{Q}}$, and
- $\frac{1}{A}\sum_{e=1}^{N_{el}} L_e t_e = v_f^g$.

To arrive at the required Equations 2.36 and 2.37 for trusses or frames, one must consider three mathematical and algorithmic operations as illustrated in Figure 2.13:

1. Calculation of the local generalized force–displacement relations for a single member or element expressed in a local coordinate system,
2. Rotation of each element into a common global coordinate system, and
3. Assembly of all local contributions into a single global system of equations, including the renumbering of degrees of freedom and stiffness components (cf. Equation 2.36).

For a full treatment of truss and frame FEA, the reader is referred to Reddy (2019) or similar. However, assuming some familiarity with the subject, we proceed with the following overview to both explain the ground structure methods for topology optimization and to enable the multiple frame-centric chapters to follow.

Local Frame Response

In the context of a single frame element connecting two nodes, each with three degrees of freedom (two translational, one rotational; Figure 2.13), the local system of equations relating applied generalized forces and generalized displacements is

$$\bar{\mathbf{k}}_e \bar{\mathbf{q}}^e = \bar{\mathbf{Q}}^e, \tag{2.40}$$

where the local stiffness matrix linking nodal translations and rotations to forces and moments applied to element e can be expressed as

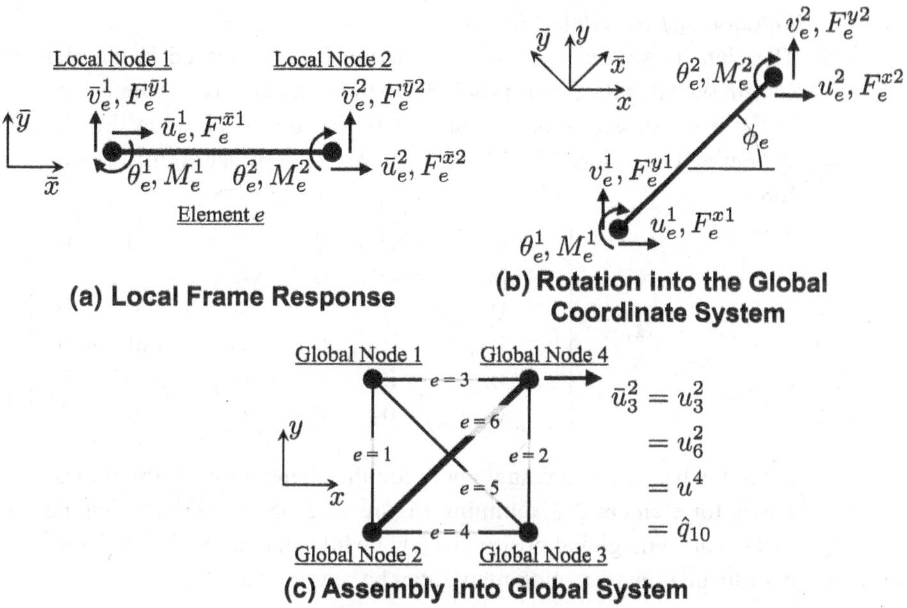

Figure 2.13 The development of frame (or truss) FEA equations must consider local response formulated in a local aligned coordinate system, rotation into the global coordinate system based on element placement within the full structure, and assembly into a globally numbered system. As an example, it is shown that the common x-displacement of Element 3, Node 2 and Element 6, Node 2 is equivalent to the x-displacement of global Node 4, which makes it the 10th overall component of the global generalized displacement vector $\hat{\mathbf{q}}$.

$$\bar{\mathbf{k}}_e = \frac{E}{L}\begin{bmatrix} A & 0 & 0 & -A & 0 & 0 \\ 0 & \frac{12I}{L^2} & \frac{6I}{L} & 0 & -\frac{12I}{L^2} & \frac{6I}{L} \\ 0 & \frac{6I}{L} & 4I & 0 & -\frac{6I}{L} & 2I \\ -A & 0 & 0 & A & 0 & 0 \\ 0 & -\frac{12I}{L^2} & -\frac{6I}{L} & 0 & \frac{12I}{L^2} & -\frac{6I}{L} \\ 0 & \frac{6I}{L} & 2I & 0 & -\frac{6I}{L} & 4I \end{bmatrix}_e . \tag{2.41}$$

Note the $[\cdot]_e$ notation invoked for brevity, indicating that each component is specific to element e (i.e., A_e, E_e, I_e, and L_e are the cross-sectional area, elastic modulus, second moment of area, and length associated with element e, respectively). The generalized element displacement vector expressed in the local coordinate system of the element is (see Figure 2.13)

$$\bar{\mathbf{q}}_e = \begin{bmatrix} \bar{u}_e^1 & \bar{v}_e^1 & \theta_e^{z1} & \bar{u}_e^2 & \bar{v}_e^2 & \theta_e^{z2} \end{bmatrix}^T,$$

and the generalized force vector, likewise in the local coordinate system, is

$$\bar{\mathbf{Q}}_e = \begin{bmatrix} F_e^{\bar{x}1} & F_e^{\bar{y}1} & M_e^{\theta_{z1}} & F_e^{\bar{x}2} & F_e^{\bar{y}2} & M_e^{\theta_{z2}} \end{bmatrix}^T.$$

Rotation into the Global Frame

The element stiffness matrix in Equation 2.41 is defined in the element coordinate system such that the $\bar{x}$-axis is parallel to the element axis and the $\bar{y}$-axis is perpendicular in the plane of action (cf. Figure 2.13). To convert between the element coordinate system and the global coordinate system, a coordinate transformation matrix of the form

$$\mathbf{T}_e^{\phi} = \begin{bmatrix} \cos\phi_e & \sin\phi_e & 0 & 0 & 0 & 0 \\ -\sin\phi_e & \cos\phi_e & 0 & 0 & 0 & 0 \\ 0 & 0 & 1 & 0 & 0 & 0 \\ 0 & 0 & 0 & \cos\phi_e & \sin\phi_e & 0 \\ 0 & 0 & 0 & -\sin\phi_e & \cos\phi_e & 0 \\ 0 & 0 & 0 & 0 & 0 & 1 \end{bmatrix}$$

is used, where ϕ_e is the angle between the element longitudinal axis $\bar{x}$ and the global x-axis for element e. Examining Figure 2.13 and considering element node 1 to be located at some global position (x_e^1, y_e^1) while element node 2 is located at (x_e^2, y_e^2), the coordinate transformation matrix can be rewritten as

$$\mathbf{T}_e^{\phi} = \frac{1}{L_e}\mathbf{T}_e^{\Delta} = \frac{1}{L_e}\begin{bmatrix} \Delta x_e & \Delta y_e & 0 & 0 & 0 & 0 \\ -\Delta y_e & \Delta x_e & 0 & 0 & 0 & 0 \\ 0 & 0 & 1 & 0 & 0 & 0 \\ 0 & 0 & 0 & \Delta x_e & \Delta y_e & 0 \\ 0 & 0 & 0 & -\Delta y_e & \Delta x_e & 0 \\ 0 & 0 & 0 & 0 & 0 & 1 \end{bmatrix}, \tag{2.42}$$

where

$$\Delta x_e = x_e^2 - x_e^1, \quad \Delta y_e = y_e^2 - y_e^1 \quad \text{and} \quad L_e = \sqrt{\Delta x_e^2 + \Delta y_e^2}.$$

The coordinate transformation of Equation 2.42 can be applied to Equations 2.41 to give the element stiffness matrix written in the global frame $\mathbf{K}_e$ via

$$\mathbf{K}_e = \mathbf{T}_e^{\phi,T}\bar{\mathbf{k}}_e\mathbf{T}_e^{\phi}. \tag{2.43}$$

This allows us to then write Equation 2.40 in the global coordinate system for each element as

$$\mathbf{K}_e\mathbf{q}^e = \mathbf{Q}^e. \tag{2.44}$$

The associated generalized forces and displacements illustrated for a single element in the global coordinate system are illustrated in Figure 2.13(b).

Note that $\mathbf{q}_e$ and $\mathbf{Q}_e$ emulate $\bar{\mathbf{q}}_e$ and $\bar{\mathbf{Q}}_e$, but in the global coordinate system. It is also essential to note that forces, moments, displacement, and rotations imposed on any frame structure are almost always considered in the global coordinate system, and thus the rotation of $\bar{\mathbf{Q}}_e$ and $\bar{\mathbf{q}}_e$ are rarely practically necessary.

Assembly into Global System

Assembly of the N_{el} systems of linear equations as per Equation 2.44 for each element in the global coordinate system into a single global system for a full structure is a matter of renumbering and summation. To maintain the compactness of this brief but comprehensive overview, we will introduce a compact summation operator $\widehat{\sum}$, defined for any general $\mathbf{B}$ as follows:

$$\widehat{\sum}(\mathbf{B}) = \begin{cases} \sum_{e=1}^{N_{el}} T_e^{\#,T} \mathbf{B}_e & \text{if } \mathbf{B} \text{ is a vector,} \\ \sum_{e=1}^{N_{el}} T_e^{\#,T} \mathbf{B}_e T_e^{\#} & \text{if } \mathbf{B} \text{ is a matrix.} \end{cases}$$

Here, $\mathbf{T}_e^{\#}$ is a transformation matrix of 0s and 1s that serves only to renumber the degrees of freedom and forces associated with element e into their associated global numbering. For example, considering Element 6 in Figure 2.13(c), we have for the transpose of this transformation

$$\mathbf{T}_6^{\#,T} = \begin{bmatrix} 0 & 0 & 0 & 0 & 0 & 0 \\ 0 & 0 & 0 & 0 & 0 & 0 \\ 0 & 0 & 0 & 0 & 0 & 0 \\ 1 & 0 & 0 & 0 & 0 & 0 \\ 0 & 1 & 0 & 0 & 0 & 0 \\ 0 & 0 & 1 & 0 & 0 & 0 \\ 0 & 0 & 0 & 0 & 0 & 0 \\ 0 & 0 & 0 & 0 & 0 & 0 \\ 0 & 0 & 0 & 0 & 0 & 0 \\ 0 & 0 & 0 & 1 & 0 & 0 \\ 0 & 0 & 0 & 0 & 1 & 0 \\ 0 & 0 & 0 & 0 & 0 & 1 \end{bmatrix},$$

since local degrees of freedom 1, 2, and 3 for this element map to 4, 5, and 6 respectively, and local degrees of freedom 4, 5, and 6 for this element map to 10, 11, and 12 respectively.

In most implementations, the calculation of all matrices from $\bar{\mathbf{k}}_e$ in Equation 2.41 to $\mathbf{T}_e^{\#}$ are based on two fundamental matrices that define the full assembly. The *assembly matrix* (or "connectivity" matrix) $\mathbf{A}$ defines which global nodes correspond to local nodes 1 and 2 for each element $e = 1 \ldots N_{el}$, while the *nodal position matrix* $\boldsymbol{P}$ defines the location of each global node. These are generated by the meshing process and can be written as

$$\mathbf{A} = \begin{bmatrix} n_1^1 & n_1^2 \\ n_2^1 & n_2^2 \\ \vdots & \vdots \\ n_e^1 & n_e^2 \\ \vdots & \vdots \\ n_{N_{el}}^1 & n_{N_{el}}^2 \end{bmatrix}, \quad \boldsymbol{P} = \begin{bmatrix} x_1 & y_1 \\ x_2 & y_2 \\ \vdots & \vdots \\ x_n & y_n \\ \vdots & \vdots \\ x_{N_{nd}} & y_{N_{nd}} \end{bmatrix},$$

where n_e^1 and n_e^2 are the global indices of local nodes 1 and 2, respectively, for element e (see Figure 2.13(a)) and (x_n, y_n) is the global location of global node n ($n = 1 \ldots N_{nd}$) (see Figure 2.13(c)). Note that the essential values Δx_e and Δx_e can be directly and systematically computed directly from **A** and ***P*** via

$$\Delta x_e = P_{A_{e2}1} - P_{A_{e1}1}, \ \ \Delta y_e = P_{A_{e2}2} - P_{A_{e1}2}.$$

Finally, by examining the forms of **A** and $\mathbf{T}_e^{\#,T}$, we note that the latter can be directly constructed from the former per the following, where the notation $[\mathbf{T}_e^{\#,T}]_{ij}$ refers to the component on the ith row at the jth column of $\mathbf{T}_e^{\#,T}$:

$$[\mathbf{T}_e^{\#,T}]_{ij} = 1 \quad \begin{array}{l} i = 3(\mathbf{A}_{e1} - 1) + j \\ j = 1 \ldots 3 \end{array}$$

and

$$[\mathbf{T}_e^{\#,T}]_{ij} = 1 \quad \begin{array}{l} i = 3(\mathbf{A}_{e2} - 1) + k \\ j = k + 3 \\ k = 1 \ldots 3 \end{array}$$

and

$$[\mathbf{T}_e^{\#,T}]_{ij} = 0 \quad \text{otherwise.}$$

Several of these concepts will be elucidated in examples to follow.

Returning to our developments, given the newly introduced compact summation operator, we require one final set of calculations to arrive at Equation 2.36:[13]

$$\hat{\mathbf{K}} = \widehat{\sum}(\mathbf{K}_e), \quad \hat{\mathbf{Q}} = \widehat{\sum}(\mathbf{Q}_e). \tag{2.45}$$

[13] If the forces applied to an assembled frame structure are discrete (i.e., nodal) and are applied in the global coordinate system, this operation on $\mathbf{Q}_e$ is not necessary. If, however, distributed forces (e.g., body forces such as gravity) are considered, the summation of elemental contributions is essential.

Application of Boundary Conditions and Calculation of Displacements and Forces
Once the global system given by Equation 2.36 has been assembled via the application of Equation 2.45, the global generalized displacement vector must be determined by application of the appropriate kinematic boundary conditions and solution of the resulting system. Just as an improperly constrained structure will exhibit meaningless rigid-body translations and rotations under the application of forces, the system in Equation 2.36 is likewise meaningless and unsolvable without the application of reasonable kinematic boundary conditions (i.e., known global translations and rotations at nodes). It is most common that a number of degrees of freedom are fixed (i.e., $\hat{q}_i = 0$ for some $i = 1 \ldots 3N_{nodes}$), though in principle nonzero kinematic boundary conditions are also acceptable.

Though multiple options exist, for both mathematical simplicity and computational efficiency here we consider the penalty method for boundary condition satisfaction[14] (Bathe 1996). To apply this approach, we let the penalty P be defined as

$$P = (1E4)\max(\hat{K}_{ij},\ \ i = 1 \ldots 3N_{nodes},\ \ j = 1 \ldots 3N_{nodes}).$$

Then, for any $\hat{q}_i,\ \ i = 1 \ldots 3N_{nodes}$ for which the translation or rotation is known or defined (i.e., a boundary condition), we set $a_i = \hat{q}_i$, modify the global stiffness matrix such that

$$\hat{K}_{ii} \leftarrow \hat{K}_{ii} + P,$$

and modify the global generalized force vector such that

$$\hat{Q}_i \leftarrow \hat{Q}_i + Pa_i.$$

Solution of the updated system $\hat{\mathbf{K}}\hat{\mathbf{q}} = \hat{\mathbf{Q}}$ will result in a vector of generalized displacements $\hat{\mathbf{q}}$ that satisfies the kinematic boundary conditions. Further, reaction forces $\hat{Q}_i$ associated with known generalized displacements $\hat{a}_i$ can be efficiently calculated in a postprocessing step via

$$\hat{Q}_i = -P(\hat{q}_i - a_i). \tag{2.46}$$

In this way, all generalized forces and displacements can be calculated.

While it is clear that the governing system could be solved by multiplying both sides of Equation 2.36 by the inverse of the global stiffness matrix, $\hat{\mathbf{K}}^{-1}$, such direct inversion of the global stiffness matrix can be computationally burdensome even for relatively simple structures, and several numerical alternatives are more commonly to increase the computational efficiency of solving the full system such as *LU* decomposition, Gaussian elimination, and others (Bathe 1996).

[14] The solution of frame FEA problems is widespread throughout this book, and some codes provided may instead make use of the slightly more cumbersome but exact "elimination approach" (Chandrupatla & Belegundu 2021) and may be limited to fixed boundary conditions only (i.e., $a_i = 0$).

2.3.2 Traditional Ground Structure Method (GSM)

As it is depicted in Figure 2.12, the configuration of the underlying ground structure is fixed throughout the optimization process; the final solution is then represented solely by the relative thicknesses of these spatially fixed members. This is known as the *traditional GSM*. Section 2.3.3 will later address an extension whereby the geometric configuration of the underlying ground structure is also permitted to improve during optimization.

Returning to the fundamental optimization upon which the ground structure methods are based, we seek to minimize $\hat{J}_c = \hat{\mathbf{Q}}^T \hat{\mathbf{q}}$ with respect to the vector of design variables, here the in-plane thicknesses $\mathbf{t}$. Given the analytical nature of this approach, this problem historically makes recourse to a wide range of gradient-based optimization algorithms and sensitivity analysis (Choi & Kim 2006). These approaches are especially efficient since the needed gradients can be obtained analytically using explicit expressions of the finite element matrix equations, as will be shown. Specifically, we require $\partial \hat{J}_c / \partial t_i \ \forall \ i = 1 \ldots N_{el}$.[15]

To begin the derivation of this essential term, we first recognize that, as with all other topology optimization methods considered in this chapter, this distribution of *loads* is generally held constant during the compliance minimization process. In such case, we have

$$\frac{\partial \hat{J}_c}{\partial t_i} = \hat{\mathbf{Q}}^T \frac{\partial \hat{\mathbf{q}}}{\partial t_i}, \tag{2.47}$$

where the determination of $\hat{\mathbf{q}}$ for any structural configuration (i.e., for any vector of thicknesses $\mathbf{t}$) is the essential goal of the finite element relations reviewed above.

To find $\partial \hat{\mathbf{q}} / \partial t_i$, we return to Equation 2.36, taking its derivative with respect to each design variable t_i while again recognizing that the components of $\hat{\mathbf{Q}}$ are defined and fixed for the structural design problem. This gives

$$\hat{\mathbf{K}} \frac{\partial \hat{\mathbf{q}}}{\partial t_i} + \frac{\partial \hat{\mathbf{K}}}{\partial t_i} \hat{\mathbf{q}} = \mathbf{0} \quad \Rightarrow \quad \hat{\mathbf{K}} \frac{\partial \hat{\mathbf{q}}}{\partial t_i} = -\frac{\partial \hat{\mathbf{K}}}{\partial t_i} \hat{\mathbf{q}}. \tag{2.48}$$

Multiplying both sides of Equation 2.48 by the inverse of the global stiffness matrix, $\hat{\mathbf{K}}^{-1}$, one obtains the equation for the sensitivity of the displacement vector with respect to a given structural member (element) thickness t_i,

$$\frac{\partial \hat{\mathbf{q}}}{\partial t_i} = \hat{\mathbf{K}}^{-1} \left(-\frac{\partial \hat{\mathbf{K}}}{\partial t_i} \hat{\mathbf{q}} \right). \tag{2.49}$$

Sensitivity analysis is typically performed in coordination with structural analysis, where the sensitivities are required only after the displacement solution has been determined (i.e., Equation 2.36 has already been solved) and $\hat{\mathbf{K}}^{-1}$ is known. Therefore, the only contribution to Equation 2.49 that must be calculated is $\partial \hat{\mathbf{K}} / \partial t_i$. By invoking Equations 2.43 and 2.45 and noting that the local thickness only effects the local elemental stiffness, we see that

[15] The new index i is here introduced to distinguish the required sensitivities from the summation operations over elements $e = 1 \ldots N_{el}$, as will be shown.

$$\frac{\partial \hat{\mathbf{K}}}{\partial t_i} = \frac{\partial}{\partial t_i}\left(\widehat{\sum} \mathbf{T}^{\phi,T}\bar{\mathbf{k}}_i\mathbf{T}^{\phi}\right) = \widehat{\sum}\left(\mathbf{T}_e^{\phi,T}\frac{\partial \bar{\mathbf{k}}_e}{\partial t_i}\mathbf{T}_e^{\phi}\right). \tag{2.50}$$

As an example, we can continue to assume a rectangular cross-section (cf. Equation 2.39) for which the second moment of area is $I = wt^3/12$. In such a case, Equation 2.41 can be rewritten as

$$\bar{\mathbf{k}}_e = \frac{Ewt_e}{L_e}\begin{bmatrix} 1 & 0 & 0 & -1 & 0 & 0 \\ 0 & \frac{t^2}{L^2} & \frac{t^2}{2L} & 0 & -\frac{t^2}{L^2} & \frac{t^2}{2L} \\ 0 & \frac{t^2}{2L} & \frac{t^2}{3} & 0 & -\frac{t^2}{2L} & \frac{t^2}{6} \\ -1 & 0 & 0 & 1 & 0 & 0 \\ 0 & -\frac{t^2}{L^2} & -\frac{t^2}{2L} & 0 & \frac{t^2}{L^2} & -\frac{t^2}{2L} \\ 0 & \frac{t^2}{2L} & \frac{t^2}{6} & 0 & -\frac{t^2}{2L} & \frac{t^2}{3} \end{bmatrix}_e. \tag{2.51}$$

For computational speedup, especially in the context of iterative GSM, we can rewrite the local element stiffness matrix Equation 2.51 in the following form:

$$\bar{\mathbf{k}}_e = t_e\frac{Ew}{L_e}\begin{bmatrix} 1 & 0 & 0 & -1 & 0 & 0 \\ 0 & 0 & 0 & 0 & 0 & 0 \\ 0 & 0 & 0 & 0 & 0 & 0 \\ -1 & 0 & 0 & 1 & 0 & 0 \\ 0 & 0 & 0 & 0 & 0 & 0 \\ 0 & 0 & 0 & 0 & 0 & 0 \end{bmatrix}$$

$$+\ t_e^3\frac{Ew}{L_e}\begin{bmatrix} 0 & 0 & 0 & 0 & 0 & 0 \\ 0 & \frac{1}{L_e^2} & \frac{1}{2L_e} & 0 & -\frac{1}{L_e^2} & \frac{1}{2L_e} \\ 0 & \frac{1}{2L_e} & \frac{1}{3} & 0 & -\frac{1}{2L_e} & \frac{1}{6} \\ 0 & 0 & 0 & 0 & 0 & 0 \\ 0 & -\frac{1}{L_e^2} & -\frac{1}{2L_e} & 0 & \frac{1}{L_e^2} & -\frac{1}{2L_e} \\ 0 & \frac{1}{2L_e} & \frac{1}{6} & 0 & -\frac{1}{2L_e} & \frac{1}{3} \end{bmatrix},$$

which is compactly written as

$$\bar{\mathbf{k}}_e = \frac{Ew}{L_e}\left\{t_e\bar{\mathbf{k}}_e^1 + t_e^3\bar{\mathbf{k}}_e^2\right\}. \tag{2.52}$$

Because $\bar{\mathbf{k}}_e^1$ and $\bar{\mathbf{k}}_e^2$ are constant matrices, taking the derivative of Equation 2.52 with respect to t_i is straightforward, giving

$$\frac{\partial\bar{\mathbf{k}}_e}{\partial t_i} = \frac{Ew}{L_e}\begin{cases} \bar{\mathbf{k}}_e^1 + 3t_e^2\bar{\mathbf{k}}_e^2, & i = e, \\ \mathbf{0}, & i \neq e, \end{cases} \tag{2.53}$$

Once this matrix is evaluated and populated for each element, the full global sensitivity can be calculated via Equation 2.50. However, because a local element stiffness $\bar{\mathbf{k}}_e$ and it alone is sensitive to changes in local element thickness t_i if $i = e$, the summation of Equation 2.50 becomes superfluous. The final full sensitivity to t_i is then found by invoking $i = e$ in Equation 2.53 and then computing[16]

$$\frac{\partial\hat{\mathbf{K}}}{\partial t_i} = \mathbf{T}_e^{\#,T}\mathbf{T}_e^{\phi,T}\frac{\partial\bar{\mathbf{k}}_e}{\partial t_i}\mathbf{T}_e^{\phi}\mathbf{T}_e^{\#}. \tag{2.54}$$

To return to the motivating optimization problem, $\partial\hat{\mathbf{K}}/\partial t_i$ is used to compute the sensitivities of the global compliance $\hat{J}_c$ to all components of $\mathbf{t}$ via Equations 2.47 and 2.49. Further, the sensitivities of the volume fraction with respect to $\mathbf{t}$ can be simply calculated as $\partial\hat{v}_f/\partial t_i = L_i/A$, where A is the area of the two-dimensional design domain. The existence of exact relations for $\partial\hat{J}_c/\partial t_i$ and $\partial\hat{v}_f/\partial t_i$ enables a designer to employ one of many efficient gradient-based optimization approaches. In this work, we select the sequential least squares programming (SLSQP) algorithm (Nocedal & Wright 2006) implemented in Python's *SciPy* module, which is capable of considering constrained, bounded objective functions and user-defined sensitivity calculations (i.e., Equation 2.47).

To summarize, the general implementation scheme for topology optimization via the traditional GSM is given in Algorithm 5. Note that while it is clearly necessary that $t_i \leq t_{max}$ (t_{max} being user-selected and problem-specific), it may also be necessary to enforce $t_i \geq t_{min}$ for numerical stability, especially in truss problems.

[16] Because $\mathbf{T}_e^{\phi}$ is constant for each element, analytical equations in the style of Equation 2.53 can be directly written for $\mathbf{K}_e$ instead of for $\bar{\mathbf{k}}_e$ by premultiplying all the Δx and Δy terms (cf. Equation 2.43), resulting in a much more efficient implementation that directly considers

$$\frac{\partial\hat{\mathbf{K}}}{\partial t_i} = \mathbf{T}_e^{\#,T}\frac{\partial\mathbf{K}_e}{\partial t_i}\mathbf{T}_e^{\#}.$$

Algorithm 5 General implementation algorithm for the traditional GSM of topology optimization; utilization of a preexisting gradient-based optimizer is assumed

Require: A preconfigured statically stable ground structure composed of truss or frame elements such that $\hat{\mathcal{B}}_m = \bigcup_{e=1}^{N_{el}} \mathcal{E}_e$ (i.e., full definition of $\mathbf{A}$ and $\boldsymbol{P}$) and associated boundary conditions, the Young's modulus and out-of-plane thickness for all elements E and w, respectively, and algorithmic parameters t_{min}, t_{max}, v_f^g, and `tol`

Initialize $k = 0$, $t_i = t_{max}\ \forall\ i = 1 \ldots N_{el}$
for $e = 1 \ldots N_{el}$ **do**
  Compute and store $\bar{\mathbf{k}}_e^1$ and $\bar{\mathbf{k}}_e^2$ per the definition of Equation 2.52
end for
while Convergence metric > `tol` **do** ▷ Begin optimization loop
  $k \leftarrow k + 1$
  Compute $\hat{\mathbf{K}}$ per Equation 2.45 using Equation 2.52 and the stored matrices $\bar{\mathbf{k}}_e^1$ and $\bar{\mathbf{k}}_e^2$
  Considering boundary conditions, solve $\hat{\mathbf{K}}\hat{\mathbf{q}} = \hat{\mathbf{Q}}$
  Compute $\hat{J}_c$ per Equation 2.37
  Compute $\hat{v}_f$ per Equation 2.39
  for $i = 1 \ldots N_{el}$ **do**
    Compute $\partial\hat{J}_c/\partial t_i$ per Equations 2.47, 2.49, and 2.54
    Compute $\partial\hat{v}_f/\partial t_i = L_i/A$
  end for
  By providing magnitudes $\hat{J}_c$ and $\hat{v}_f$ in addition to vectors of gradient terms $\partial\hat{J}_c/\partial\mathbf{t}$ and $\partial\hat{v}_f/\partial\mathbf{t}$ to the selected gradient-based optimizer, update all $t_{min} \leq t_i \leq t_{max}$ to minimize $\hat{J}_c$ subject to the constraint that $\hat{v}_f \leq v_f^g$; acquire convergence metric
end while

Example 2.5 Our next example compares the optimized solutions obtained using SIMP and GSM for the long cantilever benchmark problem studied in Example 2.1. While the SIMP design domain considers 160,000 pixels (corresponding to an 800 × 200 mesh), the GSM design domain consists of the 21 structural members configured as shown in Figure 2.14(a), the optimal thicknesses of which are determined using an implementation of the SLSQP algorithm (Nocedal & Wright 2006). The optimized layouts obtained using GSM and SIMP are depicted in Figures 2.14(b) and 2.14(c), respectively. As expected, the layout obtained with SIMP resembles that of a truss with the exception of the joints, where the requirement of transmitting moments and carrying the associated bending stresses produce localized reinforcements. Unlike the ground structure, however, SIMP has the advantage of having freedom to place the joints at optimized points within the design domain (see Section 2.3.3 for an alternative ground structure method where the placement of the joints is performed together with the sizing of the

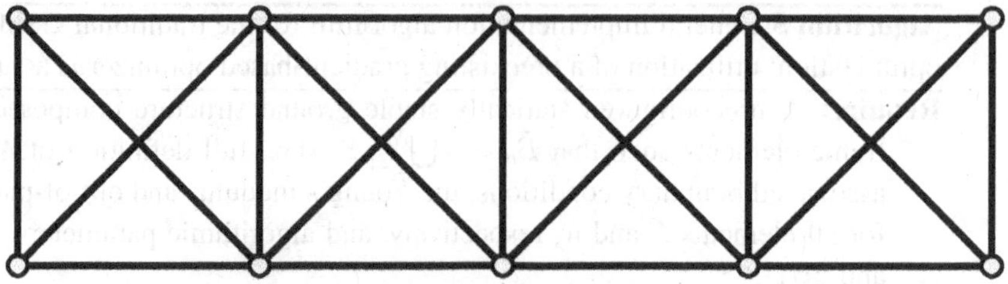

(a) Initial ground structure of 21 elements

(b) Optimized GSM solution ($v_f = 0.30$; $\hat{J}_c = 0.47$; 42 iterations)

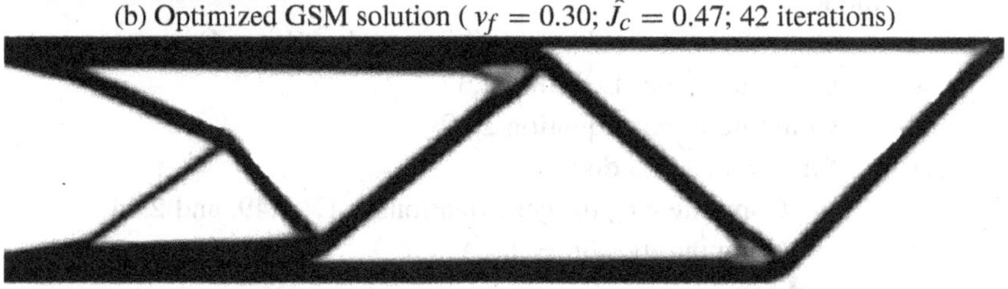

(c) SIMP solution from Figure 2.4 (800 × 200 mesh of 160,000 elements; $v_f = 0.30$; $\hat{J}_c = 0.58$; 598 iterations)

Figure 2.14 Optimized layout for the long cantilever design problem of Example 2.5 (cf. Example 2.1) using GSM as compared to the equivalent SIMP solution (cf. Section 2.2.1). The GSM solution provides a lower compliance value while requiring less functional evaluations, each of which is less computationally expensive.

cross-sectional areas). Despite the simplicity of the ground structure selected for this example the GSM design has a lower compliance than that of SIMP, and the solution for the GSM is obtained in just 42 functional evaluations as opposed to the 598 required for SIMP; this is a result of the drastically reduced number of design variables used in the ground structure (20) compared to SIMP (160,000).

However, the comparison between the compliance computed using both methods should be made with several caveats. First, the geometric sizes of the design domains are slightly different. SIMP is defined such that all pixels are located within the domain, while in the GSM only the joints are constrained to remain in the domain; thus, half of the volume of the rods along the boundary of the domain are located outside of it, providing a greater equivalent moment of inertia and thus greater stiffness. In addition, the larger effective design domain in the GSM corresponds to a smaller effective volume fraction for the optimal design and therefore (typically) an increased compliance value. Furthermore, structural members are represented in the GSM as intersecting at a single point, whereas in SIMP solutions these intersections and their effect on structural response are more accurately captured. The solution obtained with SIMP also requires a postprocessing step (not performed here) to eliminate intermediate material; this step typically increases the compliance for a given optimized layout.

2.3.3 Simultaneous Optimization of Truss/Frame Geometry and Topology

The ground structure method described in Section 2.3.2 suffers from a severe limitation; it uses a fixed layout (nodal positions and element connectivity) for the sizing optimization. However, it is intuitive that a truly optimized structure would be associated with nodal positions *not* known at the onset of the design process. To provide flexibility regarding optimal nodal position, the traditional GSM can use an exceedingly large number of joints to approximate the position of the optimized nodes in the final solution. In this section, we describe a method that circumvents that limitation by simultaneously optimizing the cross-sectional sizes of the slender members and the exact positions of their intersections. We refer to the possible removal of components (i.e., via traditional GSM) as the optimization of *topology*, while the update in nodal positions is referred to as the optimization of *geometry*.

To begin, consider the extended vector of design variables

$$\mathbf{d} = \mathbf{t} \cup \mathbf{c} = \begin{bmatrix} t_1 & t_2 & \dots & t_{N_{el}} & c_1 & c_2 & \dots & c_{2N_{Nd}} \end{bmatrix}^T,$$

where $\mathbf{t}$ remains the collection of all element thicknesses and $\mathbf{c}$ is a vector of all[17] global nodal coordinates arranged as

$$\mathbf{c} = \begin{bmatrix} x_1 & y_1 & x_2 & y_2 & \dots & x_n & y_n & \dots & x_{N_{nd}} & y_{N_{nd}} \end{bmatrix}^T,$$

where N_{nd} is the number of nodes in the entire assembly and (x_i, y_i) is the global position of global node i. Simultaneous optimization of truss geometry and topology then involves actively altering both $\mathbf{c}$ and $\mathbf{t}$ toward maximized performance. To continue mathematical development requires rigorous consideration of nodal positions, global and local, though always in the global coordinate system.

With an expanded vector of design variables, $\mathbf{d}$, we must reassess Equation 2.50. Considering the physical dissimilarity between the two types of variables contained in $\mathbf{d}$ and invoking the product rule, we have an updated form for this enabling derivative

[17] In most implementations of GSM with geometry optimization (i.e., nodal updating), the locations of some nodes will be fixed in some or all coordinate directions, such as at points of load application or fixity. The partial vector of design variables $\mathbf{c}$ will then *not* contain all nodal coordinates. However, for the sake of a general implementation whereby some coordinates x_i and/or y_i might later be algorithmically removed from $\mathbf{c}$, we here present the relations for calculating sensitivities with respect to all nodal coordinates.

$$\frac{\partial \hat{\mathbf{K}}}{\partial d_i} = \begin{cases} \mathbf{T}_e^{\#,T}\mathbf{T}_e^{\phi,T}\dfrac{\partial \bar{\mathbf{k}}_e}{\partial t_i}\mathbf{T}_e^{\phi}\mathbf{T}_e^{\#}, & i \leq N_{el}, \\ \hat{\Sigma}\left(\dfrac{\partial \mathbf{T}_e^{\phi,T}}{\partial c_j}\bar{\mathbf{k}}_e\mathbf{T}_e^{\phi} + \mathbf{T}_e^{\phi,T}\dfrac{\partial \bar{\mathbf{k}}_e}{\partial c_j}\mathbf{T}_e^{\phi} + \mathbf{T}_e^{\phi,T}\bar{\mathbf{k}}_e\dfrac{\partial \mathbf{T}_e^{\phi}}{\partial c_j}\right), & i > N_{el} \text{ and } j = i - N_{el}, \end{cases} \tag{2.55}$$

All relations required for $i \leq N_{el}$ have been previously provided in Section 2.3.2 (cf. Equation 2.54), though it is critical to note that the matrices $\bar{\mathbf{k}}_e^1$ and $\bar{\mathbf{k}}_e^2$ are no longer constant and must be updated for every change in the nodal positions (i.e., every algorithmic iteration). Here we continue with the development of needed relations for $i > N_{el}$. By invoking the property that

$$\frac{\partial}{\partial \alpha}(\mathbf{B}^T) = \left[\frac{\partial \mathbf{B}}{\partial \alpha}\right]^T,$$

for any scalar α and matrix $\mathbf{B}$, this involves only calculation of $\partial\bar{\mathbf{k}}_e/\partial c_j$ and $\partial\mathbf{T}_e^{\phi}/\partial c_j$, where $c_j = d_{(i-N_{el})}$. This endeavor of calculating the needed derivatives is largely one of index management, given the differentiation of element-level quantities with respect to global variables that is required.

The derivatives $\partial\bar{\mathbf{k}}_e/\partial c_j$ are computed by first applying the chain rule to $\bar{\mathbf{k}}_e$ (for example, as written in Equation 2.52), which gives

$$\frac{\partial \bar{\mathbf{k}}_e}{\partial c_j} = \frac{\partial \bar{\mathbf{k}}_e}{\partial L_e}\frac{\partial L_e}{\partial c_j}. \tag{2.56}$$

In the case of a two-dimensional frame constructed from members having rectangular cross-sections, the first term can be written as

$$\frac{\partial \bar{\mathbf{k}}_e}{\partial L_e} = \frac{Ewt_e}{L_e^2}\begin{bmatrix} -1 & 0 & 0 & 1 & 0 & 0 \\ 0 & -3\frac{t^2}{L^2} & -\frac{t^2}{L} & 0 & 3\frac{t^2}{L^2} & -\frac{t^2}{L} \\ 0 & -\frac{t^2}{L} & -\frac{t^2}{3} & 0 & \frac{t^2}{L} & -\frac{t^2}{6} \\ 1 & 0 & 0 & -1 & 0 & 0 \\ 0 & 3\frac{t^2}{L^2} & \frac{t^2}{L} & 0 & -3\frac{t^2}{L^2} & \frac{t^2}{L} \\ 0 & -\frac{t^2}{L} & -\frac{t^2}{6} & 0 & \frac{t^2}{L} & -\frac{t^2}{3} \end{bmatrix}_e. \tag{2.57}$$

To compute $\partial L_e/\partial c_j$, we invoke the assembly matrix $\mathbf{A}$ and nodal position matrix $\boldsymbol{P}$ previously introduced in Section 2.3.1. Given these structure-defining matrices, we can write

$$\frac{\partial L_e}{\partial c_j} = \frac{\partial}{\partial c_j}\sqrt{\Delta x_e^2 + \Delta y_e^2} = \frac{1}{L_e}\begin{cases} \Delta x_e, & (j+1)/2 = A_{e2} = n_e^2, \\ -\Delta x_e, & (j+1)/2 = A_{e1} = n_e^1, \\ \Delta y_e, & j/2 = A_{e2} = n_e^2, \\ -\Delta y_e, & j/2 = A_{e1} = n_e^1, \\ 0, & \text{otherwise.} \end{cases} \tag{2.58}$$

Recall that the essential values Δx_e and Δx_e are computed directly from $\mathbf{A}$ and $\boldsymbol{P}$ via

$$\Delta x_e = P_{A_{e2}1} - P_{A_{e1}1}, \quad \Delta y_e = P_{A_{e2}2} - P_{A_{e1}2}.$$

Finally, $\partial \mathbf{T}_e^{\phi}/\partial c_j$ must be computed. Note that if the trivial condition of Equation 2.58 is met, $\partial \mathbf{T}_e^{\phi}/\partial c_j = \mathbf{0}$ as well. Otherwise, per Equation 2.42 we have

$$\frac{\partial \mathbf{T}_e^{\phi}}{\partial c_j} = \frac{\partial}{\partial c_j}(L_e^{-1}\mathbf{T}_e^{\Delta}) = \frac{\partial L_e^{-1}}{\partial c_j}\mathbf{T}_e^{\Delta} + L_e^{-1}\frac{\partial \mathbf{T}_e^{\Delta}}{\partial c_j}, \tag{2.59}$$

where

$$\frac{\partial L_e^{-1}}{\partial c_j} = -L_e^{-2}\frac{\partial L_e}{\partial c_j} \tag{2.60}$$

and $\partial L_e/\partial c_j$ has been given in Equation 2.58. The last term to be calculated in the assessment of stiffness sensitivity is

$$\frac{\partial \mathbf{T}_e^{\Delta}}{\partial c_j} = \begin{bmatrix} \alpha & \beta & 0 & 0 & 0 & 0 \\ -\beta & \alpha & 0 & 0 & 0 & 0 \\ 0 & 0 & 0 & 0 & 0 & 0 \\ 0 & 0 & 0 & \alpha & \beta & 0 \\ 0 & 0 & 0 & -\beta & \alpha & 0 \\ 0 & 0 & 0 & 0 & 0 & 0 \end{bmatrix}, \tag{2.61}$$

where

$$\begin{aligned} &\alpha = 1, \ \beta = 0 && \text{if} \quad (j+1)/2 = A_{e2} = n_e^2, \\ &\alpha = -1, \ \beta = 0 && \text{if} \quad (j+1)/2 = A_{e1} = n_e^1, \\ &\alpha = 0, \ \beta = 1 && \text{if} \quad j/2 = A_{e2} = n_e^2, \\ &\alpha = 0, \ \beta = -1 && \text{if} \quad j/2 = A_{e1} = n_e^1. \end{aligned}$$

For many gradient-based optimizers, the sensitivities with respect to the constraints are also needed. As given in Algorithm 5,

$$\frac{\partial \hat{v}_f}{\partial t_i} = \frac{\partial \hat{v}_f}{\partial d_i} = \frac{L_i}{A}, \quad i = 1 \ldots N_{el}. \tag{2.62}$$

When considering nodal position updates, we also must consider the sum total of length changes across all elements (only some of which are associated with any nodal coordinate c_j). Specifically,

$$\frac{\partial \hat{v}_f}{\partial c_j} = \sum_{e=1}^{N_{el}} \frac{\partial \hat{v}_f}{\partial L_e} \frac{\partial L_e}{\partial c_j} = \sum_{e=1}^{N_{el}} \frac{t_e}{A} \frac{\partial L_e}{\partial c_j}, \quad j = 1 \ldots 2N_{nd}, \tag{2.63}$$

which simply requires Equation 2.58 for evaluation.

To summarize, Algorithm 6 is provided. This is a straightforward extension of Algorithm 5; however, the additional calculations required are explicitly addressed. Note that in a common truss/frame configuration such as the ×-type shown in Figure 2.13, the ratio of elements to nodes will tend toward 4:1[18] as the number of diagonally support square units increases. Given the two-dimensional nature of this problem, the additional consideration of nodal positions as design variables then increases the size of the optimization problem by a factor of 1.5–2.0 for this type of ground structure.

Example 2.6 To illustrate the concepts of assembly as they relate to truss and frame geometry optimization, we now consider the simple unit "ground structure" of Figure 2.15, where the initial x–y positions of each of four nodes are given. In the context of an iterative optimization scheme, we have

$$\mathbf{c}^{(0)} = [\ 0.0 \quad 1.0 \quad 0.0 \quad 0.0 \quad 1.0 \quad 0.0 \quad 1.0 \quad 1.0\ ]^T,$$

which, along with $\mathbf{t}^{(0)}$, is altered by the optimization process. As an example of the kinds of calculations needed to enable this process, consider element 5 in particular. We see per Equation 2.57 that

$$\frac{\partial \bar{\mathbf{k}}_5^{(0)}}{\partial L_5} = \frac{Ewt_5^{(0)}}{2} \begin{bmatrix} -1 & 0 & 0 & 1 & 0 & 0 \\ 0 & -3\frac{(t_5^{(0)})^2}{2} & -\frac{(t_5^{(0)})^2}{\sqrt{2}} & 0 & 3\frac{(t_5^{(0)})^2}{2} & -\frac{(t_5^{(0)})^2}{\sqrt{2}} \\ 0 & -\frac{(t_5^{(0)})^2}{\sqrt{2}} & -\frac{(t_5^{(0)})^2}{3} & 0 & \frac{(t_5^{(0)})^2}{\sqrt{2}} & -\frac{(t_5^{(0)})^2}{6} \\ 1 & 0 & 0 & -1 & 0 & 0 \\ 0 & 3\frac{(t_5^{(0)})^2}{2} & \frac{(t_5^{(0)})^2}{\sqrt{2}} & 0 & -3\frac{(t_5^{(0)})^2}{2} & \frac{(t_5^{(0)})^2}{\sqrt{2}} \\ 0 & -\frac{(t_5^{(0)})^2}{\sqrt{2}} & -\frac{(t_5^{(0)})^2}{6} & 0 & \frac{(t_5^{(0)})^2}{\sqrt{2}} & -\frac{(t_5^{(0)})^2}{3} \end{bmatrix}$$

[18] Any node not at a boundary is connected to eight elements, each of which is 1/2 shared with another node, hence the 4:1 ratio in the limit that boundaries become irrelevant.

and per Equation 2.58 that

$$\frac{\partial L_5^{(0)}}{\partial \mathbf{c}} = \frac{1}{\sqrt{2}}[\ -1.0 \quad 1.0 \quad 0 \quad 0 \quad 1.0 \quad -1.0 \quad 0 \quad 0\ ]^T.$$

Note that $L_5^{(0)} = \sqrt{2}$ and the various array components are all iteratively updated values (i.e., for all $k \neq 0$).

Further, we require per Equation 2.60

$$\frac{\partial L_5^{-1(0)}}{\partial \mathbf{c}} = -\frac{1}{2}\frac{\partial L_5^{(0)}}{\partial \mathbf{c}},$$

while per Equation 2.61 we have the *fixed* matrices

$$\frac{\partial \mathbf{T}_5^{\Delta}}{\partial c_1} = \frac{\partial \mathbf{T}_5^{\Delta}}{\partial c_2} = \frac{\partial \mathbf{T}_5^{\Delta}}{\partial c_3} = \mathbf{0},$$

where only the first three matrices are shown.

Finally, the sensitivities of the constraints with respect to the design variables must be calculated. Per Equations 2.62 and 2.63 we have

$$\frac{\partial \hat{v}_f^{(0)}}{\partial \mathbf{t}} = \frac{1}{A}[\ 1.0 \quad 1.0 \quad 1.0 \quad 1.0 \quad \sqrt{2}\ ]^T$$

and

$$\frac{\partial \hat{v}_f^{(0)}}{\partial \mathbf{c}} = \frac{1}{A}\left[\sum_{e=1}^{4} t_e^{(0)}\frac{\partial L_e^{(0)}}{\partial \mathbf{c}} + t_5^{(0)}\frac{1}{\sqrt{2}}[\ -1.0 \quad 1.0 \quad 0 \quad 0 \quad 1.0 \quad -1.0 \quad 0 \quad 0\ ]^T\right],$$

such that the full vector

$$\frac{\partial \hat{v}_f^{(0)}}{\partial \mathbf{d}} = \frac{\partial \hat{v}_f^{(0)}}{\partial \mathbf{t}} \bigcup \frac{\partial \hat{v}_f^{(0)}}{\partial \mathbf{c}}$$

can be populated.

Algorithm 6 General algorithm for implementation of ground structure-based topology and geometry optimization; utilization of a preexisting gradient-based optimizer is assumed

Require: A preconfigured statically stable ground structure composed of truss or frame elements such that $\hat{\mathcal{B}}_m = \bigcup_{e=1}^{N_{el}} \mathcal{E}_e$ (i.e., full definition of $\mathbf{A}$ and $\boldsymbol{P}^{(0)}$) and associated boundary conditions, the Young's modulus and out-of-plane thickness for all elements E and w, respectively, and algorithmic parameters t_{min}, t_{max}, v_f^g, and `tol`

Initialize $k = 0$, $t_i = t_{max}\ \forall\, i = 1 \ldots N_{el}$
while Convergence metric > `tol` **do** ▷ Begin optimization loop
    $k \leftarrow k + 1$
    Compute $\hat{\mathbf{K}}$ per Equation 2.45
    Considering boundary conditions, solve $\hat{\mathbf{K}}\hat{\mathbf{q}} = \hat{\mathbf{Q}}$
    Compute $\hat{J}_c$ per Equation 2.37
    Compute $\hat{v}_f$ per Equation 2.39
    for $i = 1 \ldots N_{el}$ **do**
        Compute $\partial\hat{J}_c/\partial t_i$ per Equations 2.47, 2.49, and 2.54
        Compute $\partial\hat{v}_f/\partial t_i$ per Equation 2.62
    end for
    for $i = N_{el} + 1 \ldots N_{el} + 2N_{nd}$ **do**
        $j = i - N_{el}$
        Compute $\partial\hat{J}_c/\partial c_j$ per Equations 2.47, 2.49, and 2.55 by also considering Equations 2.56 and 2.59
        Compute $\partial\hat{v}_f/\partial c_j$ per Equation 2.63
    end for
    Combine sensitivity vectors such that $\partial\hat{J}_c/\partial\mathbf{d} = \partial\hat{J}_c/\partial\mathbf{t} \cup \partial\hat{J}_c/\partial\mathbf{c}$ and $\partial\hat{v}_f/\partial\mathbf{d} = \partial\hat{v}_f/\partial\mathbf{t} \cup \partial\hat{v}_f/\partial\mathbf{c}$; eliminate sensitivities associated with fixed nodal positions (e.g., as associated with boundary conditions)
    By providing magnitudes $\hat{J}_c$ and $\hat{v}_f$ in addition to vectors of gradient terms $\partial\hat{J}_c/\partial\mathbf{d}$ and $\partial\hat{v}_f/\partial\mathbf{d}$ to the selected gradient-based optimizer, update all element thicknesses $t_{min} \leq t_i \leq t_{max}$ and nodal coordinates c_j to minimize $\hat{J}_c$ subject to the constraint that $\hat{v}_f \leq v_f^g$; acquire convergence metric
    Map the optimizer-determined vector $\mathbf{c} \rightarrow \boldsymbol{P}^{(k)}$ to update all nodal positions
end while

Example 2.7 To examine the benefits of applying both geometric and sizing optimization to a given ground structure, the benchmark problem for a long cantilever is revisited. The same initial ground structure consisting of 21 structural members and 10 nodes is considered (cf. Figure 2.16(a)); specifying that spatial coordinates of all nodes upon which boundary conditions are applied must remain constant, this results in a total of 35 design variables (21 structural member thicknesses and the (x, y) coordinates of the 7 free nodes). These variables are

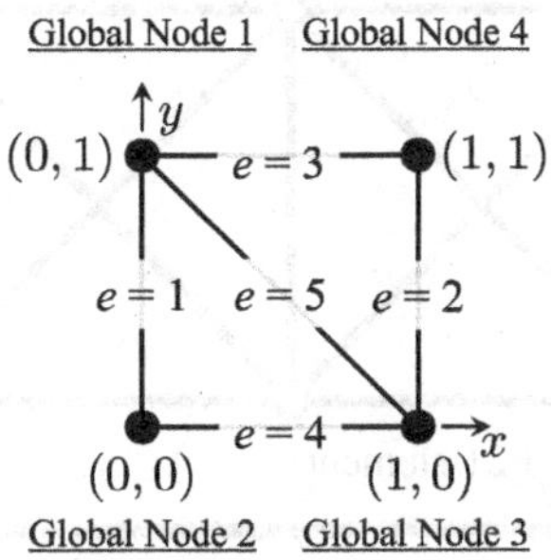

$$\mathbf{A} = \begin{bmatrix} 1 & 2 \\ 4 & 3 \\ 1 & 4 \\ 2 & 3 \\ 3 & 1 \end{bmatrix} \qquad \mathbf{P}^{(0)} = \begin{bmatrix} 0.0 & 1.0 \\ 0.0 & 0.0 \\ 1.0 & 0.0 \\ 1.0 & 1.0 \end{bmatrix}$$

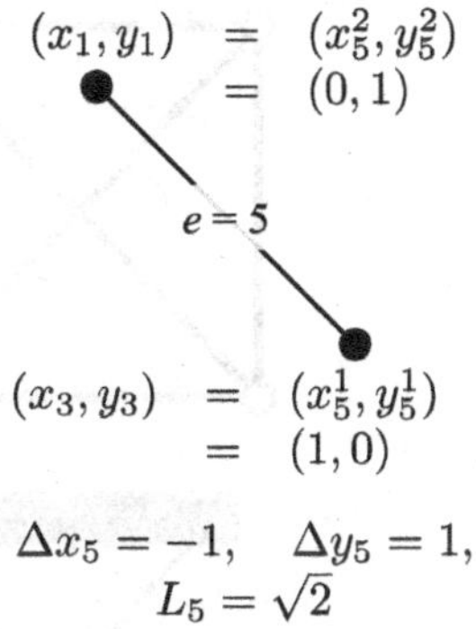

Figure 2.15 Structural layout and critical quantities associated with Example 2.3.3.

determined using the same implementation of the SLSQP algorithm with a volume constraint of $v_f = 0.30$. The result of the geometric optimization on the ground structure is shown in Figure 2.16(c); while changes in the location of many nodes appear marginal, one node in particular along the bottom of the structure is shifted upward. This serves to shorten the load path between the fixed boundary condition on the bottom-left of the structure and the applied load on the upper-right, which in turn makes additional material available for sizing optimization. This is noticeable when comparing the result in Figure 2.16(b) with that obtained without geometric optimization (cf. Figure 2.14(b)), as this additional material results in topology features and additional members that further reinforce the structure. The compliance of this solution is $\hat{J}_c = 0.46$ compared to $\hat{J}_c = 0.47$ for the traditional GSM (a reduction of 1.9%) and was obtained in 56 functional evaluations versus the 42 required for just sizing optimization.

2.4 Boundary Representation

So far, the approaches discussed have featured explicit topological representations, meaning that at least some subset of the internal variables directly defines the presence/absence of material at points throughout the design domain (cf. Section 1.1). As mentioned in Section 2.2, one drawback to these approaches is their dependence on discretization resolution quality for accurate structural response predictions. This often requires a large number of design variables such that problems can eventually become intractable. As an alternative, we recall Figure 1.1 and consider how the configuration of the *boundary* between material and void might be explicitly considered. This requires some parameterized definition of surfaces (two- or three-dimensional, depending on the design problem), where the number of parameters will generally be substantially lower than in explicit topological representations. An optimization

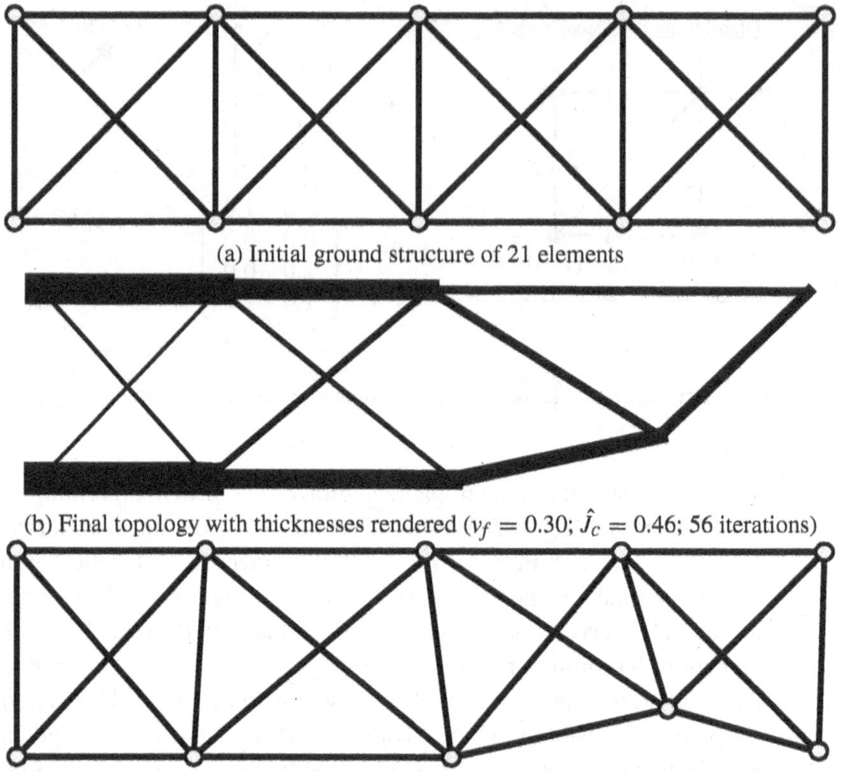

(a) Initial ground structure of 21 elements

(b) Final topology with thicknesses rendered ($v_f = 0.30$; $\hat{J}_c = 0.46$; 56 iterations)

(c) Solution rendered considering only geometric optimization (i.e., updated thicknesses not shown)

Figure 2.16 Optimized layouts for the long cantilever design problem of Example 2.7 (cf. Example 2.1) using GSM with geometric optimization. Note that compliance is reduced by 1.9% compared to traditional GSM.

over these parameters toward some objective(s) subject to some constraint(s) would comprise a boundary representation approach to topological optimization.

Among all possibilities for implementation, to date only one algorithm has received substantial attention and is reviewed herein.

2.4.1 Level Set Method

Level set methods were pioneered by Osher & Sethian (1998) in the context of curvature flow, and have since seen several variants developed for topology optimization. In the first and still most popular reference book on SIMP (Bendsøe & Sigmund 2003), the authors note the emergence of the level set method, saying "[t]he application of level set techniques for topological design has also been proposed recently ... This technique is in an initial stage of development." In the two decades since, the level set method has established itself as the leading alternative to SIMP among conventional gradient-based topology optimization methods. Here, a brief description of the level set method based on the volume expression of the shape derivative reported in Laurain (2018) is presented. This variant has the advantage of having an educational code, written in Python, readily available for experimenting at the paper cited. The reader interested in more information on the level set method in general and its variants is directed to the review papers (Gain & Paulino 2013, van Dijk et al. 2013) and the monograph (Osher & Fedkiw 2003).

Let $\mathcal{B} \in \mathbb{R}^d$, $d = 2,3$ be a fixed domain, and define on $\mathcal{B}$ an evolving level set function ϕ_t, such that

$$\mathcal{B}_m(t) = \{x \mid \phi(x,t) < 0\}, \tag{2.64}$$

which defines the layout of the structure. If $\nabla\phi(x,t) \neq 0$ on the boundary of $\mathcal{B}_{m,t}$, then

$$\partial\mathcal{B}_m(t) = \{x \mid \phi(x,t) = 0\};$$

in other words, the boundary of structure coincides with the zero level set of $\phi(x,t)$. The empty space $\mathcal{B}_v = \mathcal{B}\backslash\mathcal{B}_m$ is commonly filled with an "ersatz" or weak material; thus, we define a composite material in $\mathcal{B}$ such that

$$\mathbf{E} = \mathbf{E}_0\mathbb{I}_{\mathcal{B}_m(t)} + \epsilon\mathbf{E}_0\mathbb{I}_{\mathcal{B}_v(t)}, \tag{2.65}$$

where $\mathbf{E}_0$ is the elasticity tensor of a given material and $\epsilon > 0$ is a small number. The boundary of the design domain $\partial\mathcal{B}$ is partitioned into $\partial\mathcal{B} = \partial\mathcal{B}_D \cup \partial\mathcal{B}_N \cup \partial\mathcal{B}_F$, where the displacement is fixed on $\partial\mathcal{B}_D$, an external traction g is exerted over $\partial\mathcal{B}_N$, and $\partial\mathcal{B}_F$ is the free portion of the boundary. The equilibrium equations in $\mathcal{B}$ for small displacements are then given as

$$\nabla.(\mathbf{E}e(u)) = 0,\ x \in \mathcal{B}, \tag{2.66}$$

$$u = 0,\ x \in \partial\mathcal{B}_D, \tag{2.67}$$

$$\mathbf{E}e(u)n = g,\ x \in \partial\mathcal{B}_N, \tag{2.68}$$

$$\mathbf{E}e(u)n = 0,\ x \in \partial\mathcal{B}_F, \tag{2.69}$$

where $e(u)$ is the symmetric gradient of the displacement field u. As in previous sections, the functional of interest here is the compliance, which is given as

$$J_c = \int_{\mathcal{B}} \mathbf{E}e(u) : e(u). \tag{2.70}$$

Therefore, we can formally pose the optimization problem as below.

Find ϕ that minimizes the compliance functional, J_c, subject to:

- the satisfaction of equilibrium (cf. Equations 2.66–2.69)
- $v_f = v_f^g$,

where

$$v_f = \frac{\int \mathcal{B}_{m(t_0)}\, dx}{V_0},$$

V_0 is the total volume of the design domain $\mathcal{B}$, and v_f^g is a *goal* volume fraction to be maintained during the material location selection process.

To solve the problem, this approach performs a shape optimization of the structural layout in a manner akin to the steepest descent method, a brief description of which is provided in Appendix A. Evolving the level set function ϕ requires the computation of the cost functional; here, we consider a functional that combines both the compliance and volume:

$$\mathcal{J} = J_c + \Lambda V, \tag{2.71}$$

where J_c is the compliance functional from Equation 2.70, $V = \int \mathcal{B}_{m(t_0)}\, dx$ is the volume of the structure, and $\Lambda > 0$ is a penalizing factor (the larger the value of Λ, the smaller the volume of the optimized layout). As discussed in Appendix A, the gradient of a function can be computed as a representation of the differential of a function on an inner product space. Thus, the gradient of the cost function in Equation 2.71 can be determined by defining an inner product on the level set function, computing the differential of Equation 2.71, and using the representation equation to calculate the gradient. The differentials of both J_c and V are well-known (see Laurain (2018)) and are given by

$$dJ_c(\mathcal{B}_m(t)) \cdot \xi = \int_{\mathcal{B}} (2\nabla u^T \mathbf{E} e(u) - \mathbf{E} e(u) : e(u) I_d) : \nabla \xi, \tag{2.72}$$

and

$$dV(\mathcal{B}_m(t)) \cdot \xi = \int_{\mathcal{B}} \nabla . \xi, \tag{2.73}$$

respectively. An inner product suitable for elasticity problems, which involves both displacement and strain fields, is defined as

$$\beta(\theta, \xi) = \int_{\mathcal{B}} \alpha_1 \nabla \theta : \nabla \xi + \alpha_2 \theta \xi,$$

where $\alpha_1 > 0$ and $\alpha_2 > 0$ are constants (in Laurain (2018), $\alpha_1 = 1$ and $\alpha_2 = 1/10$). This inner product, along with the cost function differential computed using Equations 2.72–2.73, can then be substituted into the representation equation

$$\beta(\nabla J, \xi) = dJ \cdot \xi \quad \forall \xi \tag{2.74}$$

to determine ∇J. As defined, the negative of the gradient $\theta = -\nabla J$ satisfies $dJ(\mathcal{B}_m(t), \theta) = B(\theta, \theta) < 0$ (i.e., θ provides a descent direction as expected).

Having determined the search direction to update the level set function, the next step is to evolve ϕ along θ. Since the level set method is formulated in the continuum, an advection equation evolving ϕ along θ can be derived in two steps: (i) direct differentiation of $\phi(x(t), t) = 0$, where $x(t)$ is a point at the boundary of the structure and moving along θ (i.e., $\dot{x}(t) = \theta(x(t))$), and (ii) extending that equation to $\mathcal{D}$. Using the chain rule, the first step results in the following advection equation:

$$\partial_t(x(t), t) + \theta(x(t))\nabla\phi(x(t), t)) = 0.$$

In the second step, we extend this equation to $\mathcal{B}_{m,t}$ via the equation:

$$\partial_t(x, t) + \theta(x)\nabla\phi(x, t)) = 0. \tag{2.75}$$

Given an initial level set function $\phi(x, 0) = \phi_0(x)$, Equation 2.75 then prescribes the evolution of ϕ to optimize the layout shape of the structure. However, it should be noted that the equations describing this evolution, namely Equations 2.72, 2.73, 2.74, and 2.75 form a highly nonlinear system of equations: ∇J depends on the level set function ϕ, whose evolution in turn depends on ∇J in a coupled two-way manner. In addition, approximations to Equation 2.75 generate dissipation and dispersion errors that require special discretization for numerical accuracy (Vichnevetsky & Bowles 1982, Osher & Fedkiw 2003).

In this section, we consider a rectangular design domain $\mathcal{D}$ that is discretized by a uniform Cartesian mesh. To contend with the difficulties of approximating Equation 2.75, this equation is first evolved in time using a forward Euler time discretization

$$\phi_{ij}^{k+1} = \phi_{ij}^k - \Delta t \hat{H}^{LF}(p^-, p^+, q^-, q^+), \tag{2.76}$$

where $\phi_{ij}^k = \phi(k\Delta t, x_{ij})$, x_{ij} are the Cartesian nodes of the mesh, $\Delta t > 0$ is the time increment, and $\hat{H}^{LF}(p^-, p^+, q^-, q^+)$ is the following Lax–Friedrichs flux:

$$\hat{H}^{LF}(p^-, p^+, q^-, q^+) = H\left(\frac{p^- + p^+}{2}, \frac{q^- + q^+}{2}\right) - \frac{1}{2}(p^+ - p^-)\,|\theta_x| - \frac{1}{2}(q^+ - q^-)\,|\theta_y|, \tag{2.77}$$

with $\theta = (\theta_x, \theta_y)$, and

$$p^- = D_x^- \phi_{ij} = \frac{\phi_{ij} - \phi_{i-1,j}}{\Delta x}, \tag{2.78}$$

$$p^+ = D_x^+ \phi_{ij} = \frac{\phi_{i+1,j} - \phi_{i,j}}{\Delta x}, \tag{2.79}$$

$$q^- = D_y^- \phi_{ij} = \frac{\phi_{ij} - \phi_{i,j-1}}{\Delta y}, \tag{2.80}$$

$$q^+ = D_y^+ \phi_{ij} = \frac{\phi_{i,j+1} - \phi_{ij}}{\Delta y}. \tag{2.81}$$

Here $H(\nabla\phi) = \theta\nabla\phi$ is called the Hamiltonian. For numerical accuracy, a reinitialization procedure is performed every five iterations. Specifically, the following equation is solved until the steady-state is attained:

$$\partial_\tau \phi(\tau,x) + S(\phi)(\nabla\phi(\tau,x) - 1) = 0,\ (\tau,x) \in \mathbb{R}_+ \times D, \tag{2.82}$$

$$\phi(0,x) = \phi(t,x),\ x \in D, \tag{2.83}$$

where

$$S(\phi) = \frac{\phi}{\sqrt{\phi^2 + \epsilon_s^2 |\nabla\phi|^2}} \tag{2.84}$$

is an approximation of the sign function and $\epsilon_s = \min(\Delta x, \Delta y)$. The solution of equation is effected via an explicit upwind scheme as follows:

$$\phi_{ij}^{k+1} = \phi_{ij}^k - \Delta t K(p^-, p^+, q^-, q^+), \tag{2.85}$$

where

$$K(p^-, p^+, q^-, q^+) = \max(S(\phi_{ij}, 0)K^+ + \min(S(\phi_{ij}, 0)K^-, \tag{2.86}$$

and

$$K^+ = \left[\max(p^-,0)^2 + \min(p^+,0)^2 + \max(q^-,0)^2 + \min(q^+,0)^2\right]^{\frac{1}{2}}, \tag{2.87}$$

$$K^- = \left[\min(p^-,0)^2 + \max(p^+,0)^2 + \min(q^-,0)^2 + \max(q^+,0)^2\right]^{\frac{1}{2}}, \tag{2.88}$$

with p^-, p^+, q^-, q^+ computed as in Equations 2.78–2.81.

Example 2.8 For a direct comparison with SIMP, we present the optimized solution to Example 2.1 obtained using the level set method published in Laurain (2018) and presented in Algorithm 7. The material properties remain consistent with Example 2.1 and the initial layout defined by ϕ is illustrated in Figure 2.17(a); note that this represents a typical initial layout for level set methods. To satisfy the volumetric constraint, a two-step strategy was performed. First, the framework considered $\epsilon = 0.001$ and $\Lambda = 0.015$ (see Equations 2.65 and 2.71, respectively) with a reinitialization step every five iterations; this value of ϵ is recommended in Laurain (2018), and Λ was chosen by trial and error to provide a reasonable volume fraction for the optimized layout. The optimized layout obtained in this first step is shown in Figure 2.17(b) and required 235 iterations for convergence. Next, the values of ϵ and Λ were reduced to 10^{-14} and 0.014, respectively, to adjust the stiffness of the ersatz material to that of the SIMP implementation. With these updated parameters, the run

Algorithm 7 Basic algorithm for implementation of level set topology optimization

Require: A design domain $\mathcal{B}$ and associated boundary conditions, an initial level set function ϕ_{ij}^{1}, and associated line search parameters.

Initialize $k = 1$, $\Delta t = 0.5$, $N_{LS,max} = 3$, $k_{max} = 300$
while $k < k_{max}$ and $\max_{i=1,\ldots,4} |\mathcal{J}(k-i) - \mathcal{J}(k)| < (\mathcal{J}(k)/2E4)$ **do** ▷ Begin optimization loop
  Compute $\mathcal{B}_m(k)$ per Equation 2.64 and discretize
  Solve $\hat{\mathbf{K}}\hat{\mathbf{u}} = \hat{\mathbf{f}}$
  Compute $\mathcal{J}$ per Equation 2.71
  if $k > 1$, $\mathcal{J}(k) > \mathcal{J}(k-1)$, and $N_{LS} < N_{LS,max}$ **then** ▷ Reject current time step
    $N_{LS} \leftarrow N_{LS} + 1$
    Update Δt via line search method
    Update ϕ_{ij}^{k} via Equation 2.76
  else ▷ Accept current time step
    $N_{LS} \leftarrow 0$
    Choose new Δt based on convergence behavior
    Calculate θ via Equation 2.74
    Calculate $\hat{H}^{LF}$ via Equations 2.77–2.81
    Calculate ϕ_{ij}^{k+1} via Equation 2.76
    $k \leftarrow k + 1$
    if $k = 5$ **then** ▷ Reinitialize ϕ for numerical stability
      Compute S per Equation 2.84
      Compute K via Equations 2.86–2.88
      Reinitialize ϕ_{ij}^{k+1} via Equation 2.85
    end if
  end if
end while

was restarted from the solution obtained in the first step, and an optimized solution with $J_c = 0.506$ and $A = 30.0$ is obtained as shown in Figure 2.17(c). For this example, the level set method provides a lower compliance than the SIMP method, albeit with an increased number of functional and sensitivity calculations (2,779 for level set, 990 for SIMP). The level set optimized layout also results in different load paths, with a more symmetric solution obtained with SIMP for the region starting at the support up to the middle portion of the structure. Figure 2.17(d) depicts the solution obtained when applying the parameters from the second step to the initial layout and demonstrates the effect that the initial layout has on the resulting optimized structure. This dependency is an inevitable characteristic of any gradient-based method in optimization problems that are nonconvex, which almost all real engineering problems are.

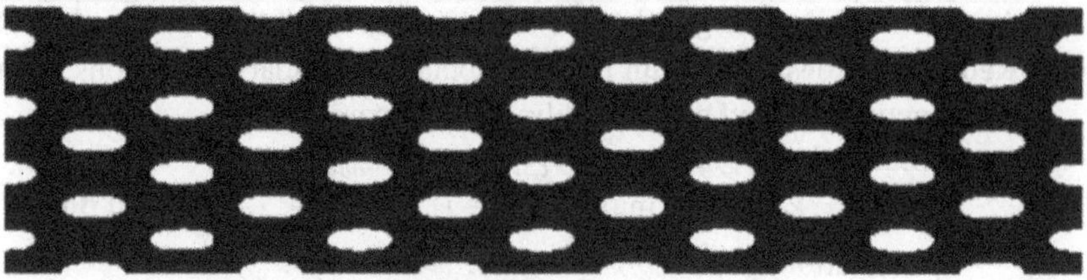

(a) Level set initial layout

(b) Step 1 solution ($v_f = 0.29$; $\hat{J}_c = 0.52$)

(c) Step 2 solution ($v_f = 0.30$; $\hat{J}_c = 0.51$)

(d) Single step solution ($v_f = 0.332$; $\hat{J}_c = 0.49$)

Figure 2.17 Optimized layouts for the long cantilever design problem of Example 2.8 (cf. Example 2.1) using the Level Set Method (600 × 150 mesh; 2,004 iterations for the combined Step 1 and Step 2 solutions).

2.5 Alternative and Emerging Approaches

Since the initial publication of the seminal text on SIMP (Bendsøe & Sigmund 2003) and the popular industrial and commercial acceptance of that method, the field of topology optimization has witnessed an explosion in its interest and publications. It would be impossible to do justice to all emerging work in the area. In this section a few prominent methods beyond those reviewed in the previous sections above are summarized and references are cited for the benefit of the interested reader.

2.5.1 Pixel/Voxel-Based

The variational growth approach operates in the context of continuum element representations (Junker & Hackl 2015). It uses an internal variable that locally quantifies the compliance of the body and derives its developmental equation via the principle of minimizing the dissipation potential. The resulting relation acts as an elasto-visco-plastic evolution equation for the internal variable. This method brings thermodynamics of material modeling into the realm of topology optimization and thus provides a physics analogue for developing solutions to issues affecting SIMP. For instance, the checkerboard pattern phenomenon has been directly addressed via this approach.

Another pixel/voxel-based approach is the geometrical projection method (GPM) (Norato, Bell & Tortorelli 2015). Similar to other alternatives presented in this book, GPM is based on collections of basic shapes to represent engineering structures. While many approaches introduced herein consider long slender trusses or frames of unspecified cross-sectional shape, GPM employs rounded rectangles in a 2-D scheme. However, GPM embeds such basic shapes in a SIMP design domain to obtain the sensitivities needed to drive the layout optimization process. The main advantages of GPM relative to the more established SIMP include a drastically reduced number of design variables, easier interpretation of the resulting structure, and better control of manufacturing parameters (e.g., member size). Figure 2.18 illustrates the application of GPM to the benchmark problem of topologically optimizing a long cantilever as considered in Example 2.1. Here a publicly available educational code (Smith & Norato 2020) with default parameters is used. For a direct comparison, the optimized layout obtained using SIMP with the same mesh resolution is shown in Figure 2.18(c). GPM requires 38% of the iterations to converge and obtains a compliance that is 34% higher relative to the SIMP solution, despite a qualitatively similar topology. The reduction in the number of iterations is a consequence of the reduction in the number of design variables (90,000 with SIMP and 96 with GPM).

2.5.2 Boundary Representation-Based and Hybrid Methods

Another important class of emerging methods that focus on the boundary between material and void regions includes the phase-field methods (Bourdin & Chambolle 2003, Wang & Zhou 2004, Wallin, Ristinmaa & Askfelt 2012). These are based on variational principles associated with phase transition in materials. The main advantages of these phase-field methods are the insights and techniques from phase transition

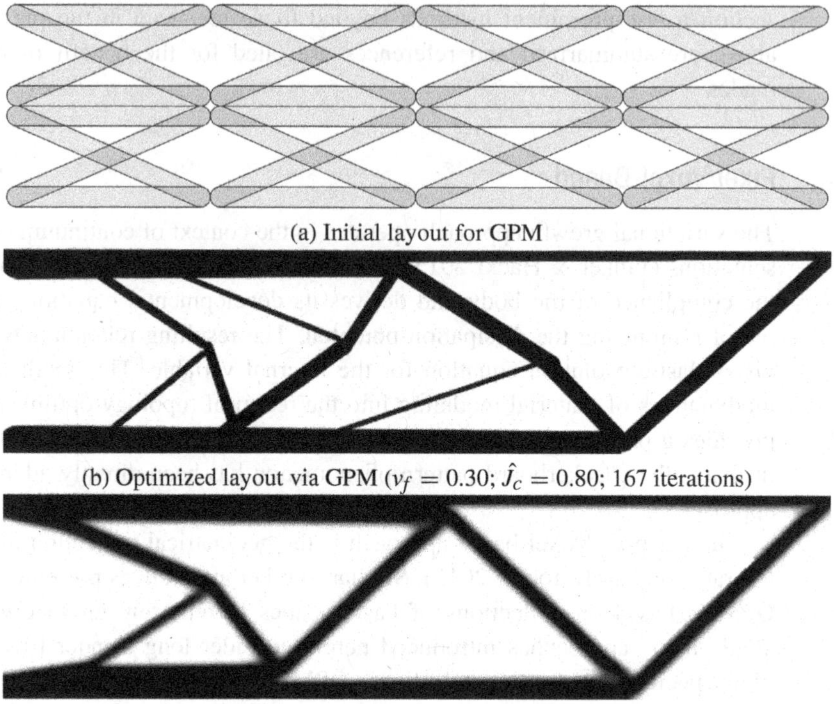

(a) Initial layout for GPM

(b) Optimized layout via GPM ($v_f = 0.30$; $\hat{J}_c = 0.80$; 167 iterations)

(c) Optimized layout via SIMP ($v_f = 0.30$; $\hat{J}_c = 0.60$; 442 iterations)

Figure 2.18 Optimized layouts for the long cantilever design problem (cf. Example 2.1) using GPM and SIMP (600 × 150 mesh).

that can be transferred to topology optimization. For instance, regularization terms can be used to simulate phase separation, where parameters are included that introduce a length scale within the topology optimization that penalizes the perimeter of all material regions in the design. Note that SIMP uses similar approaches to regularize solutions (Bendsøe & Sigmund 2003).

Also similar to SIMP, these methods define a density function $0 \leq \rho(\boldsymbol{x}) \leq 1$ over the design domain $\mathcal{B}$ to fully represent the material distribution. However, to evolve the density, instead of the compliance functional, J_c, phase-field methods employ a regularized functional Φ such that

$$\Phi(\rho, u) = \int_{\mathcal{B}} \left[\left(\frac{1}{\zeta} F(\rho(\boldsymbol{x})) + \frac{\zeta}{2} \left\| \nabla \rho(\boldsymbol{x}) \right\|^2 \right) + 2\eta J_c(\rho(\boldsymbol{x}), u(\boldsymbol{x})) \right] dV, \tag{2.89}$$

where F is a penalization function for gray material, and η and ζ are weighting parameters defining the relative importance of the structure compliance, structure perimeter, and gray material. Equilibrium conditions, volume constraint, and traction boundary

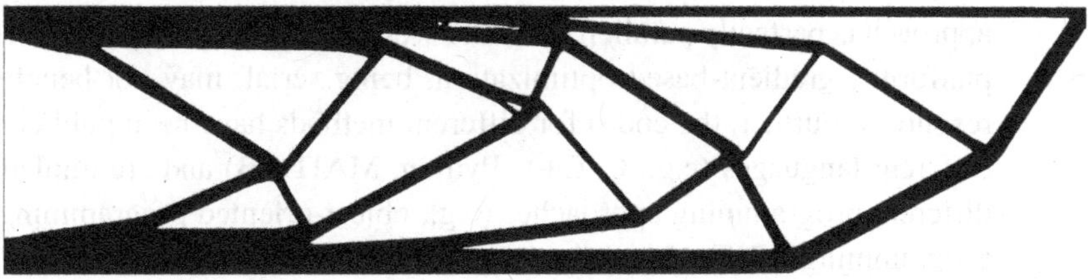

Figure 2.19 Optimized layout for the long cantilever design problem (cf. Example 2.1) using UNVARTOP (600 × 150 mesh; $v_f = 0.30$; $\hat{J}_c = 0.56$, 210 iterations).

conditions are accounted for by using a Lagrangian approach. The evolution of the density $\rho(\boldsymbol{x})$ then proceeds in the general meaning of the phase-field models (Wallin et al. 2012).

As a final emerging method to mention, we briefly discuss the UNsmooth VARiational Topology OPtimization (UNVARTOP) method (Oliver, Yago, Cante & Lloberas-Valls 2019). Similarly to the level set method, the design is defined by the region where a "discriminant function" is positive, while in the more traditional level set method the material regions are associated with negative values of a function. Further, instead of the advection equation commonly employed in Level Set Methods, a Fréchet derivative is used. An implementation of the UNVARTOP method in Matlab is publicly available (Yago, Cante, Lloberas-Valls & Oliver 2021) and is demonstrated herein. Applying it to our benchmark (long cantilever) problem, one arrives at the result shown in Figure 2.19.

2.6 Conclusions

For a much more comprehensive review of the general theory of topology optimization, the curious reader is encouraged to benefit from a range of available monographs (Rozvany 1994, Allaire 2002, Bendsøe & Sigmund 2003, Osher & Fedkiw 2003) and review papers (Eschenauer & Olhoff 2001, Rozvany 2001, Bendsøe, Lund, Olhoff & Sigmund 2005, Rozvany 2009, Gain & Paulino 2013, van Dijk et al. 2013, Deaton & Grandhi 2014, Yago et al. 2021). These references provide insightful and thorough comparative analyses of the performance and idiosyncrasies of most methods. This chapter concludes with a summary of some of the main conclusions of the recommended review papers in addition to observations provided herein.

- One must acknowledge that comparison across broad classes of methods is challenging since many factors affecting the performance of each method must be considered. For instance, a GA as employed in the Lamarckian hybrid GA/SIMP approach is perfectly parallelizable and well-suited to high-performance computing platforms; gradient-based optimization, being serial, may not benefit from these resources. Further, the codes for different methods have been publicly provided in different languages (e.g., C, C++, Python, MATLAB) and are implemented using different programming approaches (e.g., object-oriented programming, procedural programming). Different methods, including their multiple variants, are also better suited to different classes of problems, making equitable comparison difficult. The quality and complexity of an optimized solution are further dependent on chosen optimization algorithm parameters, design representation resolution, and initial layout.

- The effectiveness and efficiency of a method for a given problem depends on the properties of its primary design variable (e.g., binary versus continuous, nodal versus elemental) and enforcement of its constraints. Pixel/voxel-based methods such as SIMP, homogenization, and even variable growth use continuous design variables and must address the gray material problem. This is usually accomplished via a post-processing step that eliminates gray material throughout the domain but negatively affects the performance of the final design. Alternatively, ESO/BESO uses a binary formulation and thus computes "black-and-white" layouts directly.
- Though employing continuous design variables, boundary representation-based methods such as the level set method and UNVARTOP are naturally formulated to capture discrete segregation of material from void and thus also provide sharp "black/white" layouts. Given their use of discrete truss or frame elements to represent structure, GSM and the L-system methods that comprise the remainder of this book also require no gray material post-processing.
- SIMP, homogenization, UNVARTOP, variational growth, phase-field methods, and level set methods all display linear convergence.
- For topological designs requiring stronger, heavier structures, SIMP, UNVARTOP, and level set methods are more appropriate and produce thicker regions when compared to ESO/BESO, GSM, and GPM. Though they require more material, such solid bodies are often easier and less expensive to manufacture.
- Most of the approaches reviewed in this chapter rely on gradient-based optimization and thus are known to converge to locally optimal solutions that are highly dependent on the initial topology or representation. Restarting the optimization process multiple times using a range of initializations may be necessary to find a globally optimal solution. The Lamarckian GA/SIMP hybrid method has been especially developed to avoid this disadvantage; further, the incorporation of evolutionary algorithms into the L-system approaches in the chapters to follow makes them likewise robust against convergence to local optima.

Throughout the remainder of the book, the authors will move on from the pixel/voxel, ground structure and boundary representation-based methods to focus solely on developmental/algorithmic methods, which are not sensitive to resolution, produce no gray material, and do not converge to local optima. Though not strictly limited to *structural* network design, these new methods will be mostly demonstrated in that context with a special emphasis on frame structures (after Section 2.3.1). Fundamental concepts and simple construction algorithms will be introduced in Chapter 3 before we progress to more advanced approaches capable of outperforming all methods introduced in this chapter for some classes of problems.

3 Introduction to L-systems and Turtle Graphics Interpretation

In Chapter 1, we introduced the concepts of structural representation and topological optimization, and in Chapter 2, we provided a wide range of options regarding the latter. Further, we have addressed the motivation for a bioinspired approach to multiobjective design and introduced the EvoDevo approach and the fundamentals of genetic optimization (cf. Section 1.3). In this chapter, we begin our exploration of *developmental* topological optimization, which stands in contrast to the various subtractive approaches of Chapter 2 but rather relies on the "Genome→Development→Phenome" process shown in Figure 1.6. Specifically, we introduce a rewriting system of choice (the L-system), a means for developing over that grammar, and a simple and illustrative option for interpreting the result into a structure or network layout known as *turtle graphics*.

3.1 Introduction to L-systems

At the end of the 1960s, the eminent biologist Aristid Lindenmayer introduced a novel type of rewriting system that would go on to have important implications in a number of fields (Lindenmayer 1968*a*, Lindenmayer 1968*b*). The system, thereafter called the Lindenmayer system (or L-system for short), was initially proposed to model the development of the branched topology in plants; such branched structures are the basis of this chapter. Over the years, L-systems came to represent their own area of technical study (Prusinkiewicz & Lindenmayer 2004), with subtopics ranging from the original model for plant development to the mathematics of formal languages to the representation of virtual plants in computer graphics (Prusinkiewicz & Lindenmayer 2004) to music (Besada 2019). In this chapter, we extend the use of L-systems with their turtle interpretation to describe the topology of the structures in engineering and natural systems. In subsequent chapters, we will show that L-systems can be interpreted using techniques that lead to the representation of network topologies, whether for the purpose of designing structures, circuits, or other systems for which connectivity is critical.

Informally, L-systems are rewriting systems that iteratively generate developmental programs. These programs, when interpreted in an algorithmic manner, can produce graphic models. To begin, the pertinent definitions and objects of L-systems are briefly recalled and illustrated. The reader interested in a broader exposition of either subject should refer to Abelson and diSessa (1986a), Rozenberg and Salomaa (1997), Prusinkiewicz and Lindenmayer (2004) and the references therein.

Formally, L-systems are parallel rewriting systems consisting of an alphabet $\boldsymbol{\Sigma}$, an axiom ω^0, and a finite set of production rules $\boldsymbol{P}$, with these rules being recursively applied to an iteratively developed sequence of characters ω.

- The *alphabet* is a finite, non-empty set $\boldsymbol{\Sigma}$, whose elements Σ_i are called *characters*.[1] For example,

$$\boldsymbol{\Sigma} = \{\mathtt{A}, \mathtt{B}, \mathtt{C}, \mathtt{D}\} \quad \text{and} \quad \boldsymbol{\Sigma} = \{*, /, \#, !\} \quad \text{and} \quad \boldsymbol{\Sigma} = \{\mathtt{A}, \mathtt{B}, \#, !\}$$

  are all acceptable alphabets herein, though longer alphabets will generally be required.
- A *word* ω over $\boldsymbol{\Sigma}$ is a finite sequence of characters (or *string*) Σ_i, the empty word being denoted by $\varnothing$. The length $|\omega| \in \mathbb{N}$ of a word is defined as the integer number of non-empty characters in ω. The set of all words is denoted $\boldsymbol{\Omega}_{\Sigma}$. For example,

$$\omega = \mathtt{CCBAC} \quad \text{and} \quad \omega = *!/\#\#\# \quad \text{and} \quad \omega = !\mathtt{A}!\mathtt{B}!\#\mathtt{AAAB}$$

  are words corresponding to the three example alphabets given above and have lengths of $|\omega| = 5$, $|\omega| = 6$, and $|\omega| = 10$, respectively.
- The *axiom* is a non-empty word $\omega^0 \in \boldsymbol{\Omega}_{\Sigma}$. It will be shown to initiate the process of L-system recursion to be described. In all implementations of the L-system framework described in later sections and chapters, the axiom will have length $|\omega^0| \sim 2$ and will consist of letter characters. For example,

$$\omega^0 = \mathtt{CA}.$$

- *Production rules* are a set $\boldsymbol{P}$ of relatively compact instructions for replacing some or all of the individual characters in a word ω with new strings of characters.[2] Production rules can be *context-sensitive*, such that their application to a given target character is dependent on the surrounding characters. They can also be *context-free*, whereby their application is free of such dependencies.

  L-systems that consider context-free production rules are designated as $0L$-systems. As examples of context-free production rules over the alphabet $\boldsymbol{\Sigma} = \{\mathtt{A}, \mathtt{B}\}$, consider

$$P_1 : \mathtt{A} \rightarrow \mathtt{ABB} \quad \text{and} \quad P_2 : \mathtt{B} \rightarrow \mathtt{A},$$

  where $\boldsymbol{P} = \{P_1, P_2\}$. As given, these indicate that when $\boldsymbol{P}$ is applied to any $\omega \in \boldsymbol{\Omega}_{\Sigma}$, all A characters possibly existing in ω are replaced by the string ABB, while all B

[1] In all topological implementations described in this book we will reserve numeric characters (i.e., numbers) for use as integer and real *parameters* (as in Section 3.2); they will not be used as elements of an L-system alphabet except in the singular, simple, and classically illustrative example of the Fibonacci Series to follow shortly.

[2] It is also possible to consider production rules that replace strings of characters (having length > 1) with other strings, but this work will exclusively consider production rules as acting only on single characters at a time.

characters possibly existing in ω are replaced by the single-character string A, with these substitutions all occurring simultaneously such that a new word is formed.[3]

Without loss of generality, note that any context-free production rule P_i of a $0L$-system can then be written as

$$P_i\colon \Sigma_i \rightarrow \Lambda_i \quad \forall\, \Sigma_i \in \boldsymbol{\Sigma}, \tag{3.1}$$

where $\Lambda_i \in \boldsymbol{\Omega}_{\Sigma}$ is explicitly defined for each P_i. The possibility of identity operations $P_i\colon \Sigma_i \rightarrow \Sigma_i$ (i.e., $\Lambda_i \equiv \Sigma_i$) allows this form of the production rules to be generally applied, even for characters that are not subject to modification (to be addressed). In this work, any production rules P_i corresponding to Σ_i that are not explicitly provided are assumed to be such identity operations (i.e., characters for which rules are not defined are never replaced).

Context-sensitive production rules require a more complex notation to account for the characters adjacent to the character being replaced. There are two subclasses to be considered: production rules in $2L$-systems consider the characters both preceding (or to the left of) and following (or to the right of) the character to be replaced, while production characters in $1L$-systems consider *either* the preceding or the following character, but not both.

Expanding the notation of Equation 3.1, a $2L$ production rule is then written as

$$P_j\colon \Sigma_j^l < \Sigma_i > \Sigma_j^r \rightarrow \Lambda_j,$$

where $\Sigma_j^l \in \boldsymbol{\Sigma}$ and $\Sigma_j^r \in \boldsymbol{\Sigma}$ are fixed for each P_j such that $\Sigma_i \rightarrow \Lambda_j$ *if and only if* Σ_i is preceded by Σ_j^l *and* followed by Σ_j^r. Note that, due to the high number of options for preceding and following characters in a general alphabet (i.e., the issue of combinatorics), the length of the set $\boldsymbol{P}$ can far exceed the length of the alphabet $\boldsymbol{\Sigma}$ (cf. Equation 3.1).

Likewise, a $1L$ production rule is written in one of the following two forms:

$$P_j\colon \Sigma_j^l < \Sigma_i \rightarrow \Lambda_j$$

or

$$P_j\colon \Sigma_i > \Sigma_j^r \rightarrow \Lambda_j.$$

The comments regarding the identity operation introduced in the context of context-free production rules are likewise applicable to context-sensitive production rules.

Considering the simple alphabet $\boldsymbol{\Sigma} = \{\mathtt{A}, \mathtt{B}\}$, examples for the four possible $1L$ production rules might be

$$P_1\colon B < A \rightarrow AA, \quad P_2\colon A > B \rightarrow \varnothing, \quad P_3\colon A < B \rightarrow AB, \quad P_4\colon B > A \rightarrow ABA.$$

Note that P_2 indicates that any A followed by a B will be replaced by the null word (i.e., eliminated entirely) during rule application.

[3] This simultaneous application is in contrast to the so-called Chomsky grammars, which apply production rules sequentially (Chomsky 1959).

- Finally, *recursion* is the process by which the rules $\boldsymbol{P}$ are iteratively applied to the word ω^n such that ω^{n+1} results. This process is repeated yielding a developing set of words

$$\omega^0 \rightarrow \omega^1 \rightarrow \cdots \omega^n \rightarrow \cdots \omega^{N_r},$$

for a given number of recursions N_r.

The *axiom*, *production rules*, and process of *recursion* will now be combined such that the concept of L-system rewriting is demonstrated in several examples.

3.1.1 Introductory Examples

To begin, it is beneficial to consider simple and intuitive examples of L-systems, including those having both context-free and context-sensitive rules.

Example 3.1 Consider the Fibonacci L-system (Prusinkiewicz & Lindenmayer 2004). It is the $0L$-system $G = \{\boldsymbol{\Sigma}, \boldsymbol{P}, \omega^0\}$ defined by the binary alphabet $\boldsymbol{\Sigma} = \{0, 1\}$, the axiom $\omega^0 = 0$, and production rules $\boldsymbol{P}$ given by

$$P_1 : 0 \rightarrow 1,$$
$$P_2 : 1 \rightarrow 01.$$

Given the axiom ω^0, the associated words ω^n are then obtained by simultaneously rewriting each and all characters of the previously existing word ω^{n-1} using the production rules. The first developed word $\omega^1 = 1$ is obtained by replacing the sole character 0 in ω^0 with the character 1, as required by $P_1 : 0 \rightarrow 1$. An analogous procedure yields $\omega^2 = 01$. Given two characters 01, substitutions, $0 \rightarrow 1$ and $1 \rightarrow 01$ in ω^2, are performed simultaneously, giving $\omega^3 = 101$. Continuing this process $N_r = 6$ times, the following sequence of words is formed:

$$\begin{aligned}
\omega^0 &= 0\\
\omega^1 &= 1\\
\omega^2 &= 01\\
\omega^3 &= 101\\
\omega^4 &= 01101\\
\omega^5 &= 10101101\\
\omega^6 &= 0110110101101.
\end{aligned}$$

In this example, the length at each stage grows according to the Fibonacci series, whence its name. Indeed, $|\omega^0| = |\omega^{n-1}| + |\omega^{n-2}|$.

Retaining this binary alphabet and the same rules, a change in the axiom to $\omega = 1010$, for example, will produce a completely different series of words, though their growth in length over iterations will continue to follow a Fibonacci law.

$$
\begin{aligned}
\omega^0 &= 1010 \\
\omega^1 &= 011011 \\
\omega^2 &= 1010110101 \\
\omega^3 &= 0110110101101101 \\
\omega^4 &= 10101101011011010110101101 \\
\omega^5 &= 011011010110110101101011011010110110101101,
\end{aligned}
$$

and so on. Note that while the lengths of the iteratively developed words grew monotonically in this example, other L-systems may result in words ω^n having lengths that monotonically contract or otherwise vary in a complex fashion, as will be shown in later examples.

Example 3.2 As a second example, consider a simple L-system formulation Prusinkiewicz & Lindenmayer (2004). This consists of a binary alphabet $\boldsymbol{\Sigma} = \{\mathtt{A}, \mathtt{B}\}$, an axiom $\omega^0 = \mathtt{B}$, and production rules $\boldsymbol{P} = \{\mathtt{A} \rightarrow \mathtt{AB}, \mathtt{B} \rightarrow \mathtt{A}\}$. The iterative application of the production rules again produces a series of developmental stages (words), here and afterword defined by a string of characters. Thus, the first stage $\omega^1 = \mathtt{A}$ is obtained by substituting B with A as defined by the production rules. Similarly, the second stage $\omega^2 = \mathtt{AB}$ is obtained by simultaneously substituting $\mathtt{A} \rightarrow \mathtt{AB}$ and $\mathtt{B} \rightarrow \mathtt{A}$. Examining the first six steps of this process, one finds

$$
\begin{aligned}
\omega^0 &= \mathtt{B} \\
\omega^1 &= \mathtt{A} \\
\omega^2 &= \mathtt{AB} \\
\omega^3 &= \mathtt{ABA} \\
\omega^4 &= \mathtt{ABAAB} \\
\omega^5 &= \mathtt{ABAABABA} \\
\omega^6 &= \mathtt{ABAABABAABAAB}.
\end{aligned}
$$

Given the analogy between this L-system and that in Example 3.1 (i.e., binary alphabet and similar rules), it is not surprising that the developed words again grow monotonically in length according to a Fibonacci law.

Of course the resulting string of characters ω^6 has no inherent meaning. Indeed, to satisfy the goals of topological representation (toward optimization) on which later developments are based, graphical interpretation algorithms are required. Options for such are addressed in Section 3.2 and in Chapters 4 and 5.

Example 3.3 In this third example, we consider a context-sensitive $1L$-system given by the alphabet $\boldsymbol{\Sigma} = \{\texttt{A}, \texttt{B}, \texttt{C}\}$, axiom $\omega^0 = \texttt{ABC}$, and seven production rules:

$$
\begin{aligned}
P_1 &: \texttt{A} \to \varnothing,\\
P_2 &: \texttt{B} > \texttt{A} \to \texttt{ABC},\\
P_3 &: \texttt{B} > \texttt{B} \to \texttt{BC},\\
P_4 &: \texttt{B} > \texttt{C} \to \texttt{C},\\
P_5 &: \texttt{C} > \texttt{A} \to \texttt{A},\\
P_6 &: \texttt{C} > \texttt{B} \to \texttt{B},\\
P_7 &: \texttt{C} > \texttt{C} \to \texttt{C}.
\end{aligned}
$$

To clarify, P_2 can be stated as "if B is followed by A, then replace B with ABC." Because this context-sensitive $1L$-system considers the characters to the right of the character to be replaced, this implies two options for addressing any characters B and C at the *end* of a string:

(i) Not being explicitly associated with any of the rules in $\boldsymbol{P}$, they could be subject to the identity operation and therefore left unchanged, or
(ii) A virtual periodicity could be implemented such that the "character following the final character" is taken to be the first character of the string.

In this example, we choose the latter and extend the word to the right by periodicity. For example, when applying the production rules to character C in the axiom $\omega = \texttt{ABC}$, we activate the rule $\texttt{C} > \texttt{A} \to \texttt{A}$ as would be the case if the word ABC was periodically extended per . . . ABCABCABC. . . .

Starting with axiom $\omega^0 = \texttt{ABC}$ we obtain

$$
\begin{aligned}
\omega^0 &= \texttt{ABC},\\
\omega^1 &= \texttt{CA},\\
\omega^2 &= \texttt{A},\\
\omega^3 &= \varnothing.
\end{aligned}
$$

and $\omega^n = \varnothing$, $n > 3$. Hence, this example is contrasted with Examples 3.1 and 3.2, in which the words actually contract after the first recursion and converge to the empty word $\varnothing$ after the second recursion. Context sensitivity, however, provides great generative diversity. If the alphabet and production rules are left unchanged and the characters of the axiom are merely shuffled such that $\omega^0 = \texttt{CBA}$, we obtain a different series of words that expand prior to contracting (i.e., their length develops in a nonmonotonic manner).

$$\omega^0 = \text{CBA},$$
$$\omega^1 = \text{BABC},$$
$$\omega^2 = \text{ABCCB},$$
$$\omega^3 = \text{CCBABC},$$
$$\omega^4 = \text{CBABCCC},$$
$$\omega^5 = \text{BABCCCCC},$$
$$\omega^6 = \text{ABCCCCCCB},$$
$$\omega^7 = \text{CCCCCCCABC},$$
$$\omega^8 = \text{CCCCCCACC},$$
$$\omega^9 = \text{CCCCCACC},$$

where $\omega^n = \varnothing, n > 16$.

3.1.2 Extended L-system Representations toward Graph Construction

In applying L-systems and the instructions they generate toward the representation of network and geometric topologies, it becomes important to consider various character types when defining the alphabet $\boldsymbol{\Sigma}$, as each can represent different actions with regard to topological (e.g., graph) generation. Further, it is both algorithmically advantageous and conducive to greater diversity if one allows for some characters to be variable (subject to non-identity production rules) while others are constant (subject only to identity production rules, which need not be applied in practice). In the sections and chapters that follow, we then consider an alphabet $\boldsymbol{\Sigma}$ that consists of variable characters $\boldsymbol{\alpha}$ usually represented by English letters (e.g., $\boldsymbol{\alpha} = \{\text{A}, \text{B}, \text{C}, \text{D}, \ldots\}$; $|\boldsymbol{\alpha}| = N_\alpha$) and constant characters $\boldsymbol{\beta}$ (e.g., $\boldsymbol{\beta} = \{\text{-}, \text{+}, \text{[}, \text{]}, \ldots\}$) such that[4]

$$\boldsymbol{\Sigma} = \boldsymbol{\alpha} \cup \boldsymbol{\beta}. \tag{3.2}$$

The characters in $\boldsymbol{\alpha}$ will generally be associated with the generation or modification of the fundamental components (e.g., nodes, edges, and faces) that comprise the graphs underlying the geometric and network topologies being considered. The recursive nature associated with their variability under the application of production rules drives topological complexity. The characters in $\boldsymbol{\beta}$, on the other hand, generally direct all other processes of graph modification *not* directly associated with node, edge, or face creation or property manipulation (e.g., they may alter the state of the graph creation agent, but not of the graph itself). In the service of generative action, the axiom will always consist of characters in $\boldsymbol{\alpha}$ (i.e., $\omega_i \in \boldsymbol{\alpha}$).

Most topological representation options described in this book consider a context-free approach with rules exclusively defined so as to act on single characters. Production rules for each of the N_α characters in $\boldsymbol{\alpha}$ can then be generally and compactly written as (cf. Equation 3.1)

[4] The consideration of numeric *parameters* that modify the interpretation of characters in $\boldsymbol{\alpha}$ and $\boldsymbol{\beta}$ will be introduced in Chapter 4.

$$P_i\colon \alpha_i \to \lambda_i^1\lambda_i^2\ldots\lambda_i^{N_\Lambda},\ i = 1\ldots N_\alpha, \tag{3.3}$$

for all $\alpha_i \in \boldsymbol{\alpha}$, where $\lambda_i^1\lambda_i^2\ldots\lambda_i^{N_\Lambda}$ is a string of N_Λ characters such that each λ_i^K is selected from the set of options $\boldsymbol{\Lambda}^K$ (i.e., $\lambda_i^K \in \boldsymbol{\Lambda}^K$). In this way, the production rules can be written in a highly systematic manner conducive to evolutionary programming, as will be detailed in Section 3.3. As an additional means of clarification, consider the following example:

Example 3.4 Let us apply this notion of variable and constant characters to the generation of an instruction string to be considered in Section 3.2. Specifically, consider $N_\alpha = 4$, where $\boldsymbol{\alpha} = \{\texttt{A},\texttt{B},\texttt{C},\texttt{D}\}$ and a two-character axiom $\omega^0 = \texttt{DB}$. Assuming $\boldsymbol{\beta} = \{\texttt{-},\texttt{+},\texttt{[},\texttt{]},*,\texttt{/},\texttt{\#},\texttt{!},\varnothing\}$, let us choose $N_\Lambda = 14$ and the following options for the creation of the production rules:

$$\boldsymbol{\Lambda}^1 = \{\texttt{[},\varnothing\},\ \boldsymbol{\Lambda}^2 = \{\texttt{-},\varnothing,\texttt{+}\},\ \boldsymbol{\Lambda}^3 = \{*,\varnothing,\texttt{/}\},\ \boldsymbol{\Lambda}^4 = \{\texttt{\#},\varnothing,\texttt{!}\},\ \boldsymbol{\Lambda}^5 = \boldsymbol{\alpha},\ \boldsymbol{\Lambda}^6 = \boldsymbol{\alpha},$$
$$\boldsymbol{\Lambda}^8 = \{\texttt{[},\varnothing\},\ \boldsymbol{\Lambda}^9 = \{\texttt{-},\varnothing,\texttt{+}\},\ \boldsymbol{\Lambda}^{10} = \{*,\varnothing,\texttt{/}\},\ \boldsymbol{\Lambda}^{11} = \{\texttt{\#},\varnothing,\texttt{!}\},\ \boldsymbol{\Lambda}^{12} = \boldsymbol{\alpha},\ \boldsymbol{\Lambda}^{13} = \boldsymbol{\alpha}.$$

Thus, each of the characters λ_i^j, $j = 1\ldots 6$ and $j = 8\ldots 13$ for rule P_i is exclusively chosen from among the options in set $\boldsymbol{\Lambda}^j$. The assignments for λ_i^7 and λ_i^{14} are algorithmically chosen such that the "open bracket" indicators "[" in $\boldsymbol{\Lambda}_i^1$ and $\boldsymbol{\Lambda}_i^8$, if assigned, are closed. The four production rules shown in Figure 3.1 are then formed using random assignments $\lambda_i^K \in \boldsymbol{\Lambda}^K\ \forall K = 1\ldots 14$. The complete results of the first two recursions and partial results of the third are then shown in Figure 3.1.

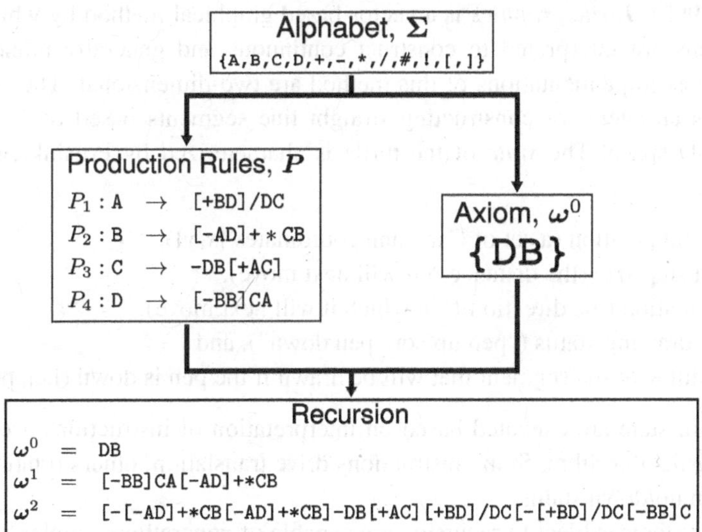

Figure 3.1 Example of recursively generated final strings using an L-system approach consisting of a two-character axiom and production rules constructed with a specific alphabet. Note that A, B, C, and D represent variable characters.

3.2 Turtle Graphics Interpretation

The L-system formulations presented in Section 3.1 are effective in encapsulating great complexity and thus diversity into a relatively straightforward encoding. This was demonstrated through various examples, and one can imagine the variety of words that could be formed by altering the alphabet, the axiom, and/or the production rules of an L-system. The user-specified number of iterations over which recursive operations are performed allows for further tuning of the information generated. However, as previously mentioned, the L-system must be coupled with a graphical interpreter for the encoded information to be given spatial significance. This is the primary theme of this book in total: to show various ways in which L-systems can be used to represent, and thus ultimately optimize, load-bearing structures and other networks. The book concerns a family of methods whereby graph-creating *agents* perform various operations to create, delete, and modify nodes, edges, faces, and even volumes. To begin, we consider a relatively simple and well-known graphical interpretation algorithm that, though limited in the representation of well-connected networks, is straightforward to describe and is highly illustrative of key concepts that will be explained in later chapters.

The topology of branched structures in nature and engineering systems may be formally described by a graph (McDonald 1983, Roth-Nebelsick, Uhl, Mosbrugger & Kerp 2001). To produce such a graph from words, an interpreter that deterministically associates a given graph to a given word generated by the L-system is required. Here we introduce a variant of the turtle interpretation of L-systems introduced in Szilard & Quinton (1979) and extended in Prusinkiewicz (1986), Prusinkiewicz (1987), Hanan (1992). *Turtle graphics* is a vector-based graphical method by which encoded instructions are interpreted to construct continuous and generally branched paths. By far, most implementations of this method are two-dimensional. The "turtle" then represents an agent for constructing straight line segments based on its movement in this 2-D space. The *state* of the turtle is characterized by Prusinkiewicz (1986) as follows:

- its current position (a set of Cartesian coordinates $\{x, y\}$),
- its next step size (the distance δ it will next move),
- its orientation (the direction θ in which it will next move),
- its line drawing status ("pen up" or "pen down"), and
- the width w of the segment that will be drawn if the pen is down (i.e., pen width).

Changes in state are executed based on interpretation of instructions according to a deterministic algorithm. Some instructions drive translation, others rotation, and still others pen up/down status.

As L-systems subject to recursion are capable of generating complex strings, they represent a powerful option for encapsulating a great diversity of possible turtle instructions. We then consider an L-system constructed for the primary purpose of providing the same, where the turtle interpretation operates over the words of $\mathbf{\Omega}_\Sigma$. That is, given any word $\omega \in \mathbf{\Omega}_\Sigma$, each character in ω is sequentially read from left to right and

Table 3.1 Summary of turtle commands comprising the alphabet Σ assumed herein.

Character	Action
	Turtle Motion & Segment Creation
α_i	Move turtle forward** by δ_{α_i} with pen down, creating segment of type "α_i"
*	Rescale δ_{α_i} $\forall\, i = 1 \ldots N_\alpha$ such that $\delta_{\alpha_i} \leftarrow (\delta_{\alpha_i}\delta_l)$
/	Rescale δ_{α_i} $\forall\, i = 1 \ldots N_\alpha$ such that $\delta_{\alpha_i} \leftarrow (\delta_{\alpha_i}/\delta_l)$
#	Rescale segment width (i.e., pen width) such that $w \leftarrow (w\delta_w)$
!	Rescale segment width (i.e., pen width) such that $w \leftarrow (w/\delta_w)$
	Turtle Orientation
+	Rotate turtle clockwise such that $\theta \leftarrow \theta + \delta_\theta$
-	Rotate turtle counterclockwise such that $\theta \leftarrow \theta - \delta_\theta$
	Branching Operators
[	Store current turtle state (location/orientation) in pushdown stack
]	Retrieve next turtle state from pushdown stack

**Note that δ_{α_i} may take a negative value, resulting in a backwards walk that may overwrite previously existing segments.

interpreted as a command to change the turtle state. Decomposing $\boldsymbol{\Sigma}$ as per Equation 3.2 into the following constant and variable characters:

$$\boldsymbol{\alpha} = \{\mathtt{A}, \mathtt{B}, \mathtt{C} \ldots\}, \quad \boldsymbol{\beta} = \{\mathtt{*}, \mathtt{/}\ \mathtt{\#}, \mathtt{!}, \mathtt{+}, \mathtt{-}, \mathtt{[}, \mathtt{]}\}$$

allows the definition of the various turtle graphics commands given in Table 3.1.[5]

The "branching operators" are essential as they save and recall states of the agent during the generative process, the set of state information having been previously defined. This same concept will be exploited in later chapters introducing more capable tools. Specifically, the branching operators work as follows: an open " [" saves a turtle state that is immediately recovered upon interpretation of the matched "] ". In this way the line-drawing agent may be returned to any point on any previously constructed segment and, from there, be instructed to generate new segments. If these new segments do not identically retrace existing ones, a branch results. Because these operators can be nested (cf. recursions 2 and 3 in Figure 3.1), hierarchical branching is common.

Given an L-system of the type described in Section 3.1 and the commands of Table 3.1, the definition of any turtle-based graph then requires the following:

- selection of N_α and the associated characters (letters) of $\boldsymbol{\alpha} = \{\mathtt{A}, \mathtt{B}, \ldots\}$,
- selection of the axiom ω^0 based on $\boldsymbol{\alpha}$,
- construction of the production rules $\boldsymbol{P}$,
- selection of number of L-system recursions,
- initialization of δ_{α_i} and θ, and
- definition of the fixed parameters δ_l, δ_w, and $\delta_\theta \geq 0$.

[5] For simplicity, we present the method for two-dimensional geometries, the extension to three dimensions being straightforward.

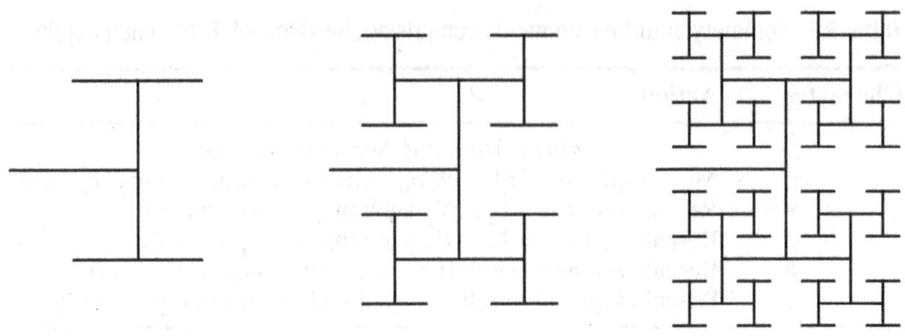

Figure 3.2 Graphs generated by the L-system $G = (\boldsymbol{\Sigma}, \boldsymbol{P}, \omega^0)$ and its turtle interpretation at given three different levels of recursion n. Here $\boldsymbol{\Sigma} = \{\texttt{A}, \texttt{B}, \texttt{C}, \texttt{*}, \texttt{/}, \texttt{+}, \texttt{-}, \texttt{[}, \texttt{]}\}$, $\omega^0 = \texttt{A}$, and $\boldsymbol{P} = \{(\texttt{A} \rightarrow \texttt{CB}), (\texttt{B} \rightarrow \texttt{[*+A][*-A]}), (\texttt{C} \rightarrow \texttt{C})\}$ while the fixed turtle parameters are $\delta_\theta = 90°$and $\delta_l = \sqrt{2}/2$.

A number of examples are now provided to exemplify the process by which L-system rewriting is used to construct turtle graphs.

Example 3.5 The first L-system/turtle example represents a simple symmetrically bifurcating tree where branches shorten at each bifurcation (Prusinkiewicz & Lindenmayer 2004). The L-system is defined by $\boldsymbol{\Sigma} = \boldsymbol{\alpha} \cup \boldsymbol{\beta}$, where $\boldsymbol{\alpha} = \{A, B, C\}$, $\boldsymbol{\beta} = \{\texttt{*}, \texttt{/}, \texttt{+}, \texttt{-}, \texttt{[}, \texttt{]}\}$, $\omega^0 = A$, and context-free production rules

$$P_1: \texttt{A} \rightarrow \texttt{CB},$$
$$P_2: \texttt{B} \rightarrow \texttt{[*+A][*-A]},$$
$$P_3: \texttt{C} \rightarrow \texttt{C}.$$

The fixed turtle parameters are chosen as $\delta_\theta = 90°$and $\delta_l = \sqrt{2}/2$, and the turtle state is initialized such that it is placed at the origin (pen down), $\theta = 0°$, and $\delta_{\alpha_i} = \sqrt{2}/2, i = 1, 2, 3$. The graphs after recursions 4, 8, and 12 are shown in Figure 3.2. It is clear that an increase in the number of recursions alone (i.e., without changes in the L-system) results in elegantly increasing topological complexity as measured by the number of bifurcations.

Example 3.6 An interesting variation on the previous tree generated in Example 3.5 corresponds to a "one-sided bifurcation" graph, which can be generated if the L-system is modified to include more variable characters $\boldsymbol{\alpha} = \{\texttt{A}, \texttt{B}, \texttt{C}, \texttt{D}, \texttt{E}, \texttt{F}\}$ with associated context-free production rules:

$$P_1: \texttt{A} \rightarrow \texttt{BC},$$
$$P_2: \texttt{B} \rightarrow \texttt{F[*+D]},$$

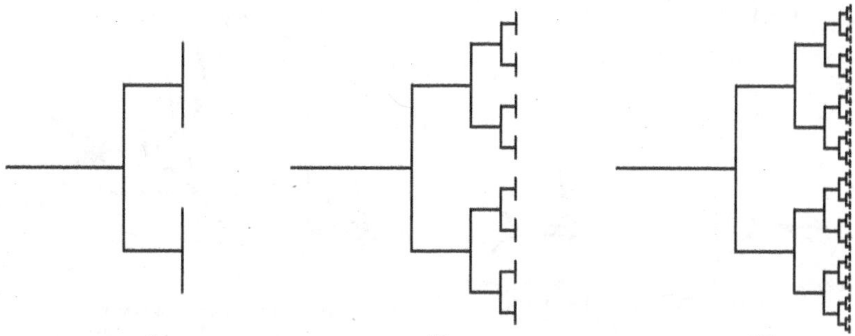

Figure 3.3 Graph generated by the L system $G = (\boldsymbol{\Sigma}, \boldsymbol{P}, \omega^0)$ and its turtle interpretation at recursions 6 (left), 12 (middle), and 18 (right), where $\boldsymbol{\alpha} = \{\texttt{A}, \texttt{B}, \texttt{C}, \texttt{D}, \texttt{E}, \texttt{F}\}$ and $\boldsymbol{P} = \{(\texttt{A} \rightarrow \texttt{BC}), (\texttt{B} \rightarrow \texttt{F[+D]}), (\texttt{C} \rightarrow \texttt{[-E]}), (\texttt{D} \rightarrow \texttt{F[*-A]}), (\texttt{E} \rightarrow \texttt{F[*+A]}), (\texttt{F} \rightarrow \texttt{F})\}$. The axiom, turtle parameters, and the initial state of the agent are unchanged from Example 3.5.

$$\begin{aligned} P_3 &: \texttt{C} \rightarrow \texttt{[*-E]}, \\ P_4 &: \texttt{D} \rightarrow \texttt{F[*-A]}, \\ P_5 &: \texttt{E} \rightarrow \texttt{F[*+A]}, \\ P_6 &: \texttt{F} \rightarrow \texttt{F}, \end{aligned}$$

while the axiom and turtle parameters are left unchanged.

The resulting graph is shown in Figure 3.3; comparing this to Figure 3.2 shows that a higher complexity in the production rules does not necessarily translate into an increased complexity of the interpreted graph. In fact, although the graphs shown in these two figures are clearly geometrically distinct, they are topologically identical.

Example 3.7 As a final example, we explicitly connect the concept of turtle graphics with Section 3.1 by considering the graph generated by the results of Example 3.4. Note that the alphabet chosen in that previous example is compatible with Table 3.1. Here we focus on the potential graph design diversity enabled by assigning unique attributes to each letter in $\boldsymbol{\alpha}$, specifically the step size δ_{α_i} and line segment color (in gray scale) associated with each α_i. Note that in the context of design, the latter could be associated with *properties* of a given line segment (e.g., material composition, sectional area, or sectional bending stiffness in the context of structures).

In addition to $G = (\boldsymbol{\Sigma}, \boldsymbol{P}, \omega^0)$ given in Example 3.4, the following turtle parameters and initial conditions are also assumed: $\delta_\theta = 40°$, $\delta_l = \delta_w = 1$, and the step sizes associated with each letter are $\delta_{\alpha_1} = 1$, $\delta_{\alpha_2} = 1/2$, $\delta_{\alpha_3} = 2$, and $\delta_{\alpha_4} = -1$ (i.e., the agent moves backwards).

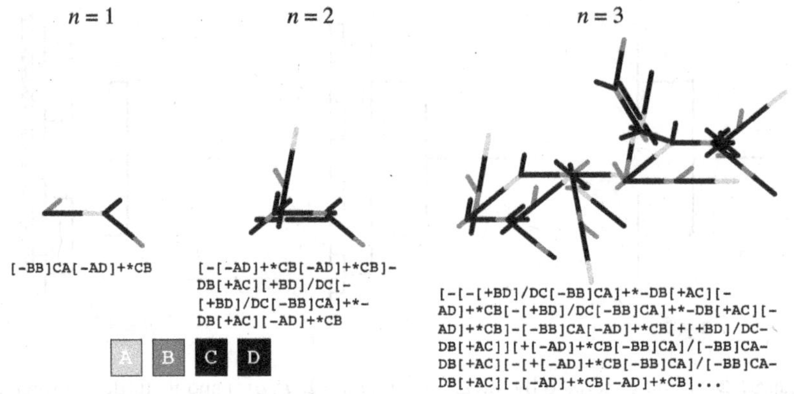

Figure 3.4 Graph generated by the L-system and first three recursions of Example 3.4. Parameters for the turtle interpretation include $\delta_\theta = 40°$ and $\delta_l = 1$. Reprinted with permission from Bielefeldt, Reich, et al. (2019).

3.3 L-system and Genetic Algorithms: Genetic Programming

Previous sections have addressed the concept of the L-systems and demonstrated how they can be interpreted into graphs via the turtle algorithm. We also showed how varying parameters defining the L-system can lead to vastly different turtle interpretations. The focus of this book, however, is the *design* of structures and networks based on underlying graphs. Here we introduce briefly how well-known evolutionary algorithms (see Section 1.3.3) are a natural choice for extending L-system-based graphing tools into actual topological optimization tools. Note that the topic of evolutionary algorithms is broad and deep; a full treatment is well beyond the scope of this work. Rather, we have provided only a high-level summary in Chapter 1 and then practically rely on open-source, widely available, and widely accepted codes. For more details on evolutionary design approaches, see, for example, Goldberg (1989), Hingston, Barone & Michalewicz (2008), Langdon & Poli (2013).

For the topological optimization problem at hand, the production of the phenotype from the genotype in the current framework requires the following four-step process:

1. Convert genetic information into a fully defined L-system,
2. Perform N_r rewriting recursions over the L-system to produce ω^{N_r} (per Section 3.1),
3. Interpret the resulting string of instructions ω^{N_r} via a graphing algorithm (e.g., turtle in the current treatment; Section 3.2),
4. Assign other physical properties as appropriate to members/regions of the resulting graph based on the design problem being considered (cf. Example 3.7, where the colors of line segments might correspond to local sectional area, bending stiffness, electrical resistance, etc.).

Given the fully defined phenotype, the design process proceeds with the calculation of the fitness of each individual design relative to the one or more performance metrics of interest to the topological designer, after which the crossover and mutation processes are applied toward cyclic evolution. All design methods and examples presented throughout this text will employ this general approach.

To continue toward case studies of the proposed design approach, we first address the particular issues of genome definition and fitness evaluation.

Design Variables. The genotype $\boldsymbol{x}$ consists of three categories of design variables: (i) those that define the L-system, (ii) those needed for the graphical interpretation (e.g., the distance and angle parameters for the turtle interpretation), and (iii) those associated with the physical properties to be assigned to the constructed graph. Let us consider these in detail.

(i) **L-system:** Fixing an alphabet $\boldsymbol{\Sigma}$ as one essential component of an L-system, genes must be used to encode the axiom $\boldsymbol{\omega}^0$, the rules $\boldsymbol{P}$, and potentially the number of recursions.

- In the approaches described, the *axiom* is usually represented by 1–3 variable letter characters of $\boldsymbol{\Sigma}$ (i.e., $\omega_i^0 \in \boldsymbol{\alpha}\ \forall\, i$). Thus, one gene or design variable must be allocated to define each of these letters of $\boldsymbol{\omega}^0$.
- The *rules* are much more involved in their representation than the axiom. While many options are available, and the curious and adventurous reader is welcome to propose many alternatives, in all cases discussed in this book, we will assume a basic context-free rule construction of the type defined in Equation 3.3 such that

$$P_i\colon \alpha_i \rightarrow \lambda_i^1 \lambda_i^2 \ldots \lambda_i^{N_\Lambda}, \quad i = 1 \ldots N_\alpha.$$

  Recall that there is one rule defined for each $\alpha_i \in \boldsymbol{\alpha}$ and that $\lambda_i^1 \lambda_i^2 \ldots \lambda_i^{N_\Lambda}$ is a string of N_Λ characters. Therefore, there may be up to $N_\alpha \times N_\Lambda$ design variables required for the full definition of all rules. (As an illustration, Example 3.4 would require $4 \times 12 = 48$ design variables for specification of the rules.) The specific manner in which these are represented within the overall set of design variables will be described shortly.
- The *number of recursions*, if included as a design variable, is simply an integer value usually between 2 and 6.

(ii) **Graphing algorithm parameters:** Once L-system rewriting has been completed, the graphical algorithm chosen to convert instructions into geometry will generally require additional variable input. For example, the turtle algorithm having the commands summarized in Table 3.1 may require the variables δ_l, δ_w, and δ_θ to be defined. Additionally, it may be desired to allow exploration of the initial values of $\delta_{\alpha_i}\ \forall\ i = 1 \ldots N_\alpha$. Therefore, the turtle graphing algorithm as presented may require up to $3 + N_\alpha$ total design variables be allocated to enable its interpretive function.

(iii) **Physical parameters:** The primary purpose of these methods is the optimization of engineering designs; therefore, the performance of some engineered topology performing some engineering function must be assessed. To that end, physical properties of constructed graphical entities must be defined. The number, type, and ranges of property values considered for each entity are highly dependent on the physical domain being considered and cannot be clearly specified here, but will be addressed in the case studies. For example, if a graph represents a *frame structure*, then each segment might have area and/or sectional bending stiffness(es) and/or various material properties that can be explored. Likewise, if a graph represents an electrical circuit, then the function of various edges (resistor vs. capacitor vs. inductor) and associated electrical properties (resistance, capacitance, inductance) must be stored and defined.

Based on the above description, a given design variable x_i might be a character, an integer, a floating-point number, or even a classification (e.g., resistor vs. capacitor). It is then important to implement a gene-mapping scheme that allows all such design variable possibilities to be considered within a single evolutionary algorithm.

Gene Representation. While it is certainly possible to develop a custom genetic algorithm (GA) or modify an existing and available one to directly consider the genes $\boldsymbol{x}$ in their various types (e.g., here string characters, real numbers, and integers) and their crossover and mutation, as an instructional resource this work will rely on widely used real number-based algorithms and their implementations. Therefore, it is useful to map the possibilities for each design variable x_i (continuous or discrete) to the real number line. Specifically, we will consider the *algorithmic* genes $\hat{\boldsymbol{x}}$ such that $0 \le \hat{x}_i \le 1 \; \forall \, i$ and further define a unique mapping $\hat{x}_i \to x_i \; \forall \, i$.

Let us first address the common case of *discrete* design variables in the definition of an L-system (e.g., the selection of first letter of the axiom or of the jth entry of the ith rule $\lambda_i^j \in \boldsymbol{\Lambda}^j$; cf. Equation 3.3). If a given discrete x_i has L_i options such that $x_i \in \boldsymbol{X}^i$ where $\text{len}(\boldsymbol{X}^i) = L_i$, then

$$x_i = X_j^i \;\; \text{if} \;\; \frac{j-1}{L_i} \le \hat{x}_i < \frac{j}{L_i}. \tag{3.4}$$

In this way, any value of $0 \le \hat{x}_i \le 1$ can be mapped to a unique selection for x_i from the options $\boldsymbol{X}^i$.

Consider an example where x_1 represents the first variable character of the axiom and $\boldsymbol{\alpha} = \{\texttt{A}, \texttt{B}, \texttt{C}\}$. Then $\boldsymbol{X}^1 = \boldsymbol{\alpha} = \{\texttt{A}, \texttt{B}, \texttt{C}\}$, $L_1 = \text{len}(\boldsymbol{X}^1) = 3$, and Equation 3.4 can be reinterpreted as

$$x_1 = \begin{cases} \texttt{A}, & 0 \le \hat{x}_i < 1/3, \\ \texttt{B}, & 1/3 \le \hat{x}_i < 2/3, \\ \texttt{C}, & 2/3 \le \hat{x}_i < 1. \end{cases}$$

In this way a specific and unique design variable can be defined based on an easily coded real number of normalized range.[6]

Further, let us address the case of *continuous* design variables, as is common in the definition of graphing algorithm parameters and physical parameters. If such a continuous x_i is being explored across the full domain $x_i^{lb} \leq x_i \leq x_i^{ub}$, then

$$x_i = x_i^{lb} + \hat{x}_i(x_i^{ub} - x_i^{lb}). \tag{3.5}$$

For example, if x_{27} represents and defines the turtle turning angle increment δ_θ such that the designer chooses to explore $5° \leq x_{27} \leq 85°$, then the implemented would evolve using $0 \leq \hat{x}_{27} \leq 1$ where, at the time of graph creation, the turning angle would be determined via $\delta_\theta = x_{27} = 5° + \hat{x}_{27}(80°)$.

To conclude, let us briefly summarize how these mappings from meaningful design variables $\boldsymbol{x}$ to algorithmic genes $\hat{\boldsymbol{x}}$ for the three essential types of variables considered in this work:

(i) **L-system:** There are three classes of design variables to be represented.
- Each character of the *axiom* ω^0 must be selected from the discrete options $x_i \in \boldsymbol{X}^i = \boldsymbol{\alpha}$ and $L_i = N_\alpha$.
- Each of the N_α *rules* requires definition of N_Λ characters, some of which are in $\boldsymbol{\alpha}$ while others are in $\boldsymbol{\beta}$ (see Section 3.1.2). Therefore, they are mapped from the continuous algorithmic genes per Equation 3.4.
- The *number of recursions* is definitively an integer, and therefore must also be mapped from the set of allowable options per Equation 3.4.

(ii) **Graphing algorithm parameters:** Most graphing parameters (and all turtle graphing parameters previously discussed) are real numbers selected from a bounded continuous domain. Therefore, Equation 3.5 is used to map any associated $\hat{x}_i \rightarrow x_i$.

(iii) **Physical parameters:** As physical parameter definition is completely driven by the physics modeled and the engineering discretion of the designer, both discrete and continuous variables might be employed. Discrete design variables might define material selection or component type (associated with a given graph segment), while continuous variables might define geometric dimensions, for example.

Complete Genetic Optimization Workflow. The complete L-system topology optimization framework employed throughout the majority of this text is summarized in Figure 3.5. The entire process is fully scripted and iterates usually until a predetermined number of evolutionary generations has been assessed. The GA (NSGA-II (Deb 2001) throughout much of this work) determines the values for the algorithmic genes ($\hat{\boldsymbol{x}}$; floating-point numbers). For each individual, these values are mapped into the design

[6] This form of Equation 3.4 is intended to provide an equal probability of selecting each $x_i \in \boldsymbol{X}^i$ (i.e., a normal distribution). These probabilities could be easily modified per designer preference to, for example, a Gaussian distribution by the selection of nonconstant intervals.

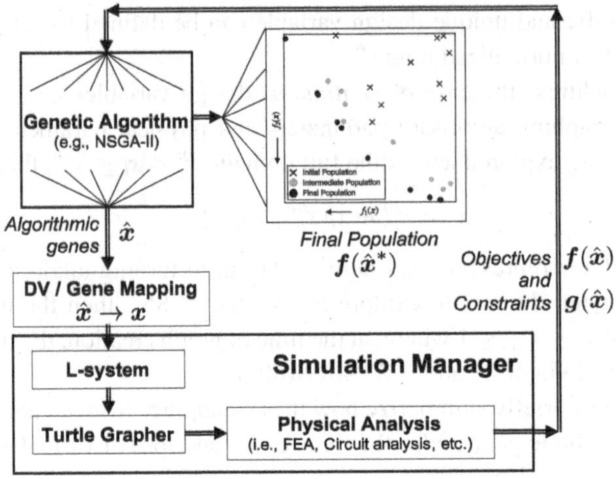

Figure 3.5 General iterative workflow employed in the topological optimization studies considered throughout (using turtle graphics as an example).

variables $\boldsymbol{x}$ for L-system determination, graphical operations, and physical property assignment per Equation 3.4 or Equation 3.5 as appropriate. Each individual topology is then created, and the associated physical system is modeled by a simulation manager. Note that each individual is independently analyzed, but these many analyses may be serial or parallel, depending on computational and licensing resources, time limitations, and computational infrastructural flexibility. Finally, the physical analysis tool returns information regarding performance objectives and constraints to the GA, which continuously improves the population toward an accurate approximation of the true Pareto frontier.

3.4 Case Studies

To demonstrate the ability of the L-system genetic design approach (cf. Figure 3.5) in determining effective topological configurations for engineered systems, two design problems are presented.[7] Both are structural in nature; designs of other types will be briefly discussed in Chapter 6. The first problem addresses load-bearing functionality in the form of a cantilevered frame configured to carry a static transverse load. The second study considers both structural and compliant regions, analogous to the functions of bone and cartilage in natural skeletal systems, for the design of an inverting compliant mechanism (Howell 2001) of the same kind used in demonstrating alternative topology optimization schemes (Bendsøe & Sigmund 2003). For both problems, the structural design approach is summarized, the optimization problem is quantified, and results are presented in the form of Pareto frontiers and associated example design configurations. Material properties for the load-bearing and compliant regions

[7] Portions reprinted with permission from Bielefeldt, Reich, et al. (2019), Copyright 2019 by Elsevier.

Table 3.2 Experimentally measured properties associated with the load-bearing and compliant materials in subsequent analyses (Bielefeldt, Reich, et al. 2019).

Material	Density (kg/m^3)	Modulus (MPa)	Poisson's Ratio
VeroWhite	1,175	1,500	0.40
TangoBlack	1,125	0.34	0.45

considered in the design analyses are shown in Table 3.2. These experimentally measured properties correspond to the VeroWhite and TangoBlack materials associated with the Stratasys Objet 500 PolyJet 3-D printer.

For each problem, two-dimensional structures are created based on the L-system representation of branched topologies graphically interpreted via the turtle algorithm (Section 3.2). However, well-behaved structural design requires three specific augmentations.

1. To prevent analysis convergence difficulties (e.g., via singular global stiffness matrices), load-carrying connectivity must be provided between boundary conditions of any type. In the present study, this is accomplished by:
   (a) Assuming an initial empty frame ("null structure") that is only sufficiently stiff to ensure problem solvability but insufficiently stiff to influence the topological optimization process, and
   (b) Specifying the turtle initialization location to be collocated with one or more boundary conditions, maximizing topological efficacy. This will be clearly demonstrated in Case Study 3.2.
2. Geometrically unbounded turtle agents will generate structures of undefined maximum size, where each new segment tends to be near the periphery, leading to networks that are spares and lack application-critical (e.g., load-transferring) intersections. Defining an initial boundary and associated algorithms that keep the agent in or on this boundary (e.g., angle-preserving path reflection operations) ensures that any structure design will not exceed a certain size and leads to far more prevalent network intersections. The turtle interpreter is then only permitted to construct new segments if such segments are interior to the boundaries. This is addressed in Appendix D.
3. Because of the importance of segment intersections and the associated connectivities across networks (structural and otherwise), segments having free ends are usually of limited utility and their existence increases solution complexity, cost, and mass, without any advantage. Algorithms that iteratively search for and eliminate free edges until none such remain provide more optimal solutions in most engineering cases of interest. This is also addressed in Appendix D.

Case Study 3.1: Transversely Loaded Frame

The first example problem considers a simple cantilevered frame composed of frame elements (cf. Section 2.3.1). Figure 3.6 illustrates the initial structure, domain

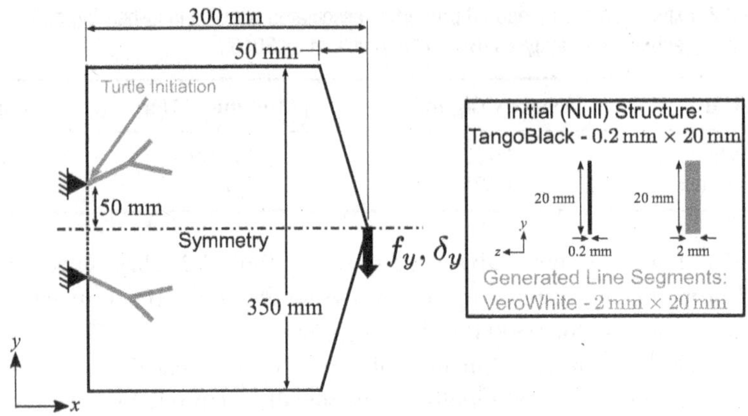

Figure 3.6 Initial (null) structure, domain boundary, turtle initialization location, and boundary conditions associated with the transversely loaded frame design study. Reprinted with permission from Bielefeldt, Reich, et al. (2019).

boundary, initial turtle agent location, and boundary conditions associated with this problem. An x–z geometric symmetry plane is assumed, allowing for topological design of only half of the structure. The initial (null) structure consists of compliant frame elements[8] that are many orders of magnitude less stiff in bending than generated internal segments, being both comprised of soft material and having an assumed cross-section of 0.2 mm × 20 mm. This initial structure serves as the domain boundary (as per issue 2 above) for each design considered. The frame boundary has a length of 300 mm and total height of 350 mm. The turtle agent is initialized 50 mm above the axis of symmetry at the minimum x position, and each line segment generated corresponds to a stiff (VeroWhite) frame element with a cross-section of 2 mm × 20 mm. A displacement boundary condition of $u_x = u_y = 0$ is enforced at the turtle initialization locations, while a concentrated force of $f_y = 0.05$ N is applied to the frame as indicated in Figure 3.6.

The full multiobjective design optimization problem with 52 design variables is defined in Table 3.3. The objective of reduced mass favors designs with a sparse topology, while the objective of increased stiffness tends to create dense topologies. The L-system follows from that of Examples 3.4 and 3.7. Each character in $\boldsymbol{\alpha}$ is assigned the same material and non-variable cross-section but is associated with a unique step size δ_{α_i}. Considering the small load and assuming small deflections for meaningful designs, a fully linear FEA with an average element length of approximately 30 mm is employed.[9]

8 Note that these compliant regions are irrelevant as load-bearing members if any internally connected segments are generated. The absence of connected internal structure results in a low-mass, low-stiffness design that serves as the baseline for all normalization in this design study.

9 As noted in Bielefeldt, Reich, et al. (2019), the FEA in this case used a "batched" approach whereby all structural designs in a given generation were arrayed as disconnected bodies in a *single FEA analysis*. In

Table 3.3 Definition of the multiobjective topology optimization problem regarding a light and stiff transversely loaded structural frame using turtle interpretation of L-system encodings

Dual Objective Optimization Problem Statement	
Minimize:	normalized mass $\hat{m}$
& Maximize:	normalized stiffness $\hat{k}$
by varying:	2 axiom characters in ω_0, 4 rule assignments (12 independent genes each), segment length factor $30 \leq f_L \leq 150$ mm, & segment turn angle $18° \leq \delta_\theta \leq 45°$
subject to:	no constraints
Alphabet and Turtle Graphics Definitions	
`A`:	VeroWhite, cross-section=2 mm×20 mm, $\delta_{\alpha_1} = f_L$
`B`:	VeroWhite, cross-section=2 mm×20 mm, $\delta_{\alpha_2} = f_L/2$
`C`:	VeroWhite, cross-section=2 mm×20 mm, $\delta_{\alpha_3} = f_L/3$
`D`:	VeroWhite, cross-section=2 mm×20 mm, $\delta_{\alpha_4} = f_L/5$
	$\delta_l = \delta_w = 1$
NSGA-II Parameters (Deb, Agrawal, Pratap & Meyarivan 2000)	
	100 members for 1,000 generations, $P_{cross} = 0.9$, $\eta_{cross} = 20$ $P_{mut} = 1/52$, $\eta_{mut} = 20$

The Pareto frontier associated with the multiobjective optimization problem of Table 3.3 is shown in Figure 3.7 along with three example Pareto-optimal topologies found by the optimizer. The mass and stiffness of each design is normalized by the derived properties of the "null" design, which is the lightest and least stiff design possible. For further assessment, we consider the utopia point (the point that maximizes or minimizes all objective functions) in the objective space. A particular topology of interest (herein referred to as the fabricated design) would then be the design on the frontier closest to the utopia point (normalized mass of 20.98, normalized stiffness of 5.72E6).

Also depicted in Figure 3.7 is the frontier associated with a series of Michell trusses (Rozvany, Bendsøe & Kirsch 1995), which have been shown mathematically to be optimal in the sense of maximizing stiffness per unit mass for linear problems where buckling is neglected (Michell 1904*b*). These structures consist of beam elements with the same material properties and cross-sections as described above. It is important to acknowledge that despite analyzing 100,000 topologies, the L-system/turtle framework is unable to match the performance of the Michell truss. The highest mass L-system/turtle designs are outperformed by a factor of two in terms of stiffness.

Many mass-efficient structural design studies such as this performed using the L-system/turtle approach have proven the inferiority of this approach for a number of reasons. Fundamentally, the generation of branched paths that only intersect in an irregular and indiscriminate manner – largely due to reflections off

this way, parallelization of the population was accomplished without multiple simultaneous calls to the FEA solver (i.e., Abaqus).

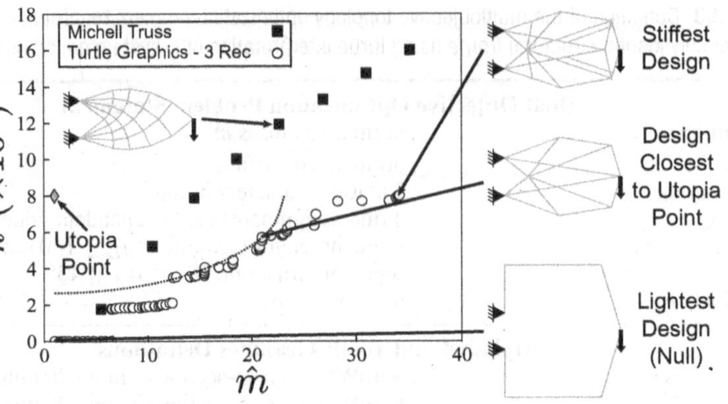

Figure 3.7 Approximated Pareto frontier of the structural frame design problem after 1,000 generations compared with Michell truss (i.e., known optimal) solutions, along with the utopia point for the L-system optimization frontier. Dashed line represents points of constant distance from this utopia point. Reprinted with permission from Bielefeldt, Reich, et al. (2019).

boundaries – simply does not lead to an efficient transfer of loads between boundary conditions (or the communication of information across a network in general). Also common are nonstraight segments as a consequence of the turning operator and/or the removal of free-ended segments. This leads to bending within segments, which is not mass-efficient since material near the neutral axis frame elements is unstressed. It is shown in Chapters 4 and 5 that topologies designed using face/cell-based algorithms are far better at emulating the structural efficiencies of a Michell truss. However, for functional objectives not associated with mass-efficient load carrying, the turtle interpretation can produce effective designs as discussed in Case Study 3.2.

Case Study 3.2: Tensile Inverter Compliant Mechanism

This case study considers a nonlinear compliant mechanism – a *tensile inverter*. Here, mass efficiency is neglected as an objective. The initial structure, domain boundary, turtle agent initialization location, and boundary conditions associated with this mechanical problem are illustrated in Figure 3.8. The initial structure consists of both compliant and stiff beam segments with an assumed rectangular cross-section of 2 mm × 10 mm. Load-bearing beams labeled A and D in this initial structure serve as part of the domain boundary. Beams B and C, on the other hand, are necessary to ensure the existence of a load path for all designs, but their bent form and soft material make them negligible in meaningful designs, and they are not considered during postprocessing. As in Case Study 3.1, a geometric symmetry is enforced about the $y = 0$ plane. This plane and another offset 100 mm above also serve as boundaries and enclose the domain within which the turtle agent may operate.

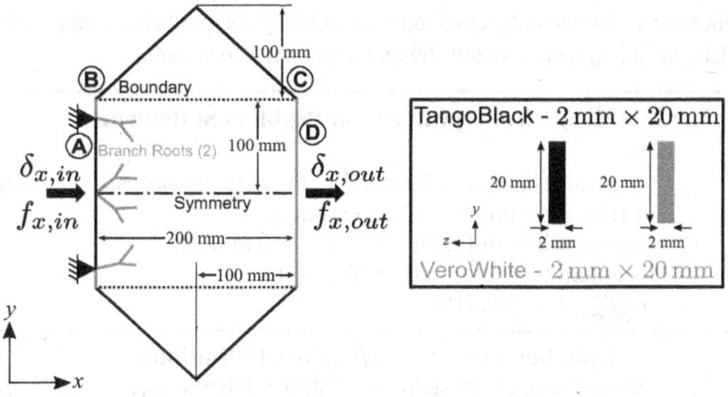

Figure 3.8 Initial structure, domain boundary, branch root location, and boundary conditions associated with the tensile inverter design study. Reprinted with permission from Bielefeldt, Reich, et al. (2019).

As an example of flexibility of this approach, a simple modification is made relative to Case Study 3.1. *Two* branched turtle paths are considered simultaneously (before accounting for symmetry). The two paths initiate from the two boundary conditions on the left side of the structure. This *multibranch* approach is implemented by simply splitting the two-character axiom into two discrete single-character strings, to which an independent rewriting process is applied as per Section 3.1. The two resulting strings are independently interpreted as separate, though often intersecting, branched paths.

To affect the "inversion" of force and motion, a displacement boundary condition of $u_x = \delta_{x,in}$ is applied to the lower branch root, while a displacement boundary condition of $u_x = u_y = 0$ is applied to the upper branch root. The goal of a well-designed tensile inverter is to maximize both the positive output displacement in the x-direction and positive output force in x-direction (i.e., $\delta_{x,out}$ and $f_{x,out}$, as in Figure 3.8) for given associated negative inputs $\delta_{x,in}$ and conjugate $f_{x,in}$. Thus, two independent structural analyses are considered in the assessment of each design. In the first analysis, a displacement is applied to the input node ($\delta_{x,in} = -5$ mm), while the resulting output displacement $\delta_{x,out}$ is measured when no load is present ($f_{x,out} = 0$), providing an indication of *free displacement inversion*. In the second analysis, the output node is forced back to its reference position ($\delta_{x,out} = 0$ mm), and the reaction forces at both the input and the output points ($f_{x,in}$ and $f_{x,out}$) are recorded, providing an indication of the *blocked force inversion*. Given the description of loading conditions associated with Figure 3.8, the free displacement inversion ratio and blocked force inversion ratio are defined respectively as

$$r_\delta = \left. \frac{-\delta_{x,out}}{\delta_{x,in}} \right|_{f_{x,out}=0} \quad \text{and} \quad r_f = \left. \frac{f_{x,out}}{f_{x,in}} \right|_{\delta_{x,out}=0}. \tag{3.6}$$

The design objective is then the simultaneous maximization of both ratios. The analyses of both load cases consider the full geometric nonlinearity of the problem

Table 3.4 Specification of the multiobjective topology optimization problem for tensile compliant inverter mechanism, including a dual-axiom-independent rewriting scheme.

	Dual Objective Optimization Problem Statement
Maximize:	r_δ, r_f
by varying:	2 axiom characters for parallel rewriting processes ($\omega_0^1 \in \alpha$, $\omega_0^2 \in \alpha$), 4 rule assignments (12 genes each), segment length factor $10 \leq f_L \leq 50$ mm, & segment turn angle $10° \leq \delta_\theta \leq 40°$
subject to:	$\sigma_{Mises}^{max} \leq 33.8$ MPa
	Alphabet and Turtle Graphics Definitions
A, B	VeroWhite, cross-section = 2 mm × 10 mm, $\delta_{\alpha_1} = \delta_{\alpha_2} = f_L$
C, D	TangoBlack, cross-section = 2 mm × 10 mm, $\delta_{\alpha_3} = \delta_{\alpha_4} = f_L/2$
	$\delta_l = \delta_w = 1$
	NSGA-II Parameters (Deb et al. 2000)
	60 members for 400 generations, $P_{cross} = 0.9$, $\eta_{cross} = 20$, $P_{mut} = 1/52$, $\eta_{mut} = 20$

(e.g., internal buckling, and large rotations). To ensure the structural feasibility of potential tensile inverter designs, a stress constraint in VeroWhite members of $\sigma_{Mises}^{max} = 33.8$ MPa is imposed via the penalty method.

Details regarding the multiobjective design optimization problem associated with the tensile inverter mechanisms are specified in Table 3.4. An *a priori* understanding of the need to localize deformation motivated that compliant members be shorter than stiff ones; this is reflected in the assignment of alphabet characters to section functionalities. As before, the FEA model consists of elements with an average length of approximately 30 mm. Due to the nonlinear behavior of the mechanism, and in particular due to the effects of local buckling, implicit dynamic analysis was required and loading increments during analysis were often small. This substantially slowed the analysis process such that the total number of designs considered was necessarily reduced to $60 \times 400 = 24,000$ (Hartl, Bielefeldt, Reich & Beran 2017).

The results of the multiobjective design optimization are shown in Figure 3.9. The family of Pareto-optimal designs is diverse and effective, showing that this approach can produce a wide range of novel designs for less intuitive problems. To provide some context regarding the quality of these results, an approximate ideal solution can be easily obtained. Neglecting nonlinear contributions and assuming displacement over a constant force, the ratio of input to output work is calculated as

$$\frac{W_{out}}{W_{in}} \approx \frac{f_{x,out}\delta_{x,out}}{f_{x,in}\delta_{x,in}}. \tag{3.7}$$

Considering perfect energy conservation per $W_{out}/W_{in} = 1$, and substituting this into Equation 3.6 leads to the simple idealized frontier,

$$r_\delta = \frac{1}{r_f}. \tag{3.8}$$

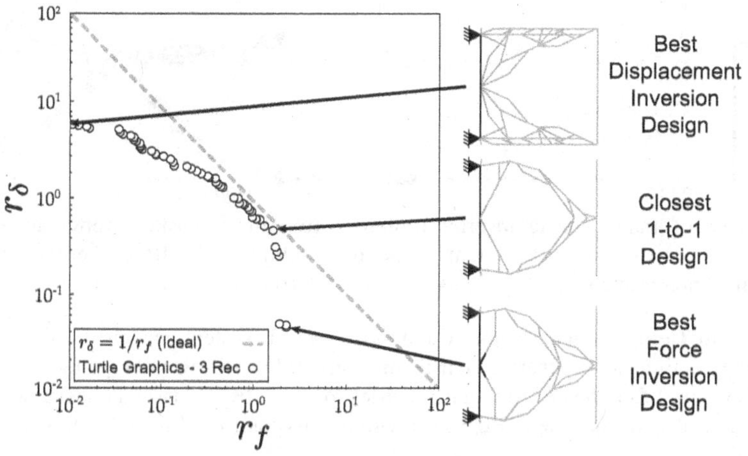

Figure 3.9 Pareto frontier of the tensile inverter design problem after 400 generations compared with an approximate analytically derived frontier. Null structure removed for clarity. Reprinted with permission from Bielefeldt, Reich, et al. (2019).

As seen in Figure 3.9, the Pareto frontier obtained using the L-system/turtle topology optimization method approaches the ideal solution but does not surpass it. A leading reason for this is that all compliant mechanism designs store some amount of the externally applied work as internal strain energy.[10]

Regarding the overall inversion effectiveness of these turtle-generated topologies, it is instructive to compare the results to known solutions, just as the results of Case Study 3.1 were compared to the Michell truss. If we consider as an example the single design closest to the ideal frontier (cf. Figure 3.10; $r_\delta = 0.484$, $r_f = 1.62$), we do observe noticeable differences in the overall topology as compared to solutions found in the literature (Sigmund 1997, Howell 2001, Tcherniak & Sigmund 2001, Bendsøe & Sigmund 2003, Hartl et al. 2017). However, the underlying geometry associated with the load path remains the same. Despite having a far different methodology, the L-system/turtle framework arrives at the same geometrical solution as traditional topology optimization methods, but without the large number of design variables and local stress concentrations associated with the SIMP topology optimization method. Furthermore, rather than a single point design, the demonstrated approach provides a wide range of diverse preliminary options from throughout the Pareto frontier from which a designer may draw inspiration. On the other hand, the turtle interpretation approach again proves ineffective in creating load-carrying structures as the blocked force inversion ratio clearly reaches a limit. This effect is not observed when a more capable interpretation is used as shown in Case Study 4.3.

[10] Note, however, that other more capable structural interpretations can generate designs that mildly surpass this idealized boundary, likely due to nonlinear effects (cf. Figure 4.29).

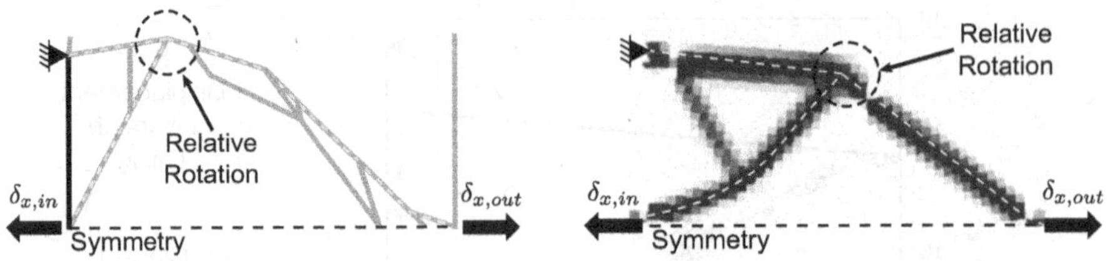

(a) One of many tensile inverter designs generated using the L-system genetic optimization framework

(b) Single tensile inverter generated using traditional SIMP (Tcherniak & Sigmund 2001)

Figure 3.10 Comparison of geometry and topology of L-system-generated and traditional topology optimization-generated tensile inverters. While the overall topologies of the mechanisms differ, the underlying geometries (denoted by the dashed lines) associated with the load paths are equivalent. Reprinted with permission from Bielefeldt, Reich, et al. (2019).

3.5 Conclusions and Extensions

The topology design approach unifying the chapters that follow is based on the idea that recursive L-systems can either generate highly complex strings of instructions for interpretation into geometric assemblies of edges and faces (Chapter 4) or can be used to directly define layouts that are themselves iteratively updated (Chapter 5). Critically, if these networks are converted into physical entities (e.g., structures and circuits) and analyzed, connections between the underlying L-system and the resulting performance can be quantified, with the ultimate goal of iterative design optimization considering multiple objectives and constraints.

The turtle algorithm demonstrated in this chapter is simple to explain and implement. It has a long history in the computer science community and can be useful in the development of new network/structure design tools. However, it does not produce the best topologies for some engineering networks, especially when load-carrying functionality is critical as below:

- Unless boundaries are introduced (and associated nonpenetration algorithms proposed), unconstrained branched structures are created. The ubiquity of free ends in such a network (cf. Figures 3.4 and 3.11) is *not* conducive to good performance in most engineered systems since the efficient sharing of information or loads between network members is usually essential and free-ended segments provide no advantage in this regard.
- Most mass-efficient structures rely on straight paths between segment intersections, maintaining a prevalence of uniaxial tensile and compressive loading as more faces become triangular (i.e., truss-like). The turtle algorithm, due to its turning behavior, is not effective at directly generating such geometric configurations but instead can create bent segments and indiscriminate intersections.

In the chapters that follow, two alternative and highly distinct classes of graphical algorithms are proposed and demonstrated that will be explicitly formulated for the

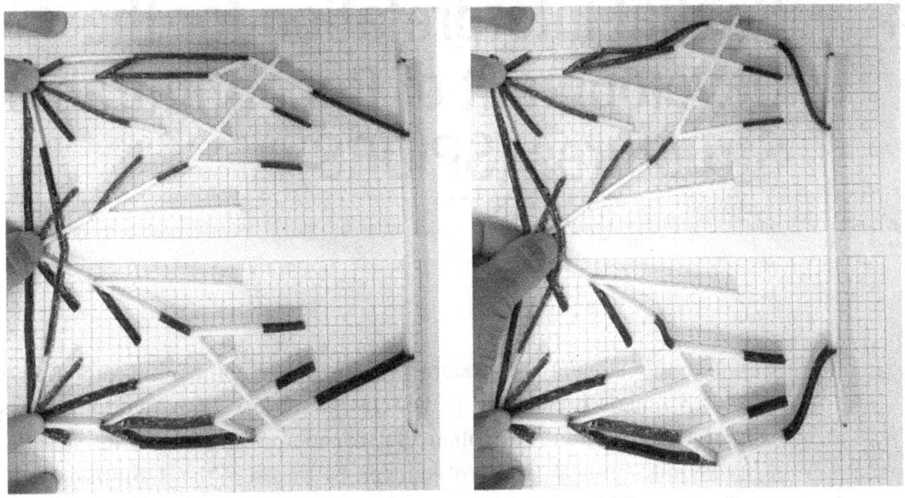

(a) Reference configuration (b) Deflected configuration

Figure 3.11 Physical demonstration of *compressive* inversion behavior comparable to that optimized in this chapter for tension, here enabled via the graph-thickening algorithms of Appendix E and dual-material additive manufacturing (Bielefeldt, Reich, et al. 2019). The failure to implement the trimming algorithm of Appendix D in this example results in a prevalence of nonfunctional free-ended segments as generated by the turtle algorithm.

generation of closed faces or cells, bounded by straight or curved edges according to user preference. A wide range of engineering design problems, including the design of electrical circuits, will be demonstrated. The intent of the introductions and demonstrations to follow is to inspire readers and prepare them to solve new problems while developing their own new design tools.

Toward the realization of real-world solutions, it is important to show how the proposed developmental approaches of this chapter, along with those of Chapters 4 and 5 to follow, can be extended from graph-based representations to volumetric three-dimensional models, computational and physical (Bielefeldt, Reich, et al. 2019). In the context of this chapter, this is captured in Figure 3.11, which physically demonstrates the performance of a *compressive inverter* analogous in operation to the tensile inverter addressed in Case Study 3.2. Appendix E provides guidance (with associated codes) regarding the algorithmic processes for converting graph-based representations into three-dimensional bodies, which can be easily realized in physical form using additive manufacturing. These workflows employ open-source software and apply to most structural topologies generated throughout the remainder of this book.

4 Spatial Interpretation for the Development of Reconfigurable Structures (SPIDRS)

Although the standard L-system with turtle graphics interpretation discussed in Chapter 3 makes it possible to generate a variety of networks or structures, from abstract fractal patterns to plant-like branching structures, its structural modeling utility is quite limited. One problem can be traced to the definition of all graph edges as a function of a single distance parameter δ_{α_i}, meaning that a given graph (e.g., structure) edge is limited to possible lengths $n\delta_{\alpha_i}$, where n is some integer. This same issue applies to orientation changes, as each change in path angle for the graph-generating turtle agent is limited to an integer multiple of the angle parameter δ_θ. Thus, while the turtle graphics interpretation approach presented in Chapter 3 can create nonintuitive self-similar structures that can solve a given multiobjective design problem, the constraints placed on distance and angle definitions limit the potential design space and can lead to L-system-generated topologies that perform well below known optimal designs.

Furthermore, the turtle graphics algorithm was actually first developed to introduce computer programming to those unfamiliar with the subject (Abelson & diSessa 1986a) and is not particularly well-suited to structural (or any network) topology development. The creation of meaningful structures specifically requires the existence of a load path between points on which displacement and/or force boundary conditions are applied. As discussed in Section 3.2, the algorithm constructs straight line segments between two points in 2-D space based on interpreted L-system instructions. Therefore, generating viable structural topologies with the turtle graphics algorithm generally relies to some extent on the intersection of line segments emanating from applied boundary conditions, where any intersections that do occur are not explicitly defined by the algorithm but rather occur by happenstance (e.g., via the intersection of two distinct branches). Additionally, the turtle graphics segment definition algorithm assigns material, section, or other segment properties in a serial manner such that the final assignment is the one considered in analysis. Thus, material and section property assignments and the formation of interesting and useful load paths, which together form the essence of topological description, occur in a coincidental manner when using the turtle graphics algorithm.

This chapter describes the development of L-system interpretation algorithm known as spatial interpretation for the development of reconfigurable structures (SPIDRS). Specifically, SPIDRS interprets strings resulting from recursion over parametric L-systems, which augment the L-system introduced in Chapter 3 with numeric parameters to greatly enhance the diversity of networks/structures that can be achieved using operations defined from the fundamentals of graph theory. These operations create and manipulate nodes and the connections between them on a graph (which may be converted into a structural or other network topology) in an efficient and precise manner.

4.1 Graph Theory: Preliminary Definitions

In mathematics, *graph theory* is the study of data structures used to model pairwise relations between objects. In general, a graph G is defined as an ordered triple $\left(N(G), E(G), \phi(G)\right)$, consisting of a nonempty set $N(G)$ of *nodes*, a set $E(G)$ of *edges* that is disjoint from $N(G)$, and an *incidence function* $\phi(G)$ that associates each edge of G with an unordered pair of (not necessarily distinct) nodes of G. If e is an edge and u and v are nodes such that $\phi(G, e) = uv$, then e is said to join u and v, and these nodes are referred to as the ends of e (Bondy & Murty 1976, Gibbons 1985, Diestel 2010).

Example 4.1 Consider, for example, the graph illustrated in Figure 4.1(a), which can be written as

$$G = \left(N(G), E(G), \phi(G)\right),$$

where

$$\begin{aligned} N(G) &= \{n_1, n_2, n_3, n_4, n_5\}, \\ E(G) &= \{e_1, e_2, e_3, e_4, e_5, e_6, e_7, e_8\}, \end{aligned}$$

and where $\phi(G)$ is defined by

$$\begin{aligned} &\phi(G, e_1) = n_1 n_2, \quad \phi(G, e_2) = n_2 n_3, \quad \phi(G, e_3) = n_3 n_3, \quad \phi(G, e_4) = n_3 n_4, \\ &\phi(G, e_5) = n_2 n_4, \quad \phi(G, e_6) = n_4 n_5, \quad \phi(G, e_7) = n_2 n_5, \quad \phi(G, e_8) = n_2 n_5. \end{aligned}$$

In a similar vein, the graph H depicted in Figure 4.1(b) can be written as

$$H = \left(N(H), E(H), \phi(H)\right),$$

where

$$\begin{aligned} N(H) &= \{n_1, n_2, n_3, n_4, n_5, n_6, n_7, n_8\}, \\ E(H) &= \{a, b, c, d, e, f, g, h, i, j, k, l\}, \end{aligned}$$

and where $\phi(H)$ is defined by

$$\begin{aligned} &\phi(H, a) = n_1 n_5, \quad \phi(H, b) = n_2 n_5, \quad \phi(H, c) = n_2 n_3, \quad \phi(H, d) = n_1 n_3, \\ &\phi(H, e) = n_3 n_5, \quad \phi(H, f) = n_3 n_4, \quad \phi(H, g) = n_4 n_5, \quad \phi(H, h) = n_2 n_4, \\ &\phi(H, i) = n_1 n_4, \quad \phi(H, j) = n_6 n_8, \quad \phi(H, k) = n_6 n_7, \quad \phi(H, l) = n_7 n_8. \end{aligned}$$

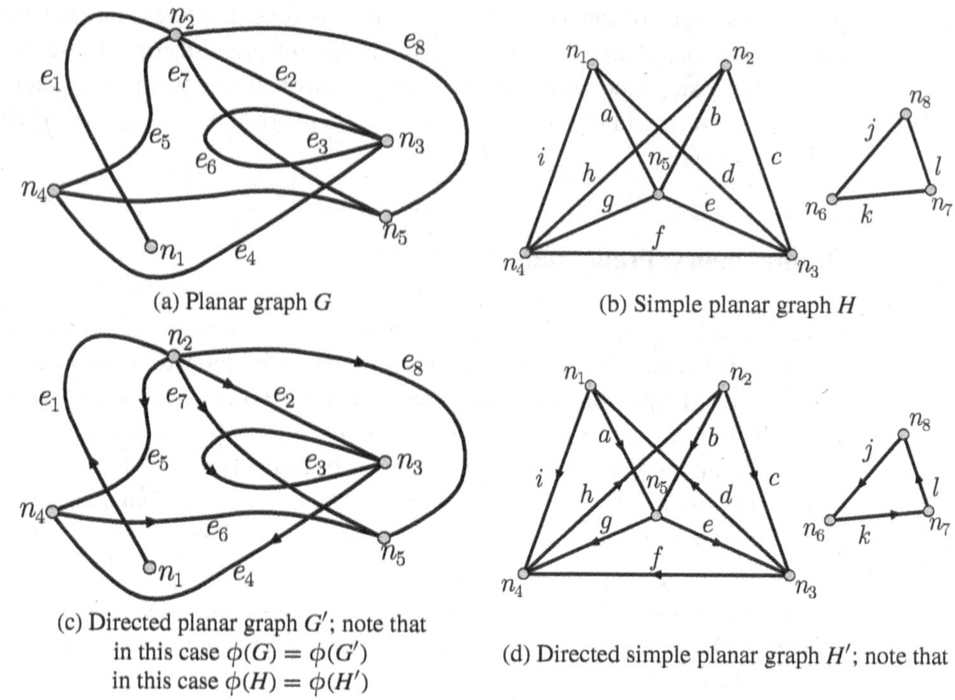

Figure 4.1 Diagrams of example graphs adapted from Bondy and Murty (1976).

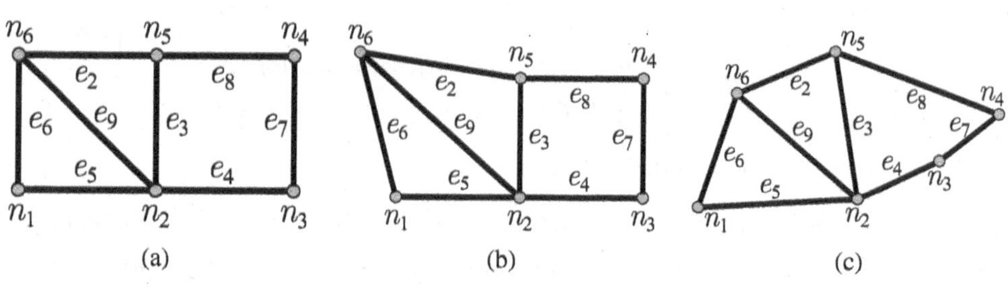

Figure 4.2 Three diagrams that all present the graphical information of graph I.

As defined, graphs are purely combinatorial entities. However, graphs are so named because they can be represented *graphically*, allowing one to understand their properties. In Figures 4.1(a)–4.1(b), each node is represented by a point and each edge by a line connecting the points that represent their ends, as defined by $\phi(G)$ or $\phi(H)$, respectively. It should be noted that there is no unique way of drawing a graph as defined, as the relative positions of points representing nodes and lines representing edges have no inherent spatial significance. For example, the diagrams shown in Figure 4.2 are all valid representations of some graph I, as each represents the same sets of nodes and edges and the same incidence function capturing their connectivity.

One can further characterize a graph based on its properties (Bondy & Murty 1976). Graphs that have a diagram whose edges intersect only at their end nodes are referred to as *planar graphs*, as such graphs can be represented in the plane in a simple manner. Though not immediately clear, both of the graphs in Figures 4.1(a)–4.1(b) are planar, as one can use the aforementioned property of graphs to reinterpret their diagrams such that their edges only intersect one another at end nodes. A graph is *simple* if it has no loops (i.e., an edge that joins a node to itself) and no two edges join the same pair of nodes. Graph G in Figure 4.1(a) is not simple, as edge e_3 forms a loop and edges e_7 and e_8 both join nodes n_2 and n_5. Graph H in Figure 4.1(b), however, is a simple graph. Graphs are referred to as *connected* if there exists a sequence of edges that link two arbitrary nodes on the graph together and *disconnected* if no such sequence exists; the idea of disconnected graphs will be revisited in Section 4.4.2. Graph G in Figure 4.1(a) represents a connected graph, as there is a path between every node associated with the graph; graph H in Figure 4.1(b) represents a disconnected graph comprised of two distinct *subgraphs*, though H could become a connected graph, for instance, by simply adding an edge between nodes n_3 and n_6.

Finally, a *directed graph* is a graph in which each edge has an assigned orientation. Directed graphs (herein referred to as G' and similar) are defined identically to general graphs with the exception of the incidence function, which no longer associates each edge of G' with an unordered pair. Instead, given $\phi(G', e) = uv$, node u is said to be the *tail* of edge e, and node v is said to be the *head* of edge e. Figure 4.1(c) depicts a directed graph G' using G (cf. Figure 4.1(a)) as its *underlying graph*. Note that in this case $\phi(G)$ correctly labels the tails and heads of each edge in $E(G')$ such that $\phi(G) = \phi(G')$. A directed graph H' based on underlying graph H (cf. Figure 4.1(b)) is shown in Figure 4.1(d). Here, $\phi(H) \neq \phi(H')$ because the edges in the set $E(H')$ have tails and heads that are not captured within $\phi(H)$, specifically $\phi(H, e)$, $\phi(H, g)$, and $\phi(H, h)$. Based on the notion that edges have an orientation, one can define the concept of a *directed walk* in a directed graph G', which is a finite non-null sequence $W(G') = (n_0, e_1, n_1, \ldots, e_k, n_k)$ whose terms are alternatively nodes and edges such that, for $i = 1, 2, \ldots, k$, the edge e_i has head n_i and tail n_{i-1}. Note that this sequence is not necessarily unique. This sequence can be further simplified to include only the sequence of nodes along the walk. For example, graphs G' and H' have directed walks of $W(G') = (n_1, n_2, n_3, n_4, n_5)$ and $W(H') = (n_1, n_5, n_3, n_1, n_4, n_2, n_5, n_4)$, respectively. In this chapter, the idea of simple, connected, planar graphs will be critical to the representation of networks/structures. The concept of a directed walk will be the key enabler of the network-generating algorithm proposed herein, which will be based on an agent executing such a walk.

4.2 Parametric L-systems

As mentioned in the introduction of this chapter, there are limiting issues with a naive use of the turtle graphics interpretation of L-systems as introduced in Chapter 3, namely with regard to the distance and angle parameters. To address these problems, Lindenmayer and others have proposed a *parametric L-system*, which enables numerical parameters to be associated with certain L-system characters (Lindenmayer 1974,

Prusinkiewicz & Hanan 1990). Parametric L-systems consist of an alphabet $\boldsymbol{\Sigma}$, a set of formal parameters $\boldsymbol{\Psi}$, and a finite set of production rules $\boldsymbol{P}$, these rules being recursively applied to form a parametric word μ.

- The *alphabet* is a finite, non-empty set $\boldsymbol{\Sigma}$ of *characters* Σ_i as defined in Section 3.1.
- *Formal parameters* $\boldsymbol{\Psi}$ are a set of arguments that are used in the specification of the parametric L-system productions to be discussed below. For example,

  $$\boldsymbol{\Psi} = \{t\} \quad \text{and} \quad \boldsymbol{\Psi} = \{x, y\} \quad \text{and} \quad \boldsymbol{\Psi} = \{h, i, k, s, t\}$$

  are all acceptable sets of formal parameters.
- A *parametric word* μ is a finite string of *modules* each consisting of a character and associated *actual* parameters; note that these numerical parameters differ from formal parameters, and their relationship will be discussed below. A module with a character $\Sigma_i \in \boldsymbol{\Sigma}$ and real-valued actual parameters $a_1, a_2, \ldots, a_n \in \mathbb{R}$ is denoted by

  $$\Sigma_i(a_1, a_2, \ldots, a_n).$$

  The set of all parametric words is denoted as $\boldsymbol{\Omega}_{\Sigma,\Psi} = \boldsymbol{\Sigma} \times \mathbb{R}$. The length $|\mu| \in \mathbb{N}$ is defined as the integer number of non-empty modules in μ, and an empty parametric word is denoted by $\varnothing$.
- The *axiom* is a non-empty parametric word $\omega^0 \in \boldsymbol{\Omega}_{\Sigma,\Psi}$ that, as described in Section 3.1, initiates the process of parametric L-system recursion.
- *Production rules* are a set $\boldsymbol{P}$ of rewriting rules. In a parametric L-system, production rules consist of three components: the *predecessor*, the *condition*, and the *successor*. For example, a context-free (cf. Section 3.1) production rule with a predecessor $\mathtt{A}(t)$, condition $t < 3$, and successor $\mathtt{B}(t+3)\mathtt{CD}(t^2, t-1)$ is written as

  $$\mathtt{A}(t)|_{t<3} \rightarrow \mathtt{B}(t+3)\mathtt{CD}(t^2, t-1).$$

  Note that in this example, all three components are a function of a formal parameter $t \in \boldsymbol{\Psi}$, which can be combined with numerical constants using the arithmetic operations $+, -, *, /$; the exponential operator $\wedge$; the relational operations $<, >, =$; the logical operators $!, \&, |$; and parentheses () to further define the production rule.[1] A production rule is applied to a module in a parametric word if the following conditions are met:
  - the character in the module and the character in the predecessor are the same (as in Section 3.1),
  - the number of actual parameters in the module is equal to the number of formal parameters in the predecessor, and
  - the condition evaluates to *true* if the actual parameters in the module are substituted for the formal parameters in the production rule.

  For example, the production rule above would be applicable to the module $\mathtt{A}(2)$, as the character $\mathtt{A}$ in the module is the same as that of the predecessor, there is one

[1] Standard rules for constructing syntactically correct expressions and for operator precedence are observed.

actual parameter in the module A(2) and one formal parameter in the predecessor A(t) and the logical expression $t < 3$ is true for $t = 2$. The resulting parametric word when applying the production rule to the module A(2) is B(5)CD(4, 1).

In general, a context-free $0L$ parametric production rule can be written as

$$P_i\colon \Sigma_i(\boldsymbol{\Psi})|_{C_i(\boldsymbol{\Psi})} \rightarrow \Lambda_i \quad \forall\, \Sigma_i \in \boldsymbol{\Sigma}, \tag{4.1}$$

where one $\Lambda_i \in \boldsymbol{\Omega}_{\Sigma,\Psi}$ is explicitly defined for each P_i. Here, $\Sigma_i(\boldsymbol{\Psi})$ is the predecessor, $C_i(\boldsymbol{\Psi})$ is a condition based on the actual parameters of the module, and Λ_i is the successor. Note that the presence of a condition $C_i(\boldsymbol{\Psi})$ allows for multiple parametric production rules to be written for the same predecessor $\Sigma_i(\boldsymbol{\Psi})$. For example, the module A(t) discussed above could have an additional production rule for the condition $t \geq 3$ with a different distinct successor Λ_i. As discussed in Section 3.1, any production rule P_i corresponding to $\Sigma_i(\boldsymbol{\Psi})$ and a valid $C_i(\boldsymbol{\Psi})$ that is not explicitly defined is assumed to be an identity operation (i.e., $P_i\colon \Sigma_i(\boldsymbol{\Psi})|_{C_i(\boldsymbol{\Psi})} \rightarrow \Sigma_i(\boldsymbol{\Psi})$) unless otherwise defined.

Like traditional L-systems, parametric L-systems can also have context-sensitive production rules. Expanding upon Equation 4.1, a parametric $2L$ production rule is given as

$$P_i\colon \Sigma_i^l(\boldsymbol{\Psi}) < \Sigma_i(\boldsymbol{\Psi}) > \Sigma_i^r(\boldsymbol{\Psi})|_{C_i(\boldsymbol{\Psi})} \rightarrow \Lambda_i,$$

where $\Sigma_i^l \in \boldsymbol{\Sigma}$ and $\Sigma_i^r \in \boldsymbol{\Sigma}$. Here, the predecessor consists of three components: the *left context* $\Sigma_i^l(\boldsymbol{\Psi})$, the *strict predecessor* $\Sigma_i(\boldsymbol{\Psi})$, and the *right context* $\Sigma_i^r(\boldsymbol{\Psi})$. In addition to those described above for context-free parametric production rules, the applicability of a given $2L$ parametric production rule is also based on the condition that $\Sigma_i(\boldsymbol{\Psi})$ is immediately preceded by $\Sigma_i^l(\boldsymbol{\Psi})$ and immediately followed by $\Sigma_i^r(\boldsymbol{\Psi})$. As an example, consider the $2L$ parametric production rule defined as

$$\text{A}(x) < \text{B}(y) > \text{C}(z)|_{x+y+z>12} \rightarrow \text{E}((x+y)/4)\text{F}((y-z)/4).$$

This production rule can be applied to the module B(5) that appears in a parametric word

$$\ldots\text{A}(7)\text{B}(9)\text{C}(3)\ldots$$

as the sequence of characters A, B, C in the production rule and parametric word are the same, the numbers of actual parameters and formal parameters coincide, and the condition $7 + 9 + 3 > 12$ is valid. As a result, the module B(9) is replaced by the successor E(4)F(1.5) in the parametric word; the modules A(7) and C(3) may be replaced by other production rules in the same derivation step.

- Finally, *recursion* is the process by which the set of production rules $\boldsymbol{P}$ are iteratively applied to the parametric word μ^n to form μ^{n+1}. This process is repeated N_r times to yield a developing set of parametric words,

$$\mu^0 \rightarrow \mu^1 \rightarrow \cdots \mu^n \rightarrow \cdots \mu^{N_r}.$$

To better elucidate the rewriting process of a parametric L-system, an example $0L$-system as presented by Prusinkiewicz and Lindenmayer (2004), is given below.

Example 4.2 Consider the parametric 0L-system $G = \{\boldsymbol{\Sigma}, \boldsymbol{\Psi}, \boldsymbol{\omega}^0, \boldsymbol{P}\}$ defined by the alphabet $\boldsymbol{\Sigma} = \{\texttt{A}, \texttt{B}, \texttt{C}\}$, formal parameters $\boldsymbol{\Psi} = \{x, y\}$, axiom $\boldsymbol{\omega}^0 = \texttt{B}(2)\texttt{A}(4,4)$, and production rules $\boldsymbol{P}$ given by

$$\begin{aligned} P_1 &: \texttt{A}(x,y)|_{y\leq 3} \rightarrow \texttt{A}(x*2, x+y), \\ P_2 &: \texttt{A}(x,y)|_{y>3} \rightarrow \texttt{B}(x)\texttt{A}(x/y, 0), \\ P_3 &: \texttt{B}(x)|_{x<1} \rightarrow \texttt{C}, \\ P_4 &: \texttt{B}(x)|_{x\geq 1} \rightarrow \texttt{B}(x-1). \end{aligned}$$

Given the axiom $\boldsymbol{\omega}^0$, the associated parametric words $\boldsymbol{\mu}^n$ are obtained by substituting each module from the previously existing parametric word $\boldsymbol{\mu}^{n-1}$ with its associated production rule (if one exists). Regarding the resulting word $\boldsymbol{\mu}^1$ from the first recursion, we see that the first module $\texttt{B}(2)$ corresponds to production rule P_4, given that $C_4(x)$ is satisfied when $x = 2$; substituting $x = 2$ into $\boldsymbol{\lambda}_4$ results in the module $\texttt{B}(1)$. Next, we see that the second module $\texttt{A}(4,4)$ corresponds with production rule P_2 given that $C_2(x,y)$ is satisfied when $y = 4$; substituting $x = y = 4$ into $\boldsymbol{\lambda}_2$ results in the series of modules $\texttt{B}(4)\texttt{A}(1,0)$. Thus, we find that $\boldsymbol{\mu}^1 = \texttt{B}(1)\texttt{B}(4)\texttt{A}(1,0)$. For the second recursion, we follow an analogous procedure to find that the modules $\texttt{B}(1) \rightarrow \texttt{B}(0)$, $\texttt{B}(4) \rightarrow \texttt{B}(3)$, and $\texttt{A}(1,0) \rightarrow \texttt{A}(2,1)$, leading to $\boldsymbol{\mu}^2 = B(0)\texttt{B}(3)\texttt{A}(2,1)$. Continuing this process $N_r = 4$ times, the following sequence of parametric words is formed:

$$\begin{aligned} \boldsymbol{\mu}^0 &= \texttt{B}(2)\texttt{A}(4,4), \\ \boldsymbol{\mu}^1 &= \texttt{B}(1)\texttt{B}(4)\texttt{A}(1,0), \\ \boldsymbol{\mu}^2 &= \texttt{B}(0)\texttt{B}(3)\texttt{A}(2,1), \\ \boldsymbol{\mu}^3 &= \texttt{CB}(2)\texttt{A}(4,3), \\ \boldsymbol{\mu}^4 &= \texttt{CB}(1)\texttt{A}(8,7). \end{aligned}$$

Note that in $\boldsymbol{\mu}^3$ the character C has no applicable production rule, and therefore when generating $\boldsymbol{\mu}^4$ (or any subsequent parametric words) this character is subjected to an identity production rule (i.e., $\texttt{C} \rightarrow \texttt{C}$).

In this chapter, we consider a *modified* parametric L-system to reduce the complexity of the L-system and limit the number of variables required for network and geometric topology generation. The alphabet $\boldsymbol{\Sigma}$ is similar to that defined in Equation 3.2 but altered to consist of three sets of characters: a set of variable characters $\boldsymbol{\alpha}$ represented by English letters (e.g., $\boldsymbol{\alpha} = \{\texttt{A}, \texttt{B}, \texttt{C}, \texttt{D}, \dots\}$; $|\boldsymbol{\alpha}| = N_\alpha$), a set of constant characters $\boldsymbol{\beta} = \{+, -\}$, and a set of constant characters $\boldsymbol{\gamma} = \{\,[\,,\,]\,\}$ such that

$$\boldsymbol{\Sigma} = \boldsymbol{\alpha} \cup \boldsymbol{\beta} \cup \boldsymbol{\gamma}.$$

The sets of constant characters are defined based on whether a module consisting of those characters requires an associated parameter (as in $\boldsymbol{\beta}$) or not (as in $\boldsymbol{\gamma}$). Following after Section 3.1, the variable characters in $\boldsymbol{\alpha}$ will generally be associated with operations that create or modify the nodes, edges, and faces of a graph (cf. Section 4.1) and by extension the network or geometric topology, while the constant characters in $\boldsymbol{\beta}$ and $\boldsymbol{\gamma}$ are associated with operations that alter the state of the graph creation agent but not the graph itself (cf. Section 4.3.2). Regarding formal parameters, we consider the set $\boldsymbol{\Psi} = \{\psi_{\alpha_1}, \psi_{\alpha_2}, \psi_{\beta_1}\}$, where ψ_{α_1} and ψ_{α_2} are parameters associated with a variable character from $\boldsymbol{\alpha}$ (i.e., $\alpha_i(\psi_{\alpha_1}, \psi_{\alpha_2})$, $\alpha_i \in \boldsymbol{\alpha}$), and ψ_{β_1} is a parameter associated with a constant character from $\boldsymbol{\beta}$ (i.e., $\beta_i(\psi_{\beta_1})$, $\beta_i \in \boldsymbol{\beta}$). The axiom $\boldsymbol{\omega}^0$ will always consist of two characters from $\boldsymbol{\alpha}$ (i.e., $\boldsymbol{\omega}^0 = \omega_1\omega_2, \omega_i \in \boldsymbol{\alpha}$) to ensure generative action, regardless of the number of recursions N_r specified.

Regarding the set of parametric production rules $\boldsymbol{P}$, we continue a context-free approach and define the condition associated with each production rule P_i to be the same (i.e., $C_i(\boldsymbol{\Psi}) = C(\boldsymbol{\Psi})$); specifically, each production rule is applied if

$$0 \le \psi_{\alpha_1} < 1 \quad \text{and} \quad 0 \le \psi_{\alpha_2} < 1.$$

Defining each actual parameter a in the parametric L-system to be bounded by the semi-open interval $[0, 1)$, we see that $C(\boldsymbol{\Psi})$ is automatically satisfied and is therefore unnecessary in the production rule definition. This allows us to specify a single production rule for each variable character, limiting both the complexity of the parametric L-system and the number of variables required to define it.[2] Thus, production rules for each of the N_α characters in $\boldsymbol{\alpha}$ for this parametric L-system formulation can be generally written as (cf. Equation 4.1)

$$P_i\colon \alpha_i \to \lambda_i^1(\psi_i^1)\lambda_i^2(\psi_i^2)\ldots\lambda_i^{N_\Lambda}(\psi_i^{N_\Lambda}),\ i = 1\ldots N_\alpha \tag{4.2}$$

for all $\alpha_i \in \boldsymbol{\alpha}$, where $\lambda_i^1\lambda_i^2\ldots\lambda_i^{N_\Lambda}$ is a string of N_Λ characters such that each λ_i^K is selected from a set of options $\boldsymbol{\Lambda}^K$ (i.e., $\lambda_i^K \in \boldsymbol{\Lambda}^K$) and $\boldsymbol{\psi}^K$ are the set of parameters associated with λ_i^K. To provide clarification, consider the below example.

Example 4.3 Let us apply the concepts introduced above to generate a parametric word $\boldsymbol{\mu}$ that will be interpreted using the approach detailed in Section 4.3. Specifically, consider $N_\alpha = 5$, where $\boldsymbol{\alpha} = \{\texttt{A}, \texttt{B}, \texttt{C}, \texttt{D}, \texttt{E}\}$, and a two-character axiom $\boldsymbol{\omega}^0 = \texttt{C(0.13,0.55)A(0.37,0.92)}$. Assume $\boldsymbol{\beta} = \{\texttt{+}, \texttt{-}\}$ and $\boldsymbol{\gamma} = \{\texttt{[}, \texttt{]}\}$. For the remainder of this chapter, we choose $N_\Lambda = 10$ and the following options for the creation of the production rules:

$$\boldsymbol{\Lambda}^1 = \{\texttt{[}, \varnothing\},\ \boldsymbol{\Lambda}^2 = \boldsymbol{\alpha},\ \boldsymbol{\Lambda}^3 = \boldsymbol{\alpha},\ \boldsymbol{\Lambda}^4 = \boldsymbol{\beta},\ \boldsymbol{\Lambda}^6 = \{\texttt{[}, \varnothing\},\ \boldsymbol{\Lambda}^7 = \boldsymbol{\alpha},\ \boldsymbol{\Lambda}^8 = \boldsymbol{\alpha},\ \boldsymbol{\Lambda}_9 = \boldsymbol{\beta}.$$

Each character λ_i^K, $K = 1\ldots 10$ for rule P_i is exclusively chosen from among the options in set $\boldsymbol{\Lambda}^K$, and the assignments for $\boldsymbol{\Lambda}_i^5$ and $\boldsymbol{\Lambda}_i^{10}$ are algorithmically chosen

[2] While a driving factor behind the adoption of the parametric L-system approach was the ability to apply parameters to certain operations to add greater modeling power to the standard L-system, an adept reader will notice that by eliminating the conditions associated with production rules we have also admittedly limited the parametric L-system by allowing only a finite number of modules to exist (i.e., if each A(t) module has the same production rule A(t) → B(0.5), the impact of the parameter in a module is seemingly diminished). However, it will be demonstrated in Sections 4.3.2–4.3.4 that the way in which each module is interpreted will vary depending on the state of the underlying graph (i.e., one B(0.5) operation does not have the same result as the next) and that the modeling power of the parametric L-system is preserved.

such that the "begin bracket" indicators " [" in $\mathbf{\Lambda}_i^1$ and $\mathbf{\Lambda}_i^6$, if assigned, are closed. Additionally, using this definition of $\mathbf{\Lambda}_K$ and recalling the number of parameters needed for each character in $\mathbf{\Sigma}$ requires that $\left|\psi_i^K\right| = 0$ for $K = 1,5,6,10$; $\left|\psi_i^K\right| = 1$ for $K = 4,9$; and $\left|\psi_i^K\right| = 2$ for $K = 2,3,7,8$. The five production rules required for each character in $\boldsymbol{\alpha}$ are then formed using random assignments, and the resulting parametric word from the first recursion and partial parametric word from the two and three recursions are shown in Figure 4.3.

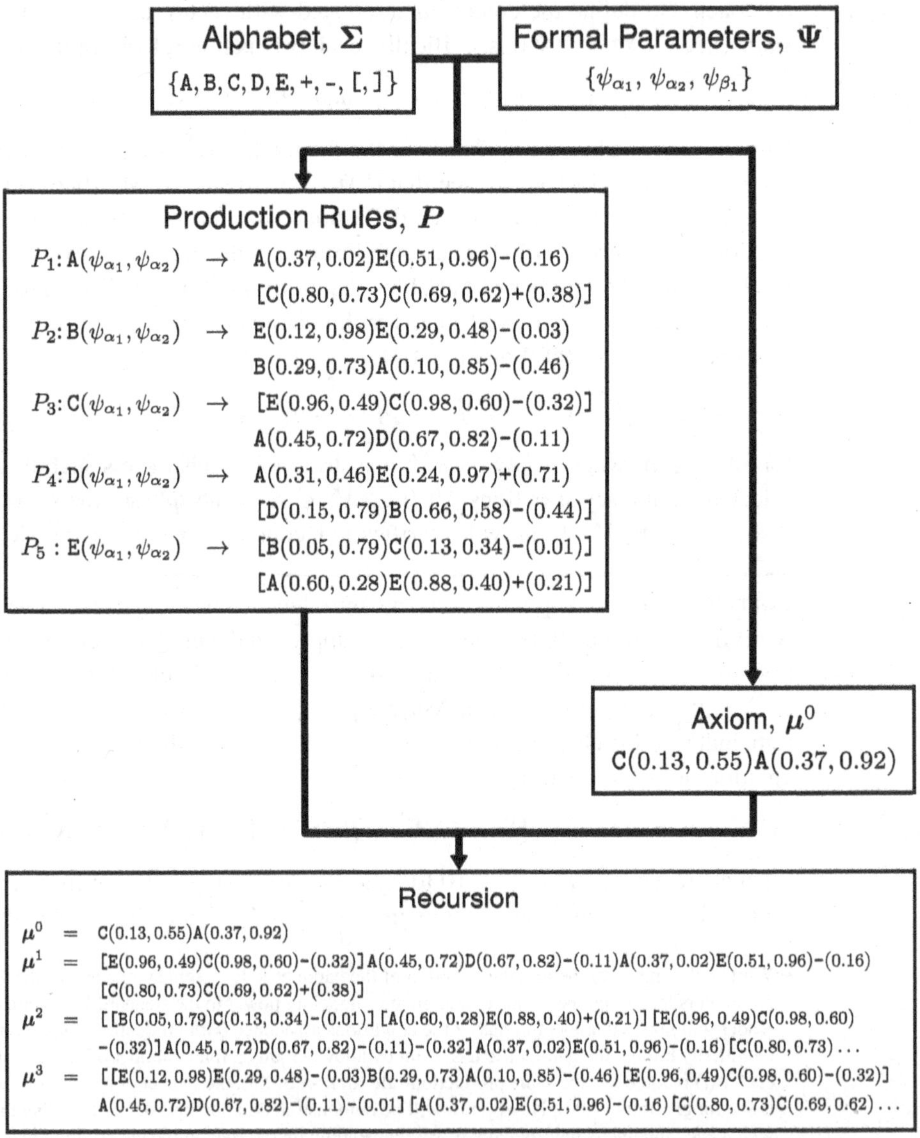

Figure 4.3 Example of recursively generated parametric words using a modified parametric L-system. Note that the symbols + and - do not denote mathematical operations, but are characters from $\boldsymbol{\beta}$ to which parameters have been assigned.

4.3 SPIDRS Algorithm

4.3.1 Half-Edge Data Structure

The concepts from graph theory introduced in Section 4.1 are implemented within the SPIDRS algorithm using a *half-edge data structure* (HEDS), also known as a *doubly connected edge list*, which was first proposed by Muller and Preparata (1978) and represents a directed simple planar graph G' in a manner that enables simple traversal and manipulation (Berg, Cheong, Kreveld & Overmars 1997). The data structure maintains a record of each node, edge, and face of the graph, which are described as follows:

- Nodes are defined in a manner consistent with Section 4.1;
- Edges are defined based on the notion that each connection between two nodes can be represented using two *half-edges* that are oriented in opposite directions and assumed to belong to two distinct faces. Thus, for an edge e that joins nodes u and v, there exist two half-edges e^+ and e^- such that $\varphi_{G'}(e^+) = uv$ and $\varphi_{G'}(e^-) = vu$. Herein, for simplicity, half-edges will be denoted as e_{uv}, where u and v are the tail and head of the half-edge, respectively. Because the HEDS assumes a planar graph, edges can only intersect one another at defined nodes.
- Faces are defined as enclosed sets of half-edges oriented such that one can start at a given node, traverse around a portion of the graph based on the orientation of adjacent half-edges, and return to that node.

Additionally, the HEDS can maintain a record of *attribute information*, which is additional information assigned to specific nodes, edges, or faces that further defines the graph (Berg et al. 1997).

To better introduce this concept, consider the directed simple planar graph J' shown in Figure 4.4. The sets of nodes, edges, and faces that define this graph are given as

$$N(J') = \{1,2,3,4,5,6\},$$

$$\boldsymbol{E}(J') = \{e_{12}, e_{21}, e_{23}, e_{32}, e_{34}, e_{43}, e_{41}, e_{14}, e_{25}, e_{52}, e_{56}, e_{65}, e_{63}, e_{36}\},$$

$$\boldsymbol{F}(J') = \{f_0, f_1, f_2\} = \{[e_{14}, e_{43}, e_{36}, e_{65}, e_{52}, e_{21}],\ [e_{12}, e_{23}, e_{34}, e_{41}],\ [e_{25}, e_{56}, e_{63}, e_{32}]\}.$$

Note that f_0 is the exterior face of the graph. From this description, the hierarchy of the information defining the graph in this data structure is immediately apparent, as $\boldsymbol{N}(J')$ is contained within $\boldsymbol{E}(J')$, and $\boldsymbol{E}(J')$ is contained within $\boldsymbol{F}(J')$. Herein we will simplify $\boldsymbol{F}(J')$ by including only the sequence of nodes defining the face such that

$$\boldsymbol{F}(J') = \{f_0, f_1, f_2\} = \{[1,4,3,6,5,2],\ [1,2,3,4],\ [2,5,6,3]\}.$$

Thus, in the HEDS, a face can simply be thought of as a directed walk $W_{J'} = (n_0, n_1, \ldots, n_{k-1}, n_k)$ (cf. Section 4.1), where $n_k = n_0$ and is omitted from the face definition with the understanding that once an observer reaches n_{k-1} they will return to the "beginning" of the face. It is assumed within the HEDS that movement around each face is always oriented in a counterclockwise direction with the exception of the face that represents the exterior of the graph (f_0 in Figure 4.4).

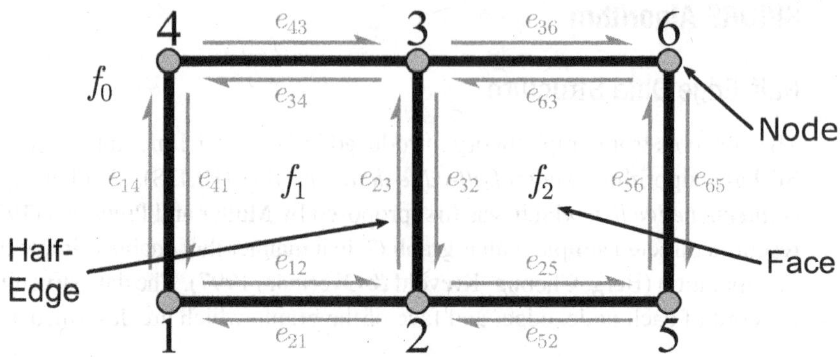

Figure 4.4 Example of a graph represented using the HEDS that serves as the foundation of the SPIDRS algorithm.

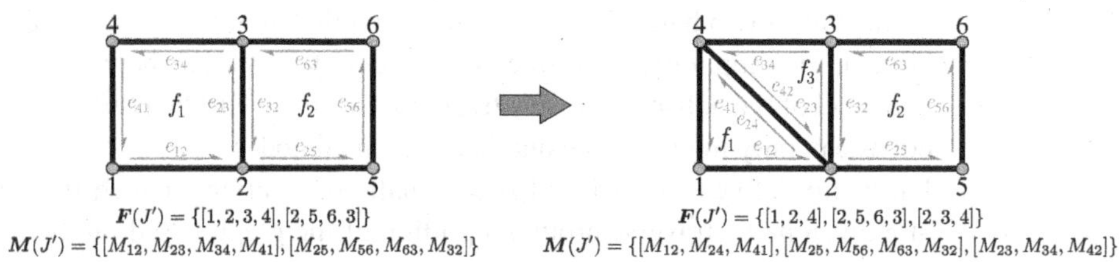

Figure 4.5 Example of how the HEDS treats topological modifications of the graph J' in Figure 4.4 (neglecting the presence of the external face f_0). Note that subdividing the face f_1 results in the creation of a new face, f_3.

HEDS formulations are used in computational geometry to describe geometric polygonal subdivisions due to the ease with which one can manipulate the topological information (which edges bound a given face, which faces are adjacent, etc.) associated with a planar graph (Berg et al. 1997). For instance, returning to the example graph J' and modifying the connectivity of the graph such that an edge is created between nodes 2 and 4 results in the subdivision of the face f_1 as illustrated in Figure 4.5. Neglecting the exterior face of the graph, graph J' can now be defined by the face set

$$\boldsymbol{F}(J') = \{[1,2,4],\ [2,5,6,3],\ [2,3,4]\}.$$

Notice that this subdivision results in the creation of a new face f_3 and a new set of twin half-edges e_{24} and e_{42}.

As discussed in Section 4.1, a graph is simply a diagram used to describe connectivity between objects and there are no inherent definitions as to the spatial locations or structural materials associated with the nodes/edges of the graph within the HEDS. Thus, at a fundamental level, the SPIDRS algorithm is simply a network design tool that defines connections between nodes and therefore has potential applications in electrical circuit design, pipeflow network design, etc. (cf. Chapter 6). For the purposes of network topology optimization, the aforementioned attribute information within the data structure is used to assign functional properties to the graph, and most often to its edges. In the case of structural design, which dominates this text, two important pieces of information are critical: (i) the actual position of each node and (ii) the meaning of each edge. Regarding the former, a coordinate function of the form $C(n) = (x_n, y_n)$, where x_n and y_n are the x- and y-coordinates for node n, is used to both define the location of each node in the graph and determine the location of newly created nodes for certain graphical operations. Regarding the latter, edges are assumed to represent high aspect ratio structural members, with the set M containing material properties and/or cross-sectional dimensions assigned to each edge. For example, the material assignments $M(J')$ of the graph shown in Figure 4.4 can be written as

$$M(J') = \{[M_{14}, M_{43}, M_{36}, M_{65}, M_{52}, M_{21}], [M_{12}, M_{23}, M_{34}, M_{41}], [M_{25}, M_{56}, M_{63}, M_{32}]\},$$

where M_{uv} is the material assignment associated with the edge e_{uv}. Note that the material associated with twin half-edges remains consistent (e.g., $M_{23} = M_{32}$), and that the set $\boldsymbol{M}(J')$ takes the same form as the face set $\boldsymbol{F}(J')$. Modifications to the material set caused by the subdividing behavior illustrated in Figure 4.5 also remain consistent with those of the face set, where after subdivision and neglecting the external face, the material set is written as

$$\boldsymbol{M}(J') = \{[M_{12}, M_{24}, M_{41}], [M_{25}, M_{56}, M_{63}, M_{32}], [M_{23}, M_{34}, M_{41}]\}.$$

Because the graph is assigned spatial significance using the coordinate function $C(n)$, the subdividing behavior and the associated modification to the set of faces and materials described above can only be validly described by this HEDS if the geometry of the domain being subdivided is convex, as the HEDS assumes that edges intersect only at nodes. As a counterexample, Figure 4.6 depicts an initial (i.e., unmodified) graph K' with an associated coordinate function defined such that the geometry of the graph is non-convex. Now consider the creation of an edge between nodes 3 and 5. From a graph theory perspective, this is valid as K' can be diagrammed such that it remains a planar graph by moving the position of node 2 (cf. Figure 4.2). However, the consideration of $C(n)$, which is essential in representing the graph as a structure, indicates that K' is uniquely diagrammed (illustrated in Figure 4.6); therefore, the creation of this new edge violates the HEDS as it will intersect the edge between nodes 1 and 2 at a point not defined by a node. This is of critical importance, as unintentional connections caused by the graphical operations to be discussed in Section 4.3.2 can (i) cause the SPIDRS algorithm to misinterpret the current state of the graph topology and (ii) alter the structural performance of the topology such that the progress of the GA is impeded as the effects of specific SPIDRS graphical operations can become nonphysical. Thus, special care must be taken when defining the coordinate function of the initial graph to ensure that its geometry is convex.

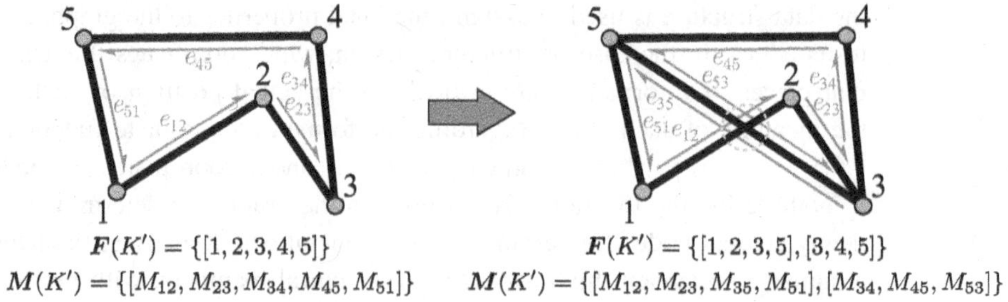

Figure 4.6 Example of how a geometrically non-convex graph K' can result in a violation of the HEDS (neglecting the presence of the external face f_0). Creating an edge between nodes 3 and 5 results in an edge intersection at a point not defined by a node (denoted by the dashed circle).

4.3.2 Graphical Operations

In an identical manner to that of Section 3.2, the parameterized L-system must be coupled with a graphical interpreter for the encoded information to be given spatial significance. Here, we introduce a series of operations that dictate the movement and actions of an edge-creating graphical agent, allowing the SPIDRS algorithm to modify and evolve the connectivity of a graph. We begin by providing a brief overview of these operations as illustrated in Figure 4.7, with a more detailed discussion as to their parameterization and function to follow in Section 4.3.3.

Two separate graphical operations govern the counterclockwise movement of the agent around a face on preexisting edges. These allow the agent to move to either preexisting nodes (Figure 4.7(a)) or newly created nodes (Figure 4.7(b)) as determined by a single associated L-system parameter (cf. Section 4.3.3). In a similar manner, separate operations dictate the creation of new edges (i.e., structural members) between existing nodes (Figure 4.7(c)) or between an existing node and a newly created node (Figure 4.7(d)); in both cases, two corresponding parameters define where edges are being created and attribute information associated with the new edge (i.e., structural material assignment). The decoupling of movement about a graph and edge creation dictates that edges connecting two nodes are constructed deliberately rather than coincidentally (cf. Section 3.5). Constraints are placed on the two edge-creating operations such that the HEDS that serves as the foundation of SPIDRS remains valid (e.g., edges cannot be created between two nodes that already share an edge). An additional operation serves to change the attribute information of the edge the agent is currently traversing, once again ensuring that the "overwriting" of material assignment occurs deliberately.

Due to the subdividing nature of how SPIDRS modifies graph connectivity, additional operations inspired by the orientation and branching operations in Section 3.2 are implemented to allow the agent to move between faces on the graph. Parameterized operations for clockwise (Figure 4.7(f)) and counterclockwise (Figure 4.7(g)) turns are used to reposition the agent in a neighboring face, thereby reducing the likelihood of graphical modifications being concentrated within the same area of the

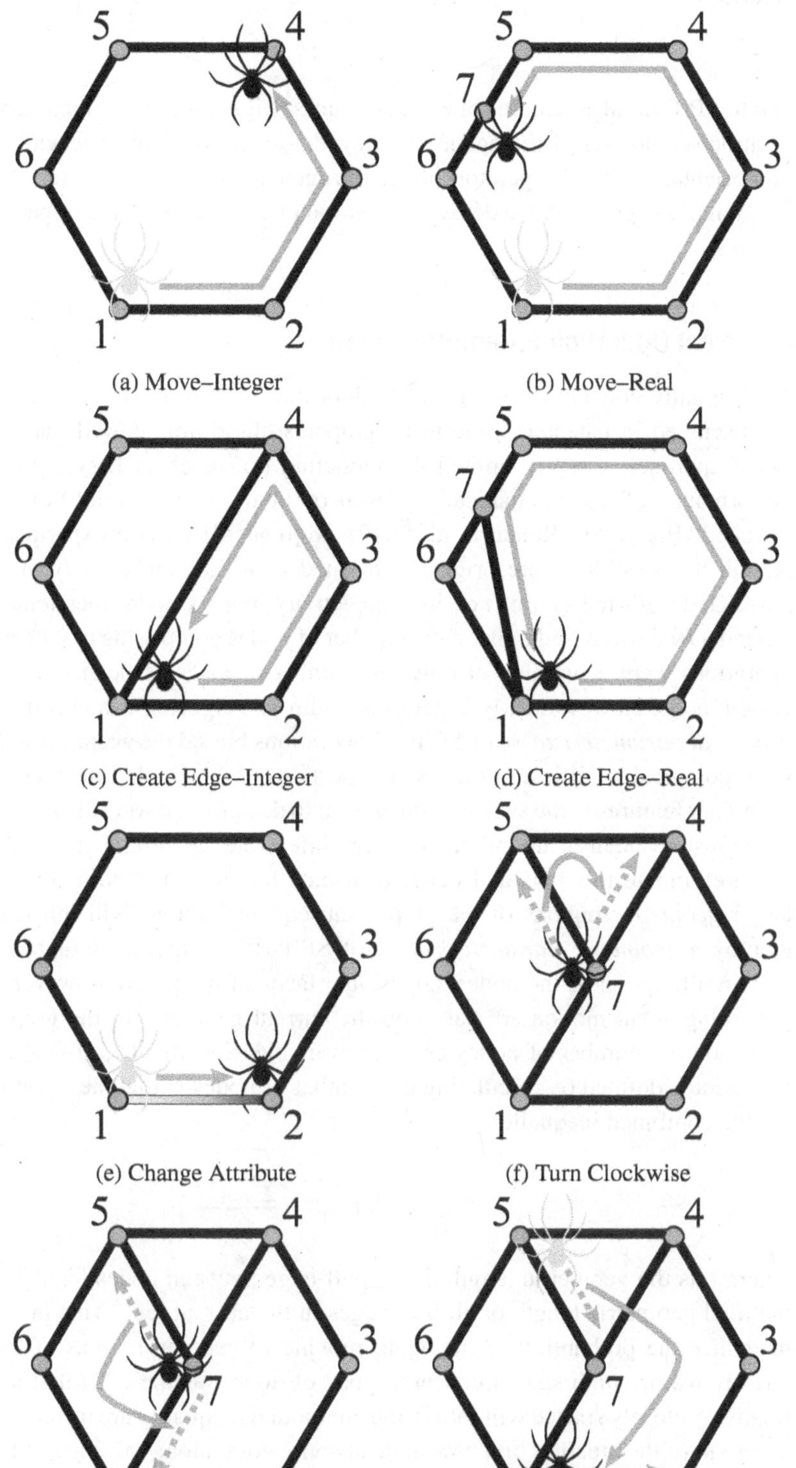

Figure 4.7 Examples of the seven SPIDRS graphical operations; turns in each direction are explicitly illustrated.

graph. SPIDRS also features bracket operators (Figure 4.7(h)), which save the current location/orientation of the agent or return the agent to its most recently saved location/orientation. These operators afford the agent the opportunity to remove itself from a region of the graph with a dense distribution of edges and navigate sparser developed regions.

4.3.3 Graphical Operation Parameterization

As originally conceived, the SPIDRS algorithm was to remain strictly graph-based. As discussed in Chapter 3, the turtle graphics algorithm showed that its geometry-based approach severely limited the modeling power of an L-system such that the performance of a mathematically proven optimal solution could not practically be matched (Bielefeldt, Reich, et al. 2019). Motivated by these experiences, SPIDRS graphical operations were originally defined as a function of *only* the information associated with the graph (i.e., its connectivity, not its nodal locations). Movement around a face was based only on the number of nodes comprising that face, and turning operations were a function of only the number of faces associated with the agent's current node. However, early experiences with this algorithm showed that this purely *graphical parameterization* of SPIDRS operations biased the agent toward concentrating topological modifications in specific portions of the graph. Fine features promoted even finer features in the same region and led to designs that were clearly geometrically (and thus physically) suboptimal and even infeasible (Bielefeldt 2020).

To eliminate this bias and better distribute topological modifications throughout larger *geometric* regions of the graph, subsequent sections will introduce and then employ a *geometric parameterization* of SPIDRS operations. Graphical operations remain a function of the nodes, edges, and faces of the graph; however, calculations governing agent motion are based on the current geometry of the graph rather than solely on the number of nodes or faces available. Specifically, in-face movement of the agent is defined by calculating the number of nodes N that the agent should move via the combined inequality

$$\frac{\sum_{i=1}^{N} l_i}{L} \leq \psi_{\alpha_1} < \frac{\sum_{i=1}^{N+1} l_i}{L}, \tag{4.3}$$

where l_i is the geometric length of the half-edge between nodes i and $i+1$, and L is the total geometric length of all half-edges in the current face. This inequality serves to reduce the probability of the agent moving to geometric areas of the graph that already feature high spatial concentrations of nodes, as the odds that nodes that are relatively closely spaced will satisfy the compound inequality are lower. For operations that require the creation of a new node at some point along the edge of the graph, the agent moves by N nodes in its current face via Equation 4.3 and then by some distance d between the current node and next node defined as

$$d = \frac{L\psi_{\alpha_1} - \sum_{i=1}^{N} l_i}{\sum_{i=1}^{N+1} l_i - \sum_{i=1}^{N} l_i}. \tag{4.4}$$

In a similar manner, out-of-face movement (turning) operations are parameterized by solving for the number of faces M the agent should move through via the combined inequality

$$\frac{\sum_{i=1}^{M} a_i}{A} \leq \psi_{\beta_1} < \frac{\sum_{i=1}^{M+1} a_i}{A}, \tag{4.5}$$

where a_i is the area of a specified face i and A is the sum of all the areas of all the faces the agent could potentially move to (i.e., not including the face the agent is currently in). Examples of how Equations 4.3–4.5 are employed for each graphical operation in the context of an initial hexagonal geometry are illustrated in Figure 4.7 and discussed below.

Move–Integer

The Move–Integer operation is represented in the parametric L-system encoding as $\mathtt{A}(\psi_{\alpha_1})$ and moves the agent to a preexisting node in the current face. In Figure 4.7(a), consider that the agent begins at node 1 and receives a command $\mathtt{A}(0.54)$. Substituting $\psi_{\alpha_1} = 0.54$ into Equation 4.3 and assuming that the length of each edge is unity such that $L = 6$, the inequality is satisfied when $N = 3$. Thus, the agent moves along the face by three nodes to node 4. Note that this command results only in the movement of the agent and does not modify the description of the graph (i.e., F and M remain unchanged).

Move–Real

The Move–Real operation is represented in the parametric L-system encoding as $\mathtt{B}(\psi_{\alpha_1})$ and moves the agent to a newly created node in the current face. This operation is illustrated in Figure 4.7(b), where the agent begins at node 1 and receives a command $\mathtt{B}(0.75)$. Substituting $\psi_{\alpha_1} = 0.75$ into Equation 4.3 and repeating the assumption that the length of each edge is unity results in the inequality being satisfied when $N = 4$, and the agent moves four nodes to node 5. Next, substituting this value and the value of ψ_{α_1} into Equation 4.4 results in $d = 0.5$, which is the distance the agent will move along the edge between nodes 5 and 6 before creating node 7. The description of the graph is accordingly modified as follows:[3]

$$\boldsymbol{F} = \{[1,2,3,4,5,6]\} \rightarrow \boldsymbol{F} = \{[1,2,3,4,5,7,6]\},$$

$$\boldsymbol{M} = \{[M_{12}, M_{23}, M_{34}, M_{45}, M_{56}, M_{61}]\} \rightarrow \boldsymbol{M} = \{[M_{12}, M_{23}, M_{34}, M_{45}, M_{57}, M_{76}, M_{61}]\},$$

and the geometric coordinates for node 7 are added to the coordinate function within the HEDS. Note that because a new node is added to $\boldsymbol{F}$, a new material attribute assignment must also be added to $\boldsymbol{M}$ such that the material attribute along the edge remains consistent (i.e., $M_{56} \rightarrow \{M_{57}, M_{76}\}; M_{57} = M_{76}$). Additionally, since a half-edge has been subdivided (in this example, e_{56} becomes e_{57}, e_{76}), the SPIDRS algorithm also subdivides the twin of that half-edge (e.g., e_{65} becomes e_{67}, e_{75}). Finally, the distance

[3] For these examples, the exterior face of the graph (e.g., f_0 in Figure 4.4) is neglected to more concisely elucidate how the various graphical operations effect the description of the graph.

d along the edge that the agent can move before creating a new node is restricted such that $0.1 \leq d \leq 0.9$ to avoid both tight node clustering (e.g., creating nodes very near the ends of edges) and the formation of smaller features; commands that result in $0.1 > d$ or $d > 0.9$ are set to 0.1 and 0.9, respectively.

Create Edge–Integer

The Create Edge–Integer operation is represented in the parametric L-system encoding as $\mathtt{C}(\psi_{\alpha_1}, \psi_{\alpha_2})$ and causes the agent to move to a preexisting node (based on the parameter ψ_{α_1}) in the current face before creating an edge back to the original location on the graph, thus subdividing the face. The parameter ψ_{α_2} defines the material attribute to be assigned to the newly created edge. Figure 4.7(c) shows an example of the agent starting at node 1 and receiving the command $\mathtt{C}(0.54, 0.81)$, where $\psi_{\alpha_2} = 0.81$ corresponds to the material attribute $M_{41} = M_{14}$ (in a manner consistent with Section 4.3.1). As with the example in Figure 4.7(a), the agent moves three nodes to node 4 before constructing a new edge back to its starting location (node 1). The description of the structure is then modified as follows:

$$\boldsymbol{F} = \{[1,2,3,4,5,6]\} \rightarrow \boldsymbol{F} = \{[1,2,3,4],[1,4,5,6]\},$$
$$\boldsymbol{M} = \{[M_{12}, M_{23}, M_{34}, M_{45}, M_{56}, M_{61}]\} \rightarrow \boldsymbol{M} = \{[M_{12}, M_{23}, M_{34}, M_{41}], [M_{14}, M_{45}, M_{56}, M_{61}]\}.$$

After subdividing the graph, the location and orientation of the agent is defined to be along the original half-edge (i.e., since the agent originated along the half-edge e_{12}, after executing this operation it will remain within the face that contains this half-edge). This operation is ignored if moving by N node(s) (cf. Equation 4.3) results in the agent stopping at a node that is collinear with its original node (e.g., nodes 2 or 6 in Figure 4.7(c)), as this can result in overlapping face and material attribute assignments that violate the HEDS. Collinearity is internally tracked using a graph-based method based on additions and/or modifications to the initial set of edges. Additionally, if the agent is currently traversing the exterior face f_0, this operation is modified to consider only agent movement (i.e., when on f_0 $\mathtt{C}(\psi_{\alpha_1}, \psi_{\alpha_2}) \rightarrow \mathtt{A}(\psi_{\alpha_1})$) to avoid unintentional intersections between edges as illustrated in Figure 4.6.

Create Edge–Real

The Create Edge–Real operation is represented in the parametric L-system encoding as $\mathtt{D}(\psi_{\alpha_1}, \psi_{\alpha_2})$ and causes the agent to move to a newly created node in the current face before creating an edge back to the original location on the graph, thus subdividing the face. The parameter ψ_{α_2} again defines the material attribute to be assigned to the newly created edge. Consider the example in Figure 4.7(d), where the agent begins at node 1 and receives a command of $\mathtt{D}(0.50, 0.39)$, where $\psi_{\alpha_2} = 0.29$ corresponds to the material attribute $M_{71} = M_{17}$. As with the example in Figure 4.7(b), the agent moves by four nodes within the face and then creates node 7 halfway between nodes 5 and

6. The agent then creates an edge from node 7 back to its original location (node 1). Thus, the description of the graph is modified per

$$\boldsymbol{F} = \{[1,2,3,4,5,6]\} \rightarrow \boldsymbol{F} = \{[1,2,3,4,5,7],[1,7,6]\},$$
$$\boldsymbol{M} = \{[M_{12},M_{23},M_{34},M_{45},M_{56},M_{61}]\} \rightarrow \boldsymbol{M} = \{[M_{12},M_{23},M_{34},M_{45},M_{57},M_{71}], [M_{17},M_{76},M_{61}]\},$$

with the geometric coordinates for node 7 added to the coordinate function within the HEDS. As before, the location and orientation of the agent is defined to be along its original half-edge. Additionally, because a half-edge has been subdivided, the SPIDRS framework also subdivides that twin of that half-edge as previously discussed. Restrictions are once again placed both on the ability to create nodes within a certain distance of existing nodes and on the creation of edges between collinear nodes. Furthermore, if the agent is currently traversing the exterior face f_0, only the movement portion of the operation is considered (i.e., when on f_0 $\mathtt{D}(\psi_{\alpha_1},\psi_{\alpha_2}) \rightarrow \mathtt{B}(\psi_{\alpha_1})$).

Change Attribute

The Change Attribute operation is represented in the parametric L-system encoding as $\mathtt{E}(\psi_{\alpha_1})$ and changes the attribute information associated with the agent's current edge to a different value specified by the parameter ψ_{α_1} before moving the agent by one node. Here, this operation will be discussed in regards to potential changes in the material assigned to the current edge, although in the context of structural topology optimization this could also be used to facilitate changes to the in-plane thickness/cross-sectional area of structural members (cf. Case Study 4.5). Figure 4.7(e) illustrates an example where the agent begins at node 1 oriented along the half-edge e_{12}, where the material attribute M_{12} associated with this edge currently corresponds to a material designated by black lines. When it receives the command $\mathtt{E}(0.13)$, the agent modifies the attribute value of M_{12} to $\psi_{\alpha_1} = 0.13$, which corresponds to a different material designated by gray lines, before moving to node 2. Because it does not explicitly modify the description of the graph, the influence of this operation is noticeable only when converting the graph into a network to be analyzed (e.g., structural representation, as will be discussed in Section 4.3.4).

Turning Operators

Turning operations are represented in the parametric L-system encodings as $+(\psi_{\beta_1})$ and $-(\psi_{\beta_1})$ for clockwise and counterclockwise turns, respectively, and allow the agent to move between neighboring faces on the graph. Examples are shown in Figures 4.7(f)–4.7(g), where the agent begins at node 7 along the half-edge e_{75}. From this position on the graph, the agent could potentially turn into the face $[7,4,5]$, which has an area of 0.43 square units (dimensions not shown), or the face $[7,1,2,3,4]$, which has an area of 1.30 square units. In Figure 4.7(f), the agent receives the command $+(0.21)$. Substituting $\psi_{\beta_1} = 0.21$ into Equation 4.5, we see that the inequality is satisfied when $M = 1$. Thus, the agent moves in the clockwise direction by one face into the face [7,4,5]. Figure 4.7(g) depicts the agent receiving the command -(0.62); substituting $\psi_{\beta_1} = 0.62$ into Equation 4.5 results in the agent moving by $M = 1$ face in the counterclockwise direction into the [7,1,2,3,4] face. After each turning operation, the agent orients itself in the new face along the half-edge that contains its original location (i.e., e_{74} in Figure 4.7(f) and e_{71} in Figure 4.7(g)).

Bracket Operator

Bracket operators are represented in the parametric L-system encoding as `[` and `]` and signify the beginning and ending of a particular string of agent commands, respectively. Inspired by similar operations within the turtle graphics algorithm of Chapter 3 (Abelson & diSessa 1986a), an open bracket stores the current position and orientation of the agent as attribute information within the HEDS before the agent continues to execute the series of commands that follow; a closed bracket returns the agent to the most recently stored location and orientation (without constructing additional edges on the graph). An example of this operation is shown in Figure 4.7(h). The agent receives the command `[` when it is located at node 1 oriented along the half-edge e_{12}. This location and orientation on the graph are stored by the agent, which then completes a series of graphical operations as defined by the parametric L-system (not shown), resulting in its positioning at node 5, oriented along the half-edge e_{57}. It then receives the command `]` and returns to the most recently stored location (node 1), orienting itself along the same half-edge. Note that these bracket operations can be nested. Due to potential modifications of the half-edges that define stored orientations by future graphical operations (e.g., creating a node 8 in between the nodes 1 and 2 would mean that the half-edge e_{12} no longer exists), after each graphical operation the algorithm updates stored locations and orientations to reflect the current state of the graph.

4.3.4 SPIDRS-Interpreted Topologies

Table 4.1 provides a summary of the graphical operations, their associated parametric L-system character, and the number of parameters required by each operation. Recall from Section 4.2 that each variable character in $\boldsymbol{\alpha}$ has two associated parameters and each constant character in $\boldsymbol{\beta}$ has one associated parameter. In the event that the algorithm encounters an operation in which one of the parameters is unused (i.e., A($\psi_{\alpha_1}, \psi_{\alpha_2}$), with A accepting a single parameter), the second parameter is simply ignored.

The SPIDRS algorithm accepts an input parametric L-system word (i.e., a string of instructions), performs a graphical operation associated with each character in the word, manipulates the topological information of the graph based on these operations, and outputs a final graph topology. The following examples discuss this process and provide additional insight into how SPIDRS graphical operations modify both the location/orientation of the agent and the description of the graph itself.

Table 4.1 Summary of SPIDRS graphical operations.

Character	Action	No. of Parameters
	In-Face Agent Motion & Edge Creation	
A	Move N nodes on current face	1
B	Move N nodes, create node	1
C	Move N nodes, create edge back to original node	2
D	Move N nodes, create new node, create edge back to original node	2
E	Change attribute of current half-edge, move by one node	1
	Out-of-Face Agent Motion	
+	Move M faces in clockwise direction	1
-	Move M faces in counterclockwise direction	1
	Bracket Operators	
[	Store current agent state (location/orientation) in pushdown stack	0
]	Retrieve most recent agent state from pushdown stack	0

Example 4.4 As an example, suppose that we are given the parametric L-system word

$$\omega = \mathtt{A}(0.65)\mathtt{C}(0.47,0.88)\mathtt{+}(0.29)\,[\mathtt{E}(0.84)\mathtt{D}(0.34,0.81)$$
$$\mathtt{-}(0.56)\mathtt{B}(0.26)\mathtt{D}(0.44,0.06)]\;\mathtt{-}(0.75)\mathtt{C}(0.98,0.51)$$

to be interpreted by the SPIDRS algorithm and applied to a geometrically hexagonal initial graph described using

$$\boldsymbol{F} = \{[1,6,5,4,3,2],[1,2,3,4,5,6]\},$$
$$\boldsymbol{M} = \{[B,B,B,B,B,B],[B,B,B,B,B,B]\},$$

where each edge of the initial graph has a length of one unit. Note that the alphabet chosen here is compatible with Table 4.1. In this example, the material attribute refers to the color assigned to each edge, with B and G representing black and gray, respectively; the parameters within each graphical operation that define changes to these attributes (i.e., ψ_{α_2} in C and D and ψ_{α_1} in E) are defined such that $\sigma \leq 0.5 \rightarrow B$ and $\sigma > 0.5 \rightarrow G$.

The SPIDRS algorithm interprets the given word and modifies the graph as illustrated in Figure 4.8. Assuming that the agent starts at node 1 oriented along the e_{12} half-edge and beginning with the first character:

1. A(0.65): Substituting $\psi_{\alpha_1} = 0.65$ into Equation 4.3, the inequality is satisfied when $N = 3$, moving the agent to node 4 oriented along the half-edge e_{45}.
2. C(0.47,0.88): Substituting $\psi_{\alpha_1} = 0.47$ into Equation 4.3, the inequality is satisfied when $N = 2$, moving the agent to node 6. The agent then creates an edge back to its original node (4), where $\psi_{\alpha_2} = 0.88$ corresponds to a G material attribute. The description of the graph is modified to

$$F = \{[1,6,5,4,3,2],[4,5,6],[1,2,3,4,6]\},$$
$$M = \{[B,B,B,B,B,B],[B,B,G],[B,B,B,G,B]\},$$

and the agent remains located at node 4 oriented along the half-edge e_{45}.

3. +(0.29): Given $\psi_{\beta_1} = 0.29$ and the state of the graph, Equation 4.5 is satisfied when $M = 1$. The agent turns clockwise onto the exterior face $f_0 = [1,6,5,4,3,2]$, remaining located at node 4 but now oriented along the e_{43} half-edge.
4. [: The agent stores its current location/orientation in the pushdown stack.
5. E(0.84): The agent moves by one node to node 3, oriented along the half-edge e_{32}. Given $\psi_{\beta_1} = 0.84$, the material attribute assigned to the edge e_{43}/e_{34} is changed to G. The graph can now be described as

$$F = \{[1,6,5,4,3,2],[4,5,6],[1,2,3,4,6]\},$$
$$M = \{[B,B,B,G,B,B],[B,B,G],[B,B,G,G,B]\}.$$

6. D(0.34, 0.81): Substituting $\psi_{\alpha_1} = 0.34$ into Equation 4.3, the inequality with the current graph is satisfied when $N = 2$, and the agent moves to node 1 oriented along the half-edge e_{16}. Next, substituting both ψ_{α_1} and N into Equation 4.4 results in $d = 0.04$. However, as discussed in Section 4.3.3, this distance is restricted such that $0.1 \leq d \leq 0.9$ to prevent tight node clustering; thus, the agent moves $d = 0.1$ along the half-edge e_{16} before creating node 7. This results in subdividing half-edge e_{16} into two distinct half-edges e_{17} and e_{76}, as well as the subdividing of its twin half-edge ($e_{61} \rightarrow e_{67}, e_{71}$). Note that the material attribute along these half-edges remains consistent; since e_{16} and e_{61} both had the assigned attribute B, their subdivided half-edges maintain that same attribute. Ordinarily, in this operation the agent would then create an edge back to its original node (3) as denoted by the dashed line in Figure 4.8, but given that the agent is traversing the exterior face this action is ignored to preserve the HEDS. Thus, the description of the graph is now

$$F = \{[1,7,6,5,4,3,2],[4,5,6],[1,2,3,4,6,7]\},$$
$$M = \{[B,B,B,B,G,B,B],[B,B,G],[B,B,G,G,B,B]\}.$$

7. -(0.56): Substituting $\psi_{\beta_1} = 0.56$ into Equation 4.5, the inequality is satisfied when $M = 1$. The agent turns counterclockwise into the face $[1,2,3,4,6,7]$, remaining at node 7 and oriented along the half-edge e_{71}.
8. B(0.26): Given $\psi_{\alpha_1} = 0.26$, Equation 4.3 is solved when $N = 2$ and the agent moves past node 1 to node 2. Next, substituting both ψ_{α_1} and N into Equation 4.4 results in $d = 0.39$, and the agent moves by that distance along the half-edge e_{23} before creating node 8. As before, this subdivides both this half-edge as well as its twin half-edge (i.e., $e_{23} \rightarrow e_{28}, e_{83}$ and $e_{32} \rightarrow e_{38}, e_{82}$, respectively) and assigns material attributes consistent with the original half-edges. The description of the graph becomes

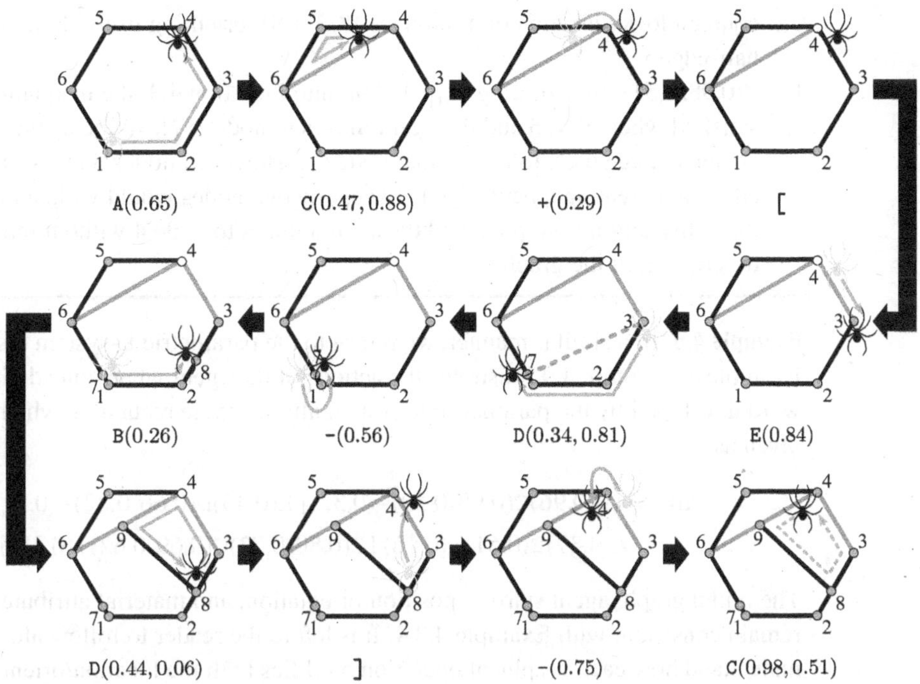

Figure 4.8 Step-by-step illustration of how the parametric L-system word given in Example 4.3.4 is interpreted by the SPIDRS algorithm and results in modifications to the initial graph.

$$\boldsymbol{F} = \{[1,7,6,5,4,3,8,2],[4,5,6],[1,2,8,3,4,6,7]\},$$
$$\boldsymbol{M} = \{[B,B,B,B,G,B,B,B],[B,B,G],[B,B,B,G,G,B,B]\}.$$

9. D(0.44,0.06): Substituting $\psi_{\alpha_1} = 0.44$, Equation 4.3 for the current graph is satisfied when $N = 2$, and thus the agent moves to node 4. Next, substituting both ψ_{α_1} and N into Equation 4.4 results in $d = 0.53$; the agent travels this distance along the half-edge e_{46} before creating node 9, thus subdividing the half-edge and its twin and assigning material attributed to these new half-edges in the same manner as previously discussed. The agent then creates an edge back to its location when it received the command (node 8), with B being assigned as this edge's material attribute via $\psi_{\alpha_2} = 0.06$. The agent is now located at node 8, oriented along the e_{83} half-edge, and the current state of the graph is described as

$$\boldsymbol{F} = \{[1,7,6,5,4,3,8,2],[4,5,6],[8,3,4,9],[1,2,8,9,6,7]\},$$
$$\boldsymbol{M} = \{[B,B,B,B,G,B,B,B],[B,B,G],[B,G,G,B],[B,B,B,G,B,B]\}.$$

10.] : The agent retrieves the latest stored location/orientation from the pushdown stack and returns to node 4, orienting itself along the half-edge e_{43} on the exterior face, without creating an edge.
11. -(0.75): Given $\psi_{\beta_1} = 0.75$, the inequality in Equation 4.5 is satisfied when $M = 2$. The agent remains at node 4 but moves by two faces in the

counterclockwise direction into face [8, 3, 4, 9], orienting itself along the e_{49} half-edge.

12. C(0.98,0.51): Substituting $\psi_{\alpha_1} = 0.98$ into Equation 4.3, the inequality is satisfied when $N = 3$ and the agent moves to node 3. Next, the agent attempts to create an edge back to its original node (4). However, nodes 3 and 4 share an edge, and creating a new edge between the two nodes would violate the HEDS; thus, this action is skipped and the agent returns to node 4 without modifying the description of the graph.

Example 4.5 In a similar manner, we return to the parametric L-system discussed in Example 4.2. Figure 4.9 illustrates the actions of the agent when considering the word developed by the parametric L-system after a single recursion, which is given as

$$\omega^1 = \texttt{[E(0.96)C(0.98,0.59)-(0.32)] A(0.45)D(0.66,0.82)-(0.11)}$$
$$\texttt{A(0.37)E(0.51)-(0.16) [C(0.80,0.73)C(0.69,0.62)+(0.38)]}\,.$$

The initial graph, agent starting position/orientation, and material attribute definitions remain consistent with Example 4.3.4. It is left to the reader to follow along and understand how each graphical operation modifies both the location/orientation of the agent and the description of the graph.

Figure 4.10 illustrates graphs generated by considering two, three, and four parametric L-system recursions. Notice that as the number of recursions increases, the complexity of the graph also tends to increase, an intuitive result considering that longer words generated by the L-system lead to more SPIDRS graphical operations and thus more opportunities for edge creation. However, this is not always the case, as it is possible for long words to (i) consist primarily of movement operations that do not modify the description of the graph or (ii) modify the graph in such a way that the majority of edge creation operations are ignored due to the constraints discussed in Section 4.3.3 and expanded upon in Example 4.3.4.

The steps toward implementing SPIDRS-interpreted parametric L-systems into a genetic topology optimization framework are similar to those described in Section 3.3 and illustrated in Figure 3.5. First, genetic information for each individual in a given generation is converted into a fully defined parametric L-system; genes defining the discrete variables that make up the axiom and production rules are evaluated using Equation 3.4, while the remaining genes associated with the continuous parameters in the set $\boldsymbol{\Psi}$ remain untouched, given that they are already bounded by the interval [0, 1]. The parametric L-system undergoes N_r recursions to produce μ^{N_r} (cf. Section 4.2), which here is interpreted using the SPIDRS algorithm rather than the turtle graphics approach in Chapter 3. As discussed in Section 4.3.1, in the case of mechanical design, the graph produced by SPIDRS can be converted into a structural topology by assuming that each edge is a high aspect ratio structural member with physical properties defined by the attribute information set $\boldsymbol{M}$. The structural topology can be analyzed

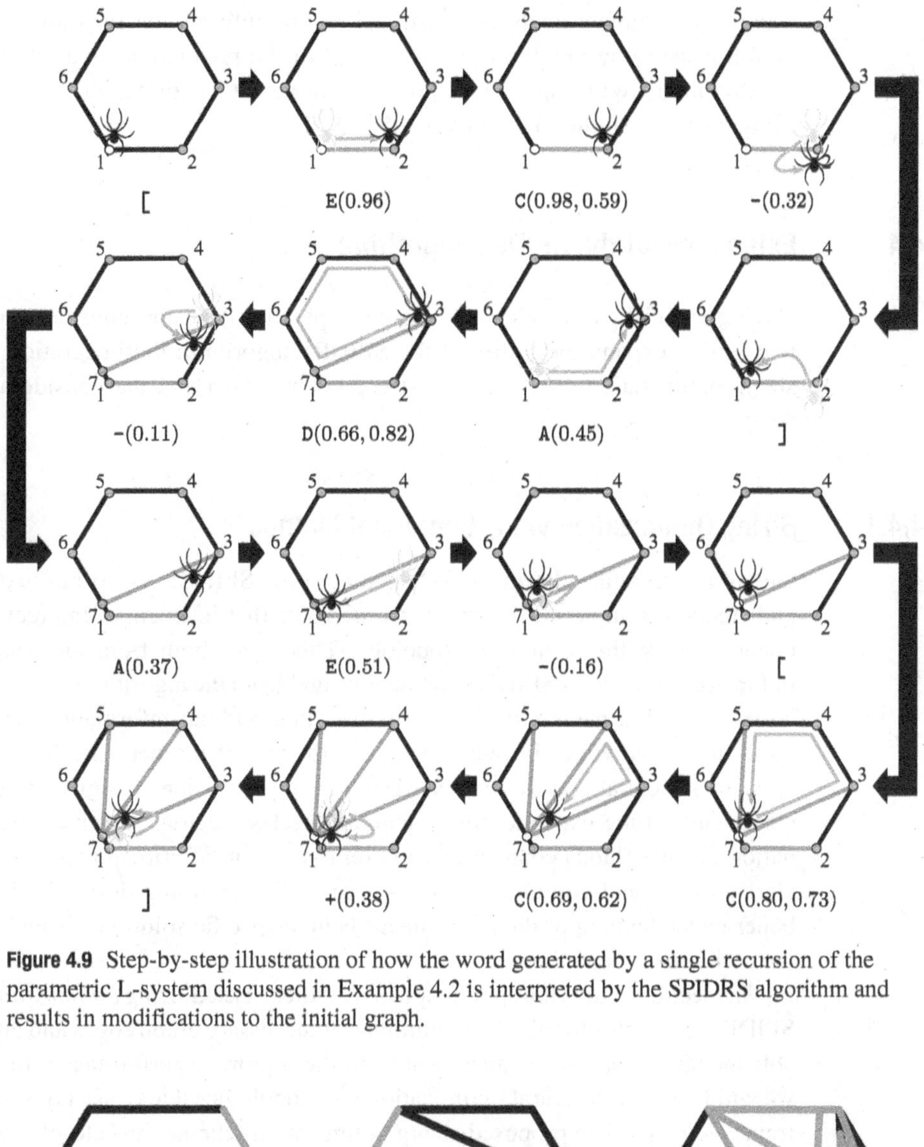

Figure 4.9 Step-by-step illustration of how the word generated by a single recursion of the parametric L-system discussed in Example 4.2 is interpreted by the SPIDRS algorithm and results in modifications to the initial graph.

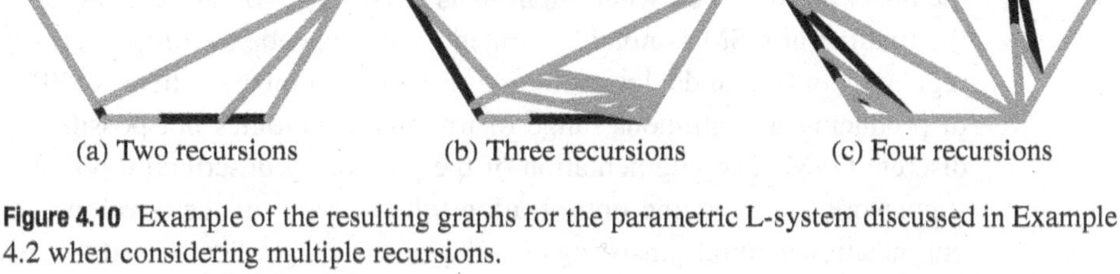

Figure 4.10 Example of the resulting graphs for the parametric L-system discussed in Example 4.2 when considering multiple recursions.

using finite element analysis (FEA), where the information regarding its structural performance (along with that from the rest of the population) is used by the GA to develop improved designs. This process will be further discussed and demonstrated using the Case Studies 4.1 and 4.3–4.5.

4.4 Extensions of the SPIDRS Algorithm

Having established an understanding of its operations, we now consider two extensions that further expand the utility of the SPIDRS algorithm: the integration of an edge-sizing optimization scheme (e.g., structural member) and the consideration of 3-D design domains.

4.4.1 Sizing Optimization via ε-Constraint Method

One of the recognized drawbacks inherent in the SPIDRS algorithm is the existence and creation of unnecessary structural members that form either indirect or auxiliary connectivity within a network topology. This arises both from the necessity of an initial graph on which SPIDRS can operate and from the algorithmic creation of edges based only on a parametric L-system encoding without information regarding how these individual edges influence the performance of the network. The existence of these nonessential connections can lead to both inefficient designs and/or increased computation time expended on potentially useless regions. Furthermore, the elimination of superfluous connections and reinforcement of critical members can further clarify trends and commonalities between well-performing designs and elucidate a better understanding of the mechanisms behind specific solutions, which is critical to the preliminary design process.

This section details the application of a gradient-based *sizing optimization* scheme to SPIDRS-generated topologies to eliminate unnecessary graph edges and allow for variable member-property enhancement to further optimize performance. In this section, we will focus on structural optimization with topologies that consist of sets of discrete frame elements. The proposed sizing optimization scheme thus closely resembles the procedure used by the GSM discussed in Section 2.3, where the cross-sectional areas of discrete members are varied using sensitivity analysis (Choi & Kim 2006), and members are removed when their areas reach zero (Bendsøe et al. 1994). However, the traditional GSM is notably limited by its dependence on predefined and usually regular layouts of nodal locations and structural members, whereas SPIDRS is capable of producing a continuous range of irregular topologies not possible via traditional discrete GSM. The augmentation of the previously described topology optimization framework with a sizing optimization scheme requires the development of a hybrid optimization method consisting of both genetic and gradient-based approaches as illustrated in Figure 4.12; such an approach has been utilized in topology optimization problems where calculated gradients are used to inform or refine topologies generated via evolutionary schemes (Shin, Todoroki & Hirano 2013, Li, Sun, Lu & Wu 2016). The major challenge in developing this hybrid framework is integrating the *single-objective* gradient-based optimization process described in Section 2.3 with a *multiobjective* genetic-based optimization framework. Here, we propose an ε-constraint approach (Miettinen 2012) to avoid the use of weighted sums, which often presupposes an intuition for the problem at hand.

To introduce this approach, consider a generalized constrained minimization design problem with two objectives, as defined in Section 1.3.3. In the context of the parametric L-system, this multiobjective optimization problem is given as

$$\begin{aligned} \boldsymbol{f}^* = & \min_{\boldsymbol{\omega},\boldsymbol{P},\boldsymbol{\Sigma}} \{f_1(\boldsymbol{\omega},\boldsymbol{P},\boldsymbol{\Sigma}), f_2(\boldsymbol{\omega},\boldsymbol{P},\boldsymbol{\Sigma})\}, \\ \text{s.t.} \quad & g_k \leq 0, \quad k = 1 \ldots N_{leq}, \end{aligned} \tag{4.6}$$

where the goal is to find a set of solutions $\boldsymbol{f}^*$ using the parametric L-system variables $\boldsymbol{\omega}$, $\boldsymbol{P}$, and $\boldsymbol{\Sigma}$. Each solution f_i^* within the set $\boldsymbol{f}^*$ is a function of the set of thicknesses $\boldsymbol{t}_i$ associated with structural members comprising the ith individual such that the objectives are Pareto optimal subject to inequality constraints g_k for $k = 1 \ldots N_{leq}$, as illustrated in Figure 4.11. The determination of $\boldsymbol{t}_i$ requires an additional distinct multiobjective optimization problem associated with the ith individual (cf. Figure 4.11) given as

$$\begin{aligned} \hat{\boldsymbol{f}}_i(\boldsymbol{w}_i,\boldsymbol{P}_i,\boldsymbol{\Sigma}_i) = & \min_{\boldsymbol{t}_i} \left\{f_1^{(i)}(\boldsymbol{t}_i), f_2^{(i)}(\boldsymbol{t}_i)\right\}, \\ \text{s.t.} \quad & g_k \leq 0, \quad k = 1 \ldots N_{leq}, \end{aligned} \tag{4.7}$$

where $\hat{\boldsymbol{f}}_i$ is a set of solutions with objective values $f_1^{(i)}$ and $f_2^{(i)}$ associated with the ith individual obtained by varying the thickness of each structural member.

The problem specified in Equation 4.7 can be efficiently analyzed using linear FEA theory and structural sensitivity analysis, which is detailed in Section 2.3.2, within a gradient-based optimization framework. However, rather than combining $f_1^{(i)}$ and

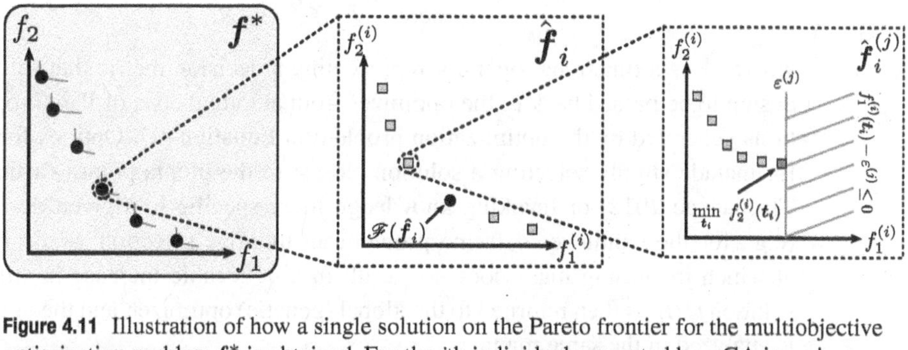

Figure 4.11 Illustration of how a single solution on the Pareto frontier for the multiobjective optimization problem $\boldsymbol{f}^*$ is obtained. For the ith individual generated by a GA in a given generation, a set of solutions is obtained using an ε-constraint approach (Miettinen 2012) that decomposes the multiobjective topology optimization problem $\hat{\boldsymbol{f}}_i$ into a series of constrained single-objective problems $\hat{f}_i^{(j)}$. A single solution selected by a function $\mathscr{F}(\hat{\boldsymbol{f}}_i)$ is then returned to the genetic algorithm.

$f_2^{(i)}$ into the single-scalar objective function required for gradient-based optimization, this framework takes an ε-constraint approach (Miettinen 2012) by decomposing the multiobjective optimization problem given in Equation 4.7 into N_ε single-objective constrained optimization problems given as

$$\begin{aligned} \hat{f}_i^{(j)}(\boldsymbol{w}_i, \boldsymbol{P}_i, \boldsymbol{\Sigma}_i) &= \min_{\boldsymbol{t}_i} f_2^{(i)}(\boldsymbol{t}_i), \\ \text{s.t.} \quad \boldsymbol{t}_{min} &\leq \boldsymbol{t}_i \leq \boldsymbol{t}_{max}, \\ f_1^{(i)}(\boldsymbol{t}_i) - \varepsilon^{(j)} &\leq 0, \\ \text{where} \quad \varepsilon_{min} &\leq \varepsilon^{(j)} \leq \varepsilon_{max}, \quad j = 1 \ldots N_\varepsilon, \end{aligned} \tag{4.8}$$

where $\boldsymbol{t}_i$ is the set of thicknesses corresponding to all members in the ith structural topology realistically bounded by $\boldsymbol{t}_{min}$ and $\boldsymbol{t}_{max}$, and ε_{min} and ε_{max} are bounds for the ε-constraint. In the most rigorous case, these are defined as

$$\begin{aligned} \varepsilon_{min} &= \min_{\boldsymbol{t}_i} f_1^{(i)}(\boldsymbol{t}_i), \\ \text{s.t.} \quad \boldsymbol{t}_{min} &\leq \boldsymbol{t}_i \leq \boldsymbol{t}_{max}, \end{aligned}$$

and

$$\begin{aligned} \varepsilon_{max} &= \max_{\boldsymbol{t}_i} f_1^{(i)}(\boldsymbol{t}_i), \\ \text{s.t.} \quad \boldsymbol{t}_{min} &\leq \boldsymbol{t}_i \leq \boldsymbol{t}_{max}, \end{aligned}$$

respectively. As illustrated in Figure 4.11, the solution to the jth ε-constrained single-objective optimization given in Equation 4.8 represents a point on the Pareto frontier for the ith individual corresponding to $f_1^{(i)} \leq \varepsilon^{(j)}$. However, the operations within the GA require a single point in objective space for a given set of genes, meaning the set of solutions $\hat{\boldsymbol{f}}_i$ resulting from Equation 4.7 must be reduced to a single solution f_i^* needed in Equation 4.6. This reduction takes the form

$$\begin{aligned} f_i^*(\boldsymbol{t}_i) &= \mathscr{F}(\hat{\boldsymbol{f}}_i), \\ \text{where} \quad \mathscr{F} &: \mathbb{R}^{N_\varepsilon} \to \mathbb{R}, \end{aligned} \tag{4.9}$$

where $\mathscr{F}$ is a functional operator representing a decision metric that selects a single design to be passed back to the optimizer from among the set of Pareto-optimal solutions generated by the optimization problem in Equation 4.7. Options for this operator include simply selecting a solution closest to the utopia point (Zavala & Flores-Tlacuahuac 2012) or invoking knowledge more specific to a given design problem (e.g., for the cantilevered frame problem an intuitive selection would be the point at which increasing mass does not result in a noticeable increase in stiffness). The solution $f_i^*(\boldsymbol{t}_i)$ is then returned to the global (genetic) optimizer, and the next individual is analyzed in the same manner.

The implementation of the proposed hybrid optimization method is illustrated in Figure 4.12. The genetic topology optimization portion of the framework remains consistent with that outlined in Section 3.3, while in this work the gradient-based portion uses the Sequential Least Squares Programming (SLSQP) algorithm (Nocedal &

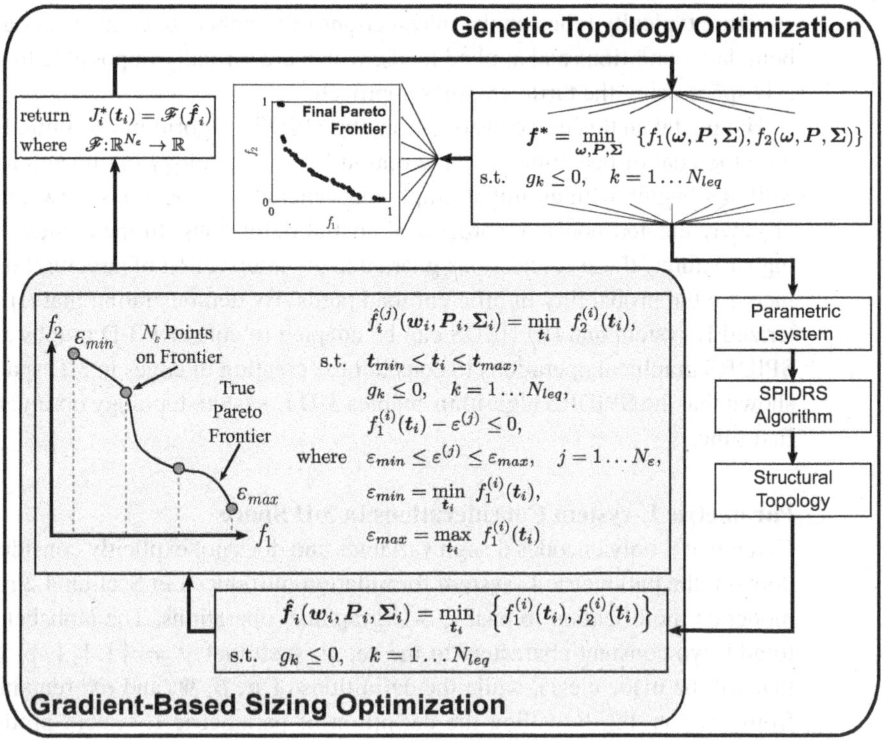

Figure 4.12 Flowchart depicting the hybrid genetic/gradient-based optimization framework developed to couple topology and sizing optimization of L-system-generated topologies.

Wright 2006) implemented in Python's *SciPy* module, which is capable of considering constrained, bounded objective functions and user-defined sensitivity calculations. The application and benefits of this approach is discussed using the Case Studies 4.1–4.3 and 4.6.

4.4.2 3-D L-system Topology Optimization

The combination of parameteric L-systems and the SPIDRS interpretation algorithm provides an effective topology optimization technique capable of considering both canonical and nonintuitive design problems, structural and otherwise (cf. Case Study 4.1–4.5). However, to this point, we have operated under the assumption that the graphs considered will be converted into 2-D networks or structures. In practice, many engineering network design problems are three-dimensional. Both density-based (Diaz & Lipton 1997, Liu & Tovar 2014, Zuo & Xie 2015) and level set (Dunning, Stanford & Kim 2015, Liu & Ma 2015, Deng & Suresh 2016) topology optimization methods have been extensively applied to 3-D structural design problems. Despite being able to render natural-looking 3-D branched designs, the turtle graphics interpretation scheme discussed in Chapter 3 is unsuitable for designing 3-D engineering structures. Such a structure typically relies on the intersections of members to create load paths between boundary conditions and applied loads, which are virtually impossible to guarantee in 3-D space using the turtle graphics approach.

Having taken this into consideration, the SPIDRS algorithm was initially developed with the goal of becoming a viable option for 3-D topology optimization. Recall that SPIDRS begins with an initial graph and generates connections between preexisting or newly created nodes, regardless of spatial definitions. In the context of engineering structures, these connections guarantee the intersection of structural members and increase the probability of efficient load paths. By demonstrating that (i) the parameterized L-system and (ii) HEDS can be adapted to consider 3-D graphs and adapting SPIDRS graphical operations to consider the creation of edges in 3-D space, it will be shown that the SPIDRS algorithm enables 3-D L-system topology optimization for the first time.

Parametric L-system Considerations in 3-D Space

Given that it only encodes design variables and does not explicitly consider the design domain, the parametric L-system formulation introduced in Section 4.2 requires only moderate modification to enable 3-D graphical operations. The alphabet is modified to add two constant characters to the set γ, such that $\gamma = \{\texttt{[},\texttt{]},\texttt{\{},\texttt{\}}\}$ (for reasons that will be made clear), while the definitions of $\boldsymbol{\alpha}$, $\boldsymbol{\beta}$, $\boldsymbol{\Psi}$, and $\boldsymbol{\omega}^0$ remain unchanged from Section 4.2. Recalling the definition of parametric L-system production rules from Equation 4.2, an update is made for 3-D design domains such that the character options become

$$\boldsymbol{\Lambda}^1 = \{\texttt{[},\varnothing,\texttt{\{}\},\ \boldsymbol{\Lambda}^2 = \boldsymbol{\alpha},\ \boldsymbol{\Lambda}^3 = \boldsymbol{\alpha},\ \boldsymbol{\Lambda}^4 = \boldsymbol{\beta},\ \boldsymbol{\Lambda}^6 = \{\texttt{[},\varnothing,\texttt{\{}\},\ \boldsymbol{\Lambda}^7 = \boldsymbol{\alpha},\ \boldsymbol{\Lambda}^8 = \boldsymbol{\alpha},\ \boldsymbol{\Lambda}^9 = \boldsymbol{\beta}.$$

As before, assignments for λ_i^5 and λ_i^{10} are made such that the indicators "[" or "{" in λ_i^1 and λ_i^6, if assigned, are closed. It should be noted that, as in Section 4.2, each production rule is still defined by only 18 independent design variables. Thus, despite shifting to the consideration of 3-D structural design domains, *there is no increase in the number of design variables required.*

SPIDRS Considerations in 3-D Space

To enable 3-D network and especially structural design problems using the SPIDRS algorithm, it must be demonstrated that both the underlying HEDS defining a graph and the SPIDRS graphical operations introduced in Section 4.3 remain valid in 3-D space. Regarding the HEDS, recall that the definition of a graph includes only sets of nodes and edges and a function that relates these two sets; these elements are given no spatial significance (cf. Section 4.1). Using this, one can say that a given graph, and therefore the HEDS that represents it, is independent of the geometric dimensionality of the structural design domain to which it is applied. Example 3-D structural design domains and their associated 2-D planar graph diagrams are illustrated in Figure 4.13. Furthermore, the HEDS discussed in Section 4.3.1 is based on the idea that each edge of a graph can be represented using two half-edges assumed to belong to different faces

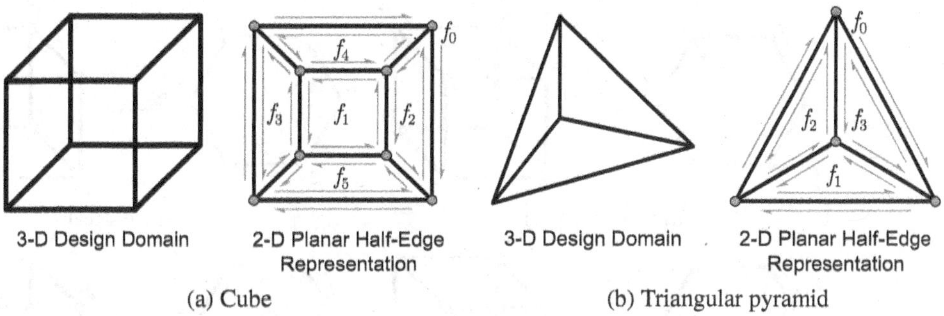

Figure 4.13 Example 3-D structural design domains and their associated 2-D planar graph diagrams using the HEDS.

and oriented in opposite directions. Edges only intersect one another at points on the graph defined by a node (i.e., the graph is planar). Thus, the HEDS will remain a viable foundation for the SPIDRS algorithm when considering 3-D design domains provided (i) consistency is maintained when defining the initial graph and its associated attribute information and (ii) assuming that one only wishes to create topological modifications within faces associated with the initial graph. It will be demonstrated that creating edges between faces in different geometric planes, which is a goal of 3-D topological design, results in graphs that are not planar but are still able to be represented using the HEDS.

With the HEDS capable of representing 3-D topologies, one must consider the graphical operations implemented by the SPIDRS algorithm (cf. Section 4.3.2). As with the graph itself, these operations are written primarily as a function of the nodes, edges, and faces of the graph and do not explicitly consider the spatial positions of these entities. Thus, the graphical operations are independent of the geometric dimensionality of the design domain. Figure 4.14 illustrates the various SPIDRS graphical operations for both 2-D and 3-D structural design domains. Note that the in-face operations (cf. Figures 4.14(a)–4.14(d)) can traverse and initiate topological modifications on any planar face of the graph as before. Out-of-face turning operations (Figures 4.14(e)–4.14(f)) allow the agent to move between faces in different planes that share a common node. Additionally, the bracket operation (Figure 4.17(b)) operates in a manner consistent with Section 4.3.2.

Having shown that the SPIDRS algorithm developed in Section 4.3 remains valid when considering 3-D space, a new graphical operation capable of facilitating the creation of edges *between* 2-D planar faces is required. Rather than the single-face subdividing behavior associated with edge-creation operations previously introduced, this operation allows for the creation of an edge between nodes contained within separate faces (i.e., there is not a face in the graph that contains both nodes). Without this operation, faces representing different planes in the initial graph could continue to be subdivided and become more complex, but no new topology *interior* to the volume would be generated.

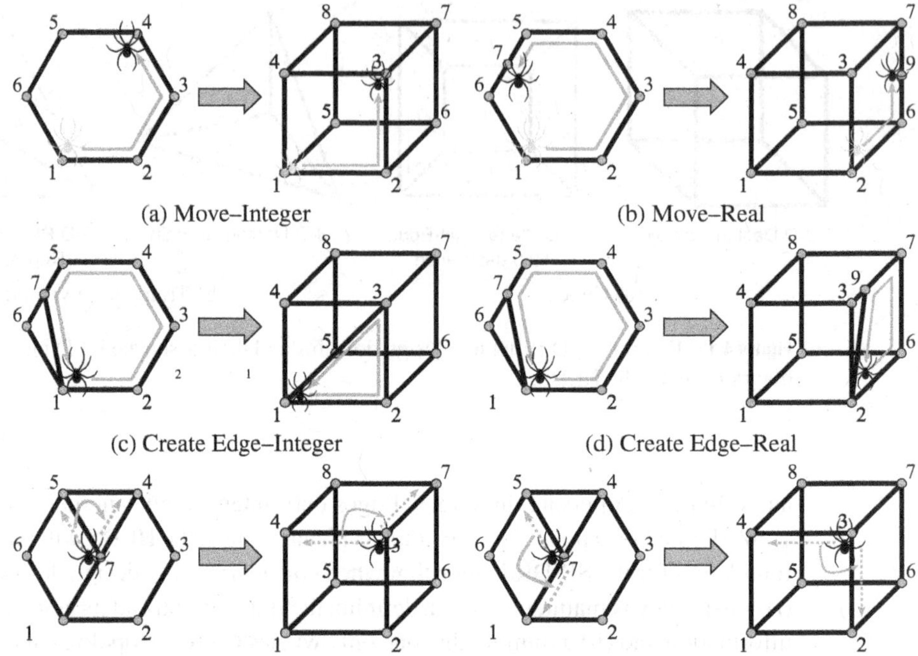

(a) Move–Integer (b) Move–Real

(c) Create Edge–Integer (d) Create Edge–Real

(e) Clockwise turning operation (f) Counterclockwise turning operation

Figure 4.14 Examples of how the SPIDRS graphical operations introduced in Section 4.3.2 remain valid when considering a 3-D structural design domain.

To elucidate how the principles of graph theory can make such an operation possible in the context of the HEDS, we first consider the simple graph illustrated in Figure 4.15(a). Using the notation introduced in Section 4.3.1, this graph can be represented by the face set

$$\boldsymbol{F} = \{[1,2,3,4],[3,8,10,9],[5,6,7,8]\}.$$

Now, consider the creation of an edge between node 2 (associated with f_1) and node 8 (associated with f_3) as illustrated in Figure 4.15(b). All edges are necessarily associated with faces, and this new edge must be associated with one or more faces as well. Recall that a face is defined in Section 4.3.1 as an enclosed set of half-edges that a graphical agent can traverse before returning to its original location on the graph. Leveraging this concept and considering the orientation of the two half-edges associated with the newly created edge, the face set can be modified such that f_1 and f_3 are merged by the addition of the new edge, giving

$$\boldsymbol{F} = \{[1,2,8,5,6,7,8,2,3,4],[3,8,10,9]\}.$$

After traversing the half-edge e_{12} and arriving at node 2 of the updated graph, an agent can move along the newly created half-edge e_{28} to node 8. From here, the agent moves about the portion of the graph previously defined as f_3 before arriving back

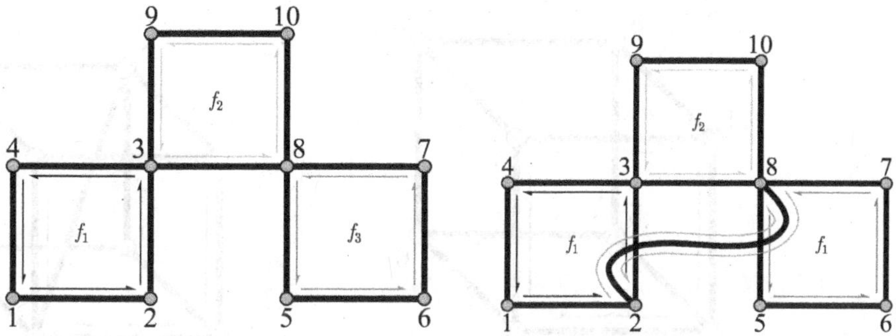

$\boldsymbol{F} = \{[1,2,3,4],[3,8,10,9],[5,6,7,8]\}$

(a) Original graph with two distinct faces

$\boldsymbol{F} = \{[1,2,8,5,6,7,8,2,3,4],[3,8,10,9]\}$

(b) Graph after the creation of an edge between two distinct faces

Figure 4.15 Example of an edge-creation operation between two distinct faces of a graph. Unlike previous operations, this can be considered a merging of two distinct faces. Note that the newly created edge is drawn here to preserve the proper motion of the agent.

at node 8. In a similar manner, moving along the half-edge e_{82} results in a return to node 2, after which the agent continues to move about the portion of the graph previously defined as f_1 before returning to node 1. The modified face set F remains consistent with the definition of a face given in Section 4.3.1. Thus, while the edge-creation operations defined in Section 4.3.2 are based on the idea of *subdividing* a given face, this proposed operation for 3-D structure creation *merges* two distinct faces together, specifically to allow a new edge connecting two previously disconnected faces.

In the same manner, this operation can be applied to 3-D design domains as illustrated in Figure 4.16(a). Consider the directed simple planar graph K', which is defined by

$$\boldsymbol{F}(K') = \{[1,2,3,4],[2,6,7,3],[5,8,7,6],[1,4,8,5],[2,1,5,6],[3,7,8,9]\}.$$

As before, assume that an edge is created between nodes 2 and 8 (Figure 4.16(b)), which in this context are associated with faces $[1,2,3,4]$ and $[5,8,7,6]$, respectively. Note that while K' is not a disconnected graph, one can say that these two faces are disconnected as they do not share a common node. Once again, leveraging the definition of a face allows for the face set describing K' to be redefined as

$$\boldsymbol{F}(K') = \{[1,2,8,7,6,5,8,2,3,4],[2,6,7,3],[1,4,8,5],[2,1,5,6],[3,7,8,9]\},$$

thus merging the faces $[1,2,3,4]$ and $[5,8,7,6]$ via the newly created edge. Most significantly, this operation allows for the creation of edges between faces defined to be in different graphical planes of the 3-D design space.

To implement this edge-creating operation in the SPIDRS algorithm, inspiration is taken from the Bracket operation first introduced in Section 4.3.3 and illustrated in Figure 4.17(a). Recall that SPIDRS interprets an open bracket in the L-system

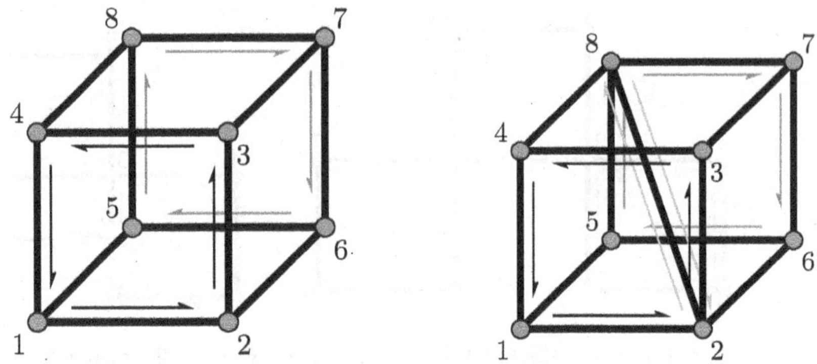

$F(K') = \{[1,2,3,4],[2,6,7,3],[5,8,7,6],$
$[1,4,8,5],[2,1,5,6],[3,7,8,4]\}$

(a) Original graph

$F(K') = \{[1,2,8,7,6,5,8,2,3,4],[2,6,7,3],$
$[1,4,8,5],[2,1,5,6],[3,7,8,4]\}$

(b) Modified graph

Figure 4.16 Example of an edge-creation operation between two disconnected faces of the directed planar graph K' associated with a 3-D structural design domain. This operation allows for edges to be created between faces defined to be in different planes of the design space. Note that only the half-edges associated with the faces of interest are denoted.

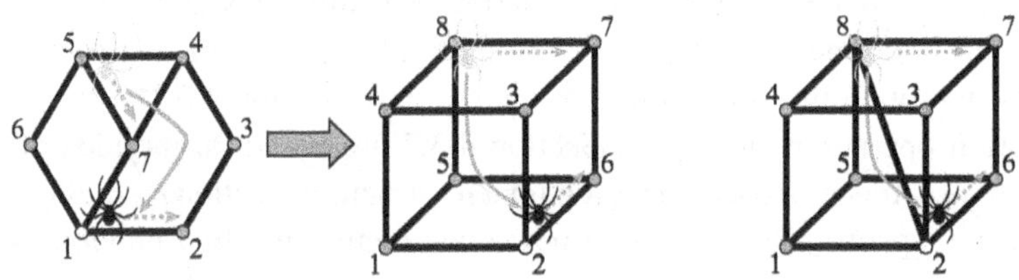

(a) Bracket operation, [. . .]   (b) Create Interfacial Edge, { . . . }

Figure 4.17 Illustration of how the Bracket operation depicted in (a) serves as inspiration for the Create Interfacial Edge operation in (b).

encoding by saving the current location of the agent and continuing to execute any commands that follow; a closed bracket returns the agent to the most recently saved location *without constructing any additional edges on the graph.* In a similar manner, the new Create Interfacial Edge operation illustrated in Figure 4.17(b) is represented by the open and closed braces added to the L-system alphabet; however, unlike the Bracket operation, as the agent returns to the most recently saved location, it constructs an edge between the two nodes and modifies the topological information of the graph accordingly. It should be noted that SPIDRS ensures that the current node and most recently saved location do not share a face; if this is the case, the Create Interfacial Edge operation is skipped as such in-plane modifications are already accomplished via the Create Edge–Integer operation.

The addition of the Create Interfacial Edge operation to the SPIDRS algorithm has several consequences. First, while not explicitly stated, it was assumed in Section 4.1

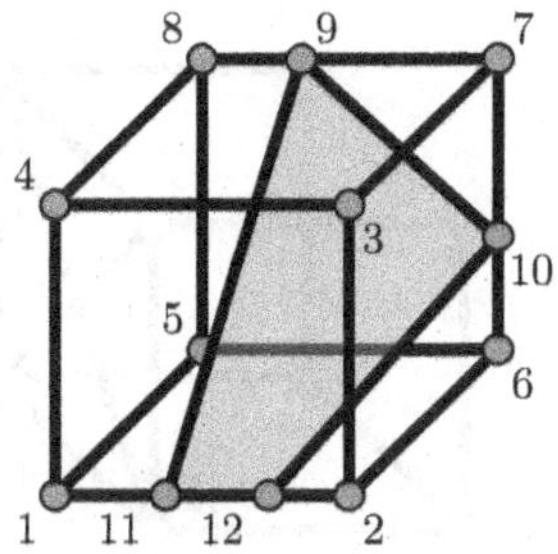

Figure 4.18 Example of a geometrically nonplanar face within a 3-D structural design domain. The possibility of such faces precludes the use of area as part of the geometric parameterization of SPIDRS turning operations.

that an edge marked a division between faces such that twin half-edges corresponded to two distinct faces. As seen in Figures 4.15 and 4.16, this is no longer guaranteed; in fact, the idea of twin half-edges corresponding to the same face is the key idea enabling this operation and does not affect the validity of the HEDS. Second, this operation leaves open the possibility for geometrically nonplanar faces (see Figure 4.18), severely complicating the area calculations required for the geometric parameterization of turning operations (cf. Section 4.3.3). To maintain computational efficiency while retaining the perceived increase in performance granted by the geometric parameterization of SPIDRS graphical operations, the parameterization of 3-D turning operations is modified to consider the perimeter of potential faces rather than area. Third, the Create Interfacial Edge operation is herein represented by characters in the set γ that do not have formal parameters assigned to them, but it is these parameters that define the material assignments of edges created using the Create Edge operations. Thus, problems herein that consider a 3-D design domain will be limited to a single material, precluding the need for the Change Attribute operation. It is left to the curious reader to implement formal parameters in connection with this command to allow such dynamic edge property assignment.

Perhaps the most impactful consequence of extending SPIDRS into 3-D space is that, in general, the edges of a graph can no longer be guaranteed to intersect only at nodes. This is because faces are no longer guaranteed to be convex and that the Create Interfacial Edge still holds when the two faces being considered lie in the same plane. For example, assume that the graph K' illustrated in Figure 4.17(b) has been further modified such that its face set is defined by

$$F(K') = \{[1,2,8,7,6,5,8,2,3,4],[2,6,3],[3,6,7],[1,4,8,5],[2,1,5,6],[3,7,8,9]\}.$$

as shown in Figure 4.19(a). Now, consider that the agent currently positioned at node 7 along the half-edge e_{76} receives the command $\}$, returns to node 2, orients itself along the half-edge e_{28}, and creates an edge between nodes 7 and 2 as it moves. This modifies the graph definition to

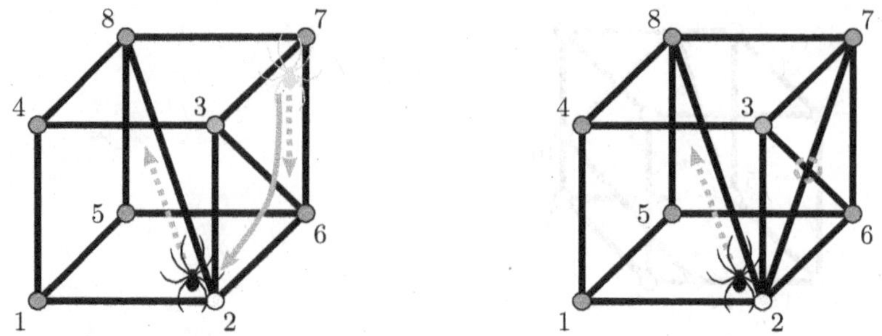

$F(K') = \{[1,2,8,7,6,5,8,2,3,4],[2,6,3],$
$[3,6,7],[1,4,8,5],[2,1,5,6],[3,7,8,4]\}$
(a) Create Interfacial Edge operation

$F(K') = \{[1,2,7,6,3,7,2,8,7,6,5,8,2,3,4],$
$[2,6,3],[1,4,8,5],[2,1,5,6],[3,7,8,4]\}$
(b) Resulting nonplanar graph K'

Figure 4.19 Example of how the introduction of the Create Interfacial Edge operation can result in the formation of a nonplanar graph due to edge intersections at locations not defined by a node, as denoted by the dashed circle in (b).

$$F(K') = \{[1,2,7,6,3,7,2,8,7,6,5,8,2,3,4],[2,6,3],[1,4,8,5],[2,1,5,6],[3,7,8,9]\}$$

This operation is valid as nodes 2 and 7 did not share a face when the operation was commanded. However, given that SPIDRS is concerned with graph connectivities and not with the spatial position of elements of the graph, the algorithm does not consider that the newly created edge between nodes 2 and 7 intersects the existing edge between nodes 3 and 6 at a point not defined by a node (denoted by a circle in Figure 4.19(b)). Depending on the topological design problem being considered and the constraints thereof, this could be remedied by calculating potential intersection points between edges and either skipping the edge-creation process or adding a node at that point. However, such a process requires explicit consideration of the geometry of the graph, which is counter to the objective of the SPIDRS algorithm. Herein, these intersections are allowed with the understanding that they must be considered during structural analysis.

Example 3-D structural topologies generated by SPIDRS using the parametric L-system are illustrated in Figure 4.20, which represents the first-ever demonstration of truly three-dimensional L-system topology optimization. As with the 2-D structural topologies in Figure 4.10, increasing the number of L-system recursions results in increased topological complexity. It is also important to stress that, as seen in the single recursion case, the introduction of the Create Interfacial Edge operation does not guarantee the formation of edges between disconnected faces in the graph. It should also be noted that, in general, the creation of one interfacial edge precipitates more due to the in-face edge-creation commands operating on a face that exists on multiple planes, as seen in the three recursion example. This is especially convenient in the creation of new and useful connectivities (e.g., load paths in a structure). The application of SPIDRS to 3-D design problems is exemplified using the Case Studies 4.6–4.7.

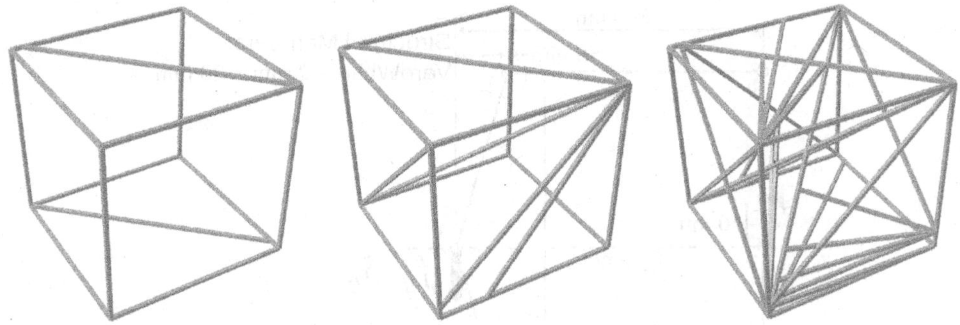

(a) One recursion (b) Two recursions (c) Three recursions

Figure 4.20 Example of 3-D structural topology generation using a SPIDRS interpretation of a parametric L-system when considering multiple recursions.

4.5 Case Studies

To demonstrate the ability of the parametric L-system and SPIDRS algorithm to generate effective structural configurations given multiple objective functions, a set of case studies is presented. These studies include design problems relating to 2-D cantilevered frames (Case Studies 4.1–4.2), compliant mechanisms (Case Studies 4.3–4.4), structures with variable stiffness responses (Case Study 4.5), and 3-D cantilevered frames subjected to various load cases (Case Studies 4.6–4.7). As in Chapter 3, results will be presented in the form of associated Pareto frontiers that are compared to resulting designs from alternative topology optimization methodologies and/or known optimal/ideal solutions. Where applicable, design problems will also be considered using the hybrid optimization approach introduced in Section 4.4.1 to demonstrate the benefits of considering both topology and sizing optimization simultaneously.

Case Study 4.1: 2-D Cantilevered Frame

Our first case study reconsiders the simple cantilevered frame discussed in Case Study 3.1 and illustrated in Figure 4.21. The structural design domain is defined by the initial graph

$$\boldsymbol{F} = \{[1,5,4,3,2],[1,2,3,4,5]\},$$

where these nodes are assigned spatial coordinates such that the domain has a maximum total length in x of 300 mm and a maximum total height in y of 350 mm. This initial graph and any subsequently generated structural members are assumed to consist of VeroWhite (cf. Table 3.2) with a rectangular cross-section of 2 mm × 20 mm (thicker dimension out of plane). Based on *a priori* knowledge of the desired result, geometric symmetry is assumed about the x–z plane such that only half of the topology is generated by the SPIDRS algorithm. A displacement boundary condition of $u_x = u_y = 0$ is

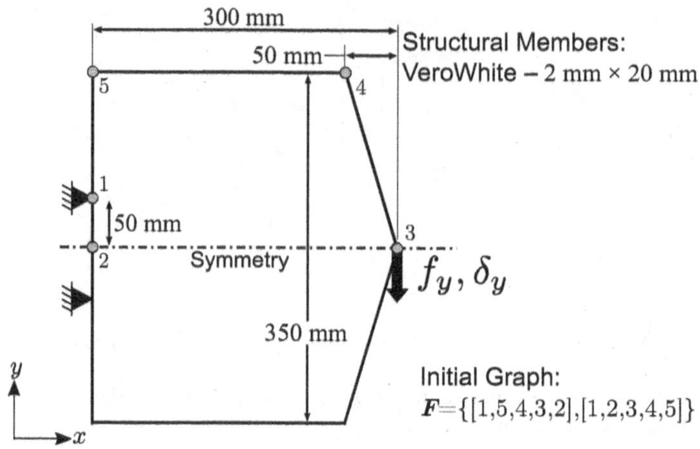

Figure 4.21 Initial graph description and structural boundary conditions associated with Case Study 4.1.

enforced at node 1 (i.e., ±50 mm above/below the axis of symmetry at the minimum x position), and a concentrated transverse force of $f_y = -0.05$ N is applied at node 3.

Details regarding the multiobjective optimization of the cantilevered frame are provided in Table 4.2. As before, the objectives in this problem are the minimization of the mass m and the maximization of the stiffness $k = f_y/\delta_y$ of the structure. To obtain nondimensional objective values, we normalize both the mass and stiffness of each topology by those associated with the "null" solution (represented by the initial graph in Figure 4.21), which is the lightest and least stiff design possible ($m_{null} = 0.57$ kg and $k_{null} = 5.98$ E-3 kN/m), such that $\hat{m} = m/m_{null}$ and $\hat{k} = k/k_{null}$. Minimization of $\hat{m}$ favors designs with a sparse topology, while maximization of $\hat{k}$ tends toward dense topologies. Given the single-material definition of the problem, the Change Attribute graphical operation (cf. Section 4.3.2) is not considered; thus, here the SPIDRS algorithm requires 74 independent design variables (four production rules with 18 variables each plus a two-character axiom) and considers four parametric L-system recursions. Assuming small deflections and rotations for meaningful designs, a fully linear structural analysis for plane strain conditions is considered (i.e., the buckling of structural members is neglected) with an average element length of approximately 30 mm. The speed of such analysis coupled with the efficient parallelization approach employed (cf. Case Study 3.1) dictates that $100 \times 1{,}000 = 100{,}000$ individuals can be analyzed in only a few hours.

The approximate Pareto frontier generated by the SPIDRS algorithm for the problem summarized in Table 3.3 is shown in Figure 4.22 along with three example topologies of interest. As expected, the lightest and least-stiff structure is simply the null solution ($\hat{m} = \hat{k} = 1.0$), while the stiffest and heaviest structure ($\hat{m} = 4.20$, $\hat{k} = 15.5$) features a comparatively dense distribution of structural members. It is also inter-

Table 4.2 Summary of the multiobjective topology optimization problem for a cantilevered frame.

Design Problem Statement	
Minimize:	normalized mass $\hat{m}$
and Maximize:	normalized stiffness $\hat{k}$
by varying:	2 axiom characters, 4 rule assignments (18 genes each)
NSGA-II Parameters (Deb et al. 2000)	
100 members for 1,000 generations, $P_{cross} = 0.9$, $\eta_{cross} = 50$, $P_{mut} = 0.1$, $\eta_{mut} = 20$	

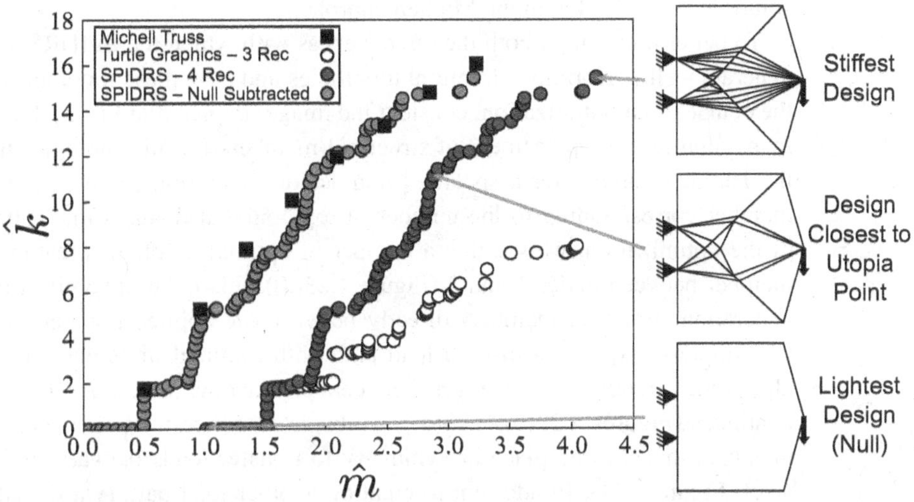

Figure 4.22 Pareto frontier associated with the cantilevered frame design problem generated using SPIDRS-interpreted L-system strings, along with comparisons to frontiers associated with a Michell truss and turtle graphics-interpreted L-system strings.

esting to observe the design on the frontier closest to the utopia point,[4] which features $\hat{m} = 2.87$ and $\hat{k} = 11.1$. Comparing the topologies associated with this frontier with those generated using a turtle graphics interpretation, we see that as mass increases SPIDRS-generated topologies demonstrate a marked improvement in performance such that for topologies with $\hat{m} > 4$, values of $\hat{k}$ are nearly double those of turtle graphics topologies.

Also shown in Figure 4.22 is the frontier associated with a series of Michell trusses (cf. Case Study 3.1) that are modeled using frame elements with the same material properties and rectangular cross-section mentioned previously. Comparing this frontier with that associated with the SPIDRS algorithm, we see that SPIDRS-generated topologies are noticeably outperformed by the Michell trusses across the entire range

[4] Defined as the point comprised of the best calculated value across each objective function that minimizes or maximizes all objective functions (here, $\hat{m}_{utp} = 1.0$ and $\hat{k}_{utp} = 15.5$).

of normalized mass values. This is attributed to the presence of the null structure in every SPIDRS-generated topology, regardless of the existence of more advantageous load paths that provide increased stiffness while requiring less total mass. Thus, by eliminating the null structure (and potentially other members that form indirect or extraneous load paths), we can expect that SPIDRS-generated topologies will demonstrate increased performances, perhaps even approaching those of a Michell truss. As a brief example, Figure 4.22 provides the frontier associated with SPIDRS-generated designs with both the mass and stiffness contributions associated with the null structure removed. This idealized set of solutions compares more favorably to those associated with the family of Michell trusses, matching their performances at higher normalized mass values and, most importantly, providing design options in between the discrete solutions associated with the Michell approach.

To better understand both the effectiveness with which the SPIDRS algorithm can generate well-performing structural topologies and how these topologies evolve over the course of an optimization, consider the images depicted in Figure 4.23. These heat maps represent the placement of structural members for all topologies that make up the Pareto frontier after a specified number of generations, with the shade of each member corresponding to the number of topologies it features in. At the beginning of the optimization, we see that a number of optimal topologies feature a structural member between nodes 1 and 3 (Figure 4.23(a)); this is intuitive given two facts: (i) generating structural members directly between the applied boundary conditions is the simplest way to construct a load path with minimal mass and (ii) the SPIDRS algorithm directly creates edges between preexisting nodes. As the optimization continues (Figures 4.23(b)–4.23(d)), we begin to notice the prevalence of structural members that use the place of symmetry to transfer loads between nodes 1, 5, and 4 (cf. Figure 4.21). In addition to creating another load path (via the edge between nodes 3 and 4) and therefore increasing stiffness, the members between the symmetry plane and node 4 will come to serve primarily as "construction lines" supporting other structural members. Specifically, Figure 4.23(e) indicates that by generation 180, Pareto optimal topologies leverage this member to create one or more load paths between nodes 1 and 3 that share a common geometry with those associated with the Michell truss and resulting designs generated by SIMP and level set implementations (cf. Figure 4.26). By the end of the optimization (Figure 4.23(f)), we see a clear underlying topology amongst the majority of the Pareto optimal topologies with the SPIDRS algorithm adjusting the placement of structural members to achieve varying performances. This kind of qualitative guidance can be very useful during preliminary design.

Motivated by the predicted performances achieved when the null structure is removed from SPIDRS-generated structures (cf. Figure 4.22), the problem is reconsidered with the sizing optimization approach introduced in Section 4.4.1 to eliminate unnecessary structural members and maximize the structural performance of a given topology. The goal of the problem remains the minimization of normalized mass $\hat{m}$ while maximizing normalized stiffness $\hat{k}$, but here the addition of a gradient-based sizing optimization scheme allows the thicknesses of all structural members $\boldsymbol{t}$ to vary.

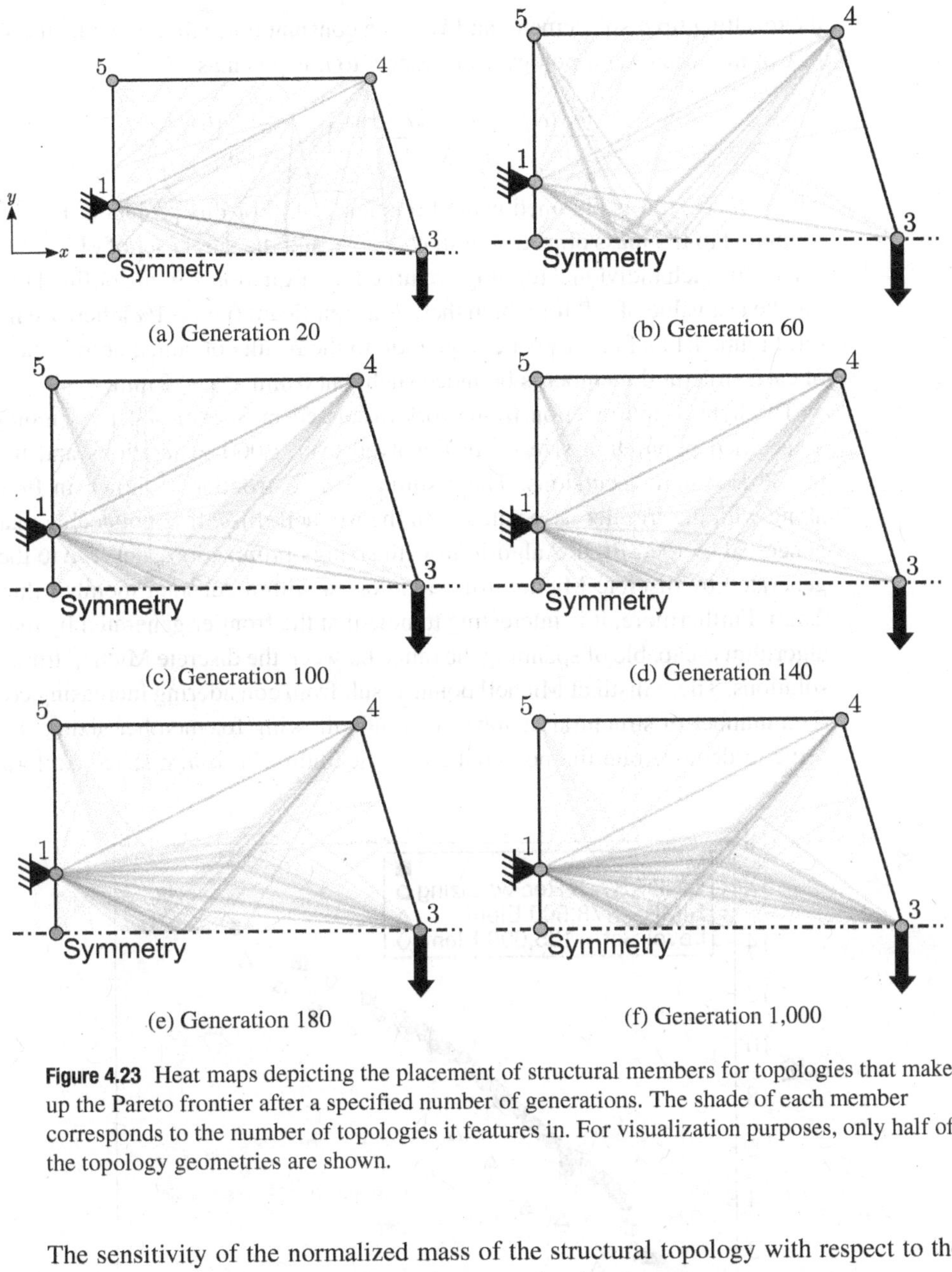

(a) Generation 20 (b) Generation 60

(c) Generation 100 (d) Generation 140

(e) Generation 180 (f) Generation 1,000

Figure 4.23 Heat maps depicting the placement of structural members for topologies that make up the Pareto frontier after a specified number of generations. The shade of each member corresponds to the number of topologies it features in. For visualization purposes, only half of the topology geometries are shown.

The sensitivity of the normalized mass of the structural topology with respect to the thickness of a given structural member t_i can be simply written as

$$\frac{\partial \hat{m}(t_i)}{\partial t_i} = \rho_i w_i L_i, \tag{4.10}$$

where ρ_i is the density of the material (cf. Table 3.2), w_i is the out-of-plane width (20 mm), and L_i is the length of that structural member. Given that normalized stiffness is defined as $\hat{k} = (f_y/\delta_y)/k_{null}$, where f_y is a constant applied force of -0.05 N, δ_y is the resulting tip displacement, and k_{null} is a constant normalizing value, the sensitivity of $\hat{k}$ of the structural topology with respect to t_i is given as

$$\frac{\partial \hat{k}(t_i)}{\partial t_i} = \frac{\partial}{\partial t_i}\left(\frac{f_y/\delta_y(t_i)}{k_{null}}\right) = -\frac{\partial \delta_y(t_i)}{\partial t_i}\frac{f_y}{{\delta_y}^2}, \tag{4.11}$$

where $\partial \delta_y(t_i)/\partial t_i$ is obtained using Equation 2.49. For this problem, the ε-constraint is applied to the mass objective with $N_\varepsilon = 15$, and the single set of objective function values for each individual topology returned (cf. Equation 4.9) are defined by calculating the last value of $\varepsilon^{(j)}$ for which there is a significant (i.e., $\geq$1%) increase in stiffness (cf. Figure 4.11). For a direct comparison to the results obtained above, the thickness of each structural member is bounded such that $0\,\text{mm} \leq t \leq 2\,\text{mm}$.

The hybrid optimization framework discussed in Section 4.4.1 is employed with a specified population size of 100 evolved over 1,000 generations and three parametric L-system recursions. The resulting Pareto frontier is shown in Figure 4.24, along with the frontier associated with the Michell truss. It is noticeable that designs generated by the SPIDRS algorithm with sizing optimization converge to the frontier generated by discrete Michell truss topologies with a constant member thickness of 2 mm. Furthermore, it is interesting to note that the frontier generated by the SPIDRS algorithm is capable of spanning the range between the discrete Michell truss topology solutions. These distinct Michell points result from considering increasing complexity (i.e., number of structural members) but not allowing for member sizing. The results serve to demonstrate the potential added flexibility in design space exploration that

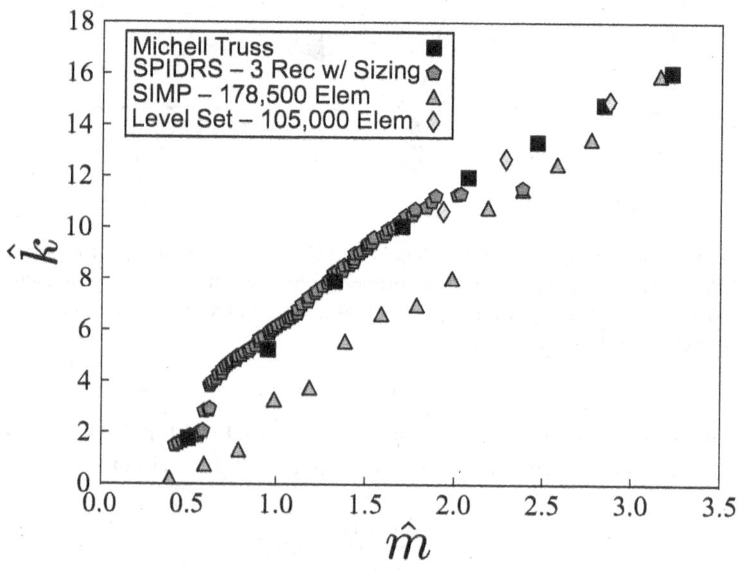

Figure 4.24 Pareto frontier associated with the 2-D cantilevered frame case study generated using the hybrid topology/sizing optimization framework (cf. Section 4.4.1) with thickness bounds of $0\,\text{mm} \leq t \leq 2\,\text{mm}$, along with comparisons to the frontiers associated with the Michell truss (constant thickness of 2 mm), a SIMP formulation, and a level set formulation.

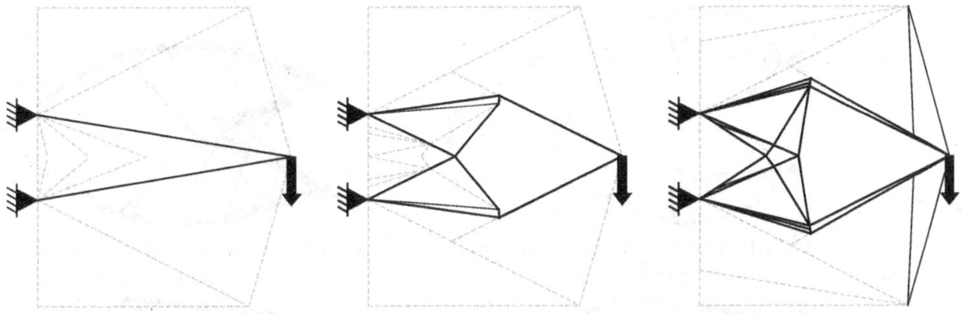

(a) Lightest design ($\hat{m} = 0.42$, $\hat{k}_{trans} = 1.51$)

(b) Design closest to the utopia point ($\hat{m} = 1.32$, $\hat{k}_{trans} = 8.26$)

(c) Stiffest design ($\hat{m} = 2.38$, $\hat{k}_{trans} = 11.5$)

Figure 4.25 Several 2-D cantilevered frame topologies of interest generated by the SPIDRS algorithm using the hybrid topology/sizing optimization framework. Structural members that do not contribute to the stiffness of the structure (represented by dashed lines) are removed, resulting in topologies that are capable of matching known optimal solutions.

using the SPIDRS algorithm can provide. Several topologies of interest generated using the hybrid topology/sizing optimization approach are shown in Figure 4.25. Here, the effect of sizing optimization can clearly be seen, as structural members that form auxiliary load paths are limited or downsized. Also eliminated are edges that serve temporary graph-based roles in which they allow the creation of critical load paths while not themselves having an structural utility, thus generating Michell-like topologies that are capable of matching known optimal solutions.

Figure 4.24 also provides the frontiers resulting from a SIMP implementation available in the Abaqus FEA suite and an open source implementation of the level set method (Otomori, Yamada, Izui & Nishiwaki 2015). These implementations consider the same design problem shown in Figure 4.21 and also consider plane strain conditions. The structural design domain in the SIMP implementation is discretized into 178,500 elements with a geometric symmetry condition (cf. Figure 4.21), while the implementation of the level set method analyzed the generated topology using 105,000 elements and no geometric symmetry condition. While their reliance on gradient-based optimization techniques would typically complicate the process, because mass is an objective we can generate these frontiers simply by increasing the permissible volume fraction. Several designs of interest found within the SIMP- and level set method-generated frontiers are shown in Figure 4.26.

Comparing these frontiers with that of the SPIDRS sizing optimization study yields several interesting conclusions. First, it is noticeable that sized structural topologies generated by the SPIDRS algorithm consistently outperform those generated using the SIMP method across nearly the entire range of normalized mass values considered. This is attributed to the fact that SIMP has less material with which to construct load paths, resulting in inefficient configurations of high aspect ratio structural members as illustrated in Figure 4.26(a). It should be noted that as the allowable volume fraction continues to increase SIMP does begin to include designs that are comparable with

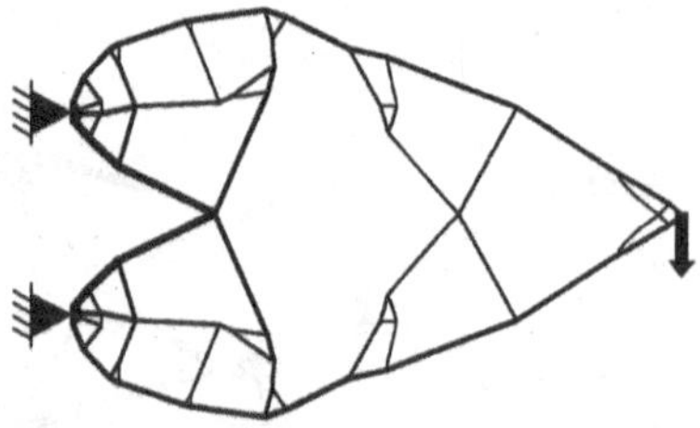

(a) SIMP-generated design with $\hat{m} = 0.79$ and $\hat{k} = 1.34$

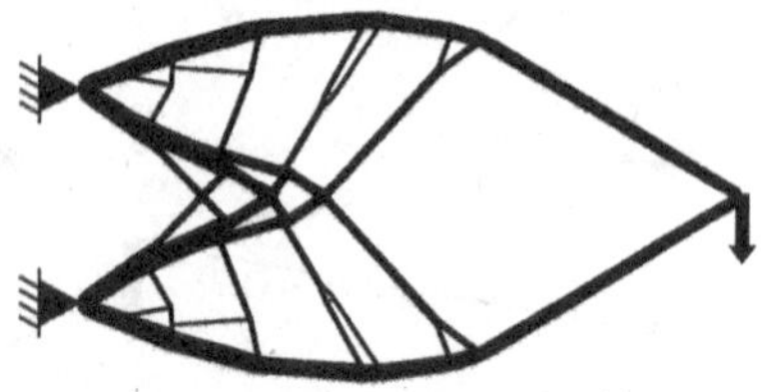

(b) SIMP-generated design with $\hat{m} = 3.16$ and $\hat{k} = 16.0$

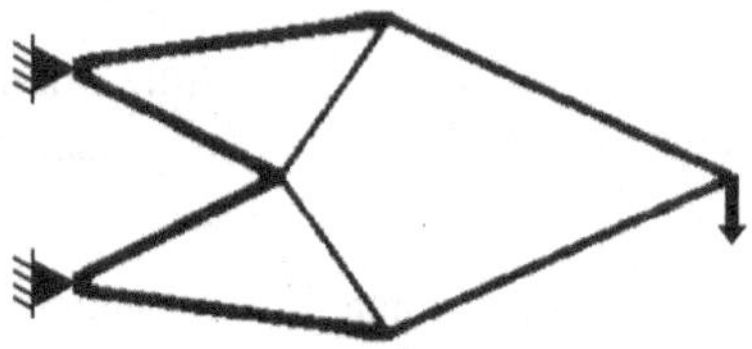

(c) Level set-generated design with $\hat{m} = 1.93$ and $\hat{k} = 10.7$

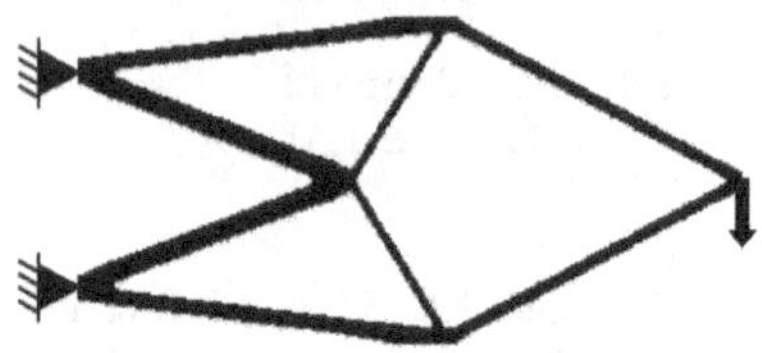

(d) Level set-generated design with $\hat{m} = 2.87$ and $\hat{k} = 15.0$

Figure 4.26 Example topologies generated using (a)–(b) an implementation of SIMP (Dassault Systèmes of America Corp. 2014) and (c)–(d) of the level set method (Otomori et al. 2015).

sized designs generated by SPIDRS (Figure 4.26(b)), though these designs tend to have structural members thicker in magnitude than the maximum allowable member thickness used in this study. Second, structural topologies obtained using the level set implementation (Figures 4.26(c)–4.26(d)) compare favorably to the performances of both the Michell truss and the sized SPIDRS-generated frontiers, but they fail to converge at lower normalized mass values, as the representation of high aspect ratio structural members becomes impractical using the underlying level set function (van Dijk et al. 2013).

Case Study 4.2: 2-D Long Cantilevered Benchmark

For completeness, we now return to the long cantilever benchmark problem introduced in Example 2.1 and discussed extensively throughout Chapter 2. Specifically, the hybrid optimization framework is adapted to also include the spatial location of nodes in the SPIDRS-generated graph using the geometry optimization approach introduced in Section 2.3.3. The initial graph for the problem is shown in Figure 4.27(a) and considers the same boundary conditions and structural member material properties as discussed in Example 2.1. Nodes where boundary conditions are applied (nodes 1, 3, and 4) remain fixed in the design domain, while the (x, y) coordinates of any remaining nodes are included in the optimization problem along with the thicknesses of subsequently generated structural members. These coordinates are constrained such that they remain within the dimensions of the initial graph, and member thicknesses are bounded along the interval $[10^{-8}, 1.5]$. Here, we consider three levels of L-system recursion, a maximum allowable volume fraction of $v_f^g = 0.3$, and a population size of 200 individuals evolved over 100 generations.

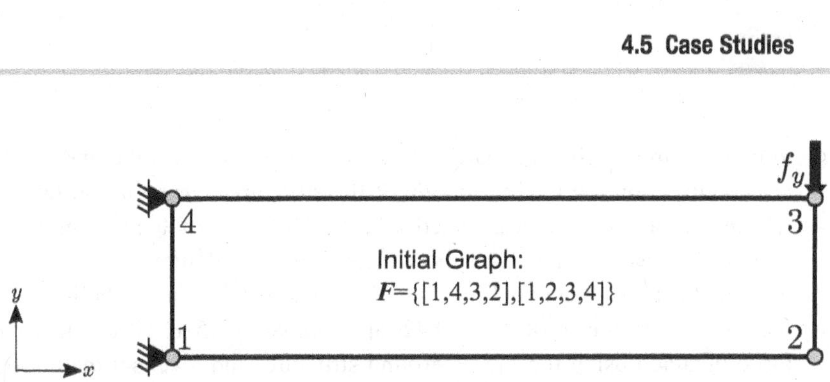

(a) Initial graph and boundary conditions associated with the long cantilever benchmark problem

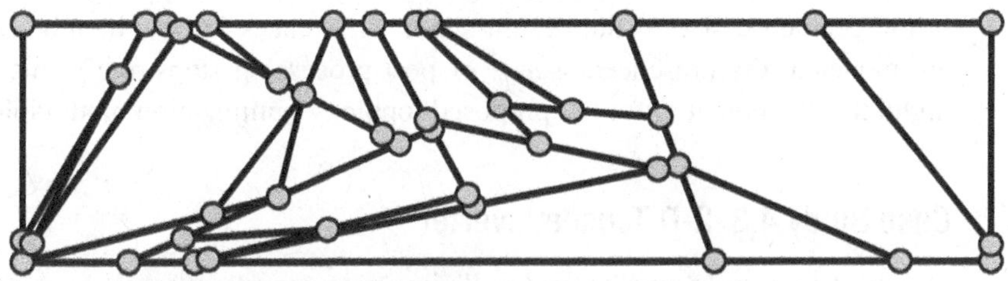

(b) Resulting optimal graph (unmodified)

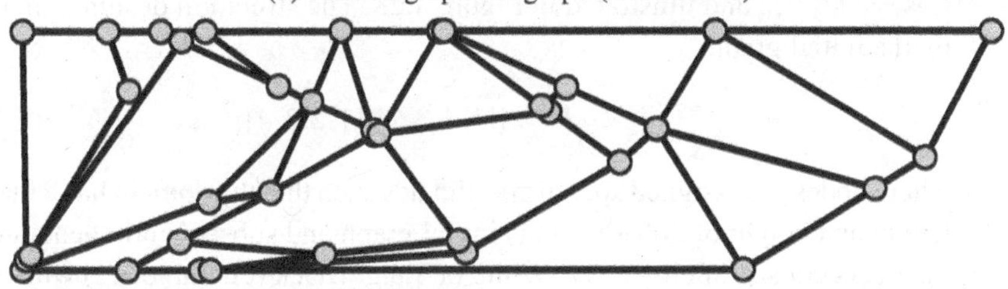

(c) Graph after geometric optimization

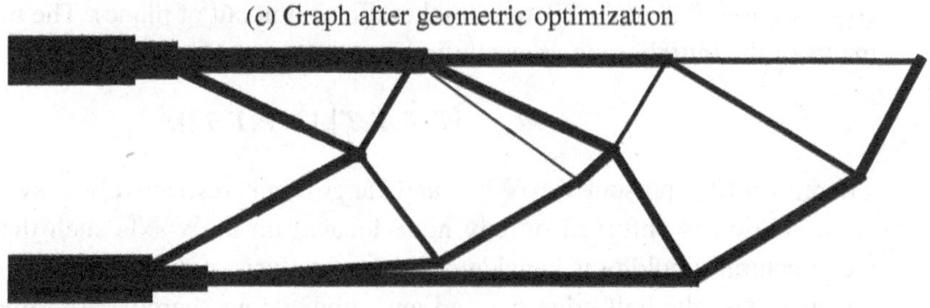

(d) SPIDRS-generated solution considering both geometric and sizing optimization; $v_f = 0.30$; $\hat{J}_c = 0.42$

Figure 4.27 Solution to the long cantilevered design problem using SPIDRS with geometric and sizing optimization, the initial graph and boundary conditions of which are illustrated in (a). The resulting optimal graph (b) is modified using geometric optimization (c) such that critical load paths are established and/or improved. Sizing optimization eliminates inefficient load paths, resulting in a design with a compliance value approximately 25%, 10%, and 18% lower than those obtained using the SIMP, ground structure, and level set methods (cf. Chapter 2), respectively.

Figure 4.27(b) illustrates the SPIDRS-generated graph that results in the lowest compliance value, which features a dense and seemingly random collection of edges. However, applying a geometry optimization step results in the layout shown in Figure 4.27(b); here, the bottom-right nodes of the graph are shifted up and to the left to create a more efficient load path between the applied force and the fixed boundary condition on the lower-left side of the structure. Furthermore, nodes in the interior of the domain are rearranged to create a series of load paths to reinforce the structure. These load paths are more clearly illustrated after considering a sizing optimization step as shown in Figure 4.27(d). The result is a Michell-like structure that features a compliance of $\hat{J}_c = 0.42$, approximately 25%, 10%, and 18% lower than those obtained using the SIMP, ground structure, and level set methods presented in Chapter 2, respectively. While it should be noted that the same caveats discussed in Section 2.3.2 apply when comparing the results of these approaches (discrepancies in the geometric size of the design space, differences in how member intersections are modeled, the presence/absence of post-processing steps, etc.), this result again highlights the potential for the proposed topology optimization methodology.

4.5.1 Case Study 4.3: 2-D Tensile Inverter

Our third case study revisits the tensile inverting compliant mechanism problem from Case Study 3.2 and illustrated in Figure 4.28. The structural design domain is defined by the initial graph,

$$\boldsymbol{F} = \{[1,4,3,2],[1,2,3,4]\},$$

where nodes are assigned spatial coordinates such that the domain has a maximum total length and height of 200 mm. This initial graph and subsequently generated structural members consist of either VeroWhite or TangoBlack (cf. Table 3.2) with a rectangular cross-section of 2 mm × 20 mm (thicker dimension out of plane). The material assignments of the initial graph are specified as

$$\boldsymbol{M} = \{[T,T,V,T],[T,V,T,T]\},$$

where V and T represent VeroWhite and TangoBlack, respectively. A symmetry boundary condition is enforced on any node located on the x-axis such that only half of the structural topology is considered during analysis; note that any edges collinear to this axis (i.e., the half-edge e_{12} and any subdivisions thereof) are removed from the final topology to ensure the construction of meaningful load paths. A displacement boundary condition of $u_x = \delta_{x,in}$ is applied to node 1, while another displacement boundary condition of $u_x = u_y = 0$ is applied to node 4.

A summary of the multiobjective optimization of the tensile inverter is shown in Table 4.3. As in Case Study 3.2, two independent load cases are considered to determine a structural topology that maximizes both the free displacement inversion ratio r_δ and the blocked force inversion ratio r_f as defined in Equation 3.6. For the first, a displacement of $\delta_{x,in} = -5$ mm is applied to node 1, and the resulting displacement $\delta_{x,out}$ at node 2 is measured. To calculate the second, node 2 is returned to its reference position (i.e., $\delta_{x,out} = 0$) while keeping $\delta_{x,in} = -5$ mm and the reaction forces at nodes 1 and 2 ($f_{x,in}$ and $f_{x,out}$, respectively) are measured. Given the use of multiple materials, this case study utilizes every graphical operation discussed in Section 4.3.2 and thus

Table 4.3 Summary of the multiobjective topology optimization problem for a tensile inverter.

Design Problem Statement	
Maximize:	r_δ, r_f
by varying:	2 axiom characters,
	5 rule assignments (18 genes each),
subject to:	$\sigma_{Mises}^{max} \leq 33.8$ MPa
NSGA-II Parameters (Deb et al. 2000)	
60 members for 400 generations,	
$P_{cross} = 0.9$, $\eta_{cross} = 50$,	
$P_{mut} = 0.1$, $\eta_{mut} = 20$	

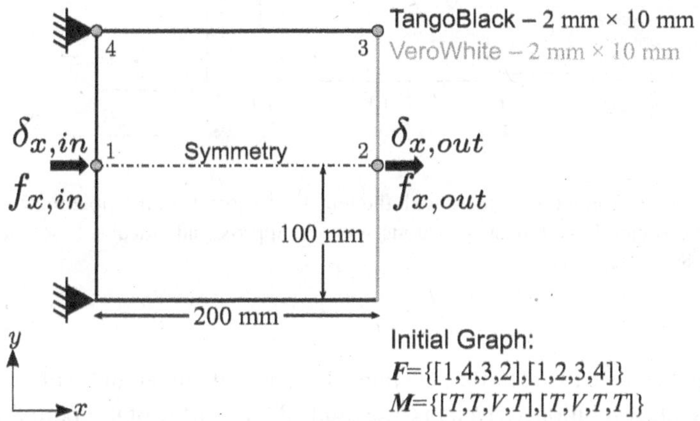

Figure 4.28 Initial graph description and boundary conditions associated with Case Study 4.2 (input displacement and forces $\delta_{x,in} < 0$ and $f_{x,in} < 0$ will be applied to assess inversion behavior).

considers a total of 92 independent design variables with three parametric L-system recursions. Both load cases described above consider the full geometric nonlinearity of the problem (i.e., internal buckling, large rotations, etc.); in particular, accounting for buckling requires the use of implicit dynamic analysis and thus small loading increments. Therefore, this study considers a population size of only 60 individuals and a total of 400 generations. Additionally, a stress constraint of $\sigma_{Mises}^{max} = 33.8$ MPa (based on experimental characterizations) is imposed to ensure the structural feasibility of resulting configurations.

The resulting Pareto frontier after considering a population of 60 individuals for 400 generations is shown in Figure 4.29, along with the Pareto frontier generated using the turtle graphics algorithm (cf. Case Study 3.2) and the approximate ideal solution defined by Equation 3.8. Comparing the two L-system topology optimization methods, the frontier associated with the SPIDRS algorithm outperforms that of the turtle graphics algorithm across all values of r_δ and r_f. It is especially interesting to note the effectiveness of solutions generated by the SPIDRS algorithm to transmit much

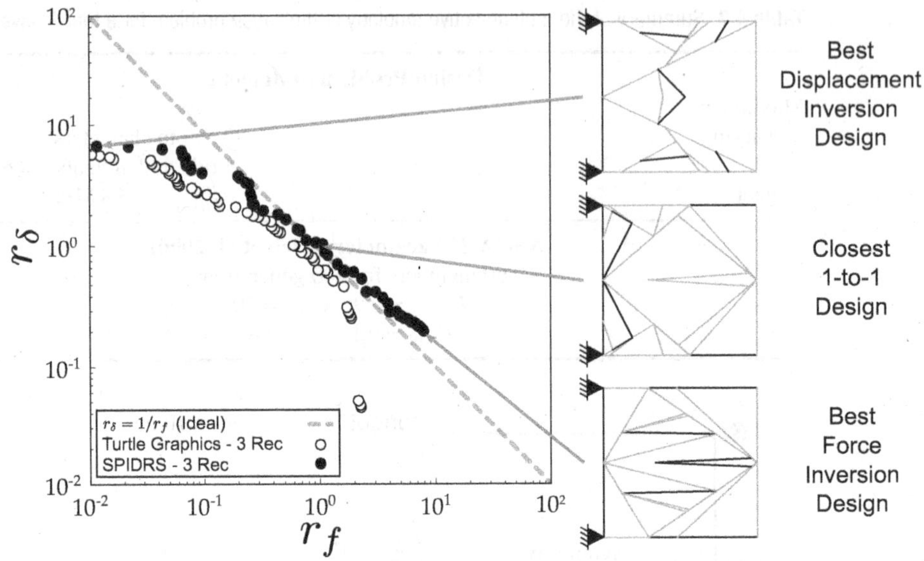

Figure 4.29 Pareto frontier associated with the tensile inverter design problem generated using SPIDRS-interpreted L-system strings along with the approximate ideal solution defined by Equation 3.8.

higher loads relative to the turtle-generated designs, with magnitudes of r_f that actually surpass the ideal solution derived in Case Study 3.2 by virtue of its approximate nature.

Additional examination of the resulting topologies offers additional insights into the response of these mechanisms with regard to the desired design objectives. Figure 4.30 shows that all the three topologies highlighted in Figure 4.29 all share the same underlying load path discussed in Case Study 3.2. However, we notice that specific changes in the geometry of this load path affect the inversion behavior of a given topology. For example, the design with the best displacement inversion behavior (Figure 4.30(a), $r_\delta = 7.69$) features a load path geometry such that the point of relative rotation is as far to the right as possible (specifically, at node 3 in Figure 4.28). For force inversion such a load path geometry performs poorly, as the resulting reaction force at node 2 when $\delta_{x,out} = 0$ acts almost entirely in the $-y$-direction; however, for free displacement inversion the mechanism uses this compressive reaction force to buckle the structural member between nodes 2 and 3 and generate a large $\delta_{x,out}$. Considering the design with the best force inversion behavior (Figure 4.30(c), $r_f = 7.95$), we see that the point of relative rotation is shifted to the left toward node 4. When $\delta_{x,out} = 0$, the reaction force at node 2 acts primarily in the $+x$-direction and thus results in a strong force inversion behavior, but the reduced compressive force in the y-direction corresponds to a reduced amount of member buckling and thus a weak displacement inversion behavior. The design that is closest to exhibiting force and displacement inversion responses equal to one (Figure 4.30(b), $r_\delta = 1.12$ and $r_f = 1.07$) has a point of relative rotation between the two, demonstrating a good trade-off between the reaction force in the $+x$-direction

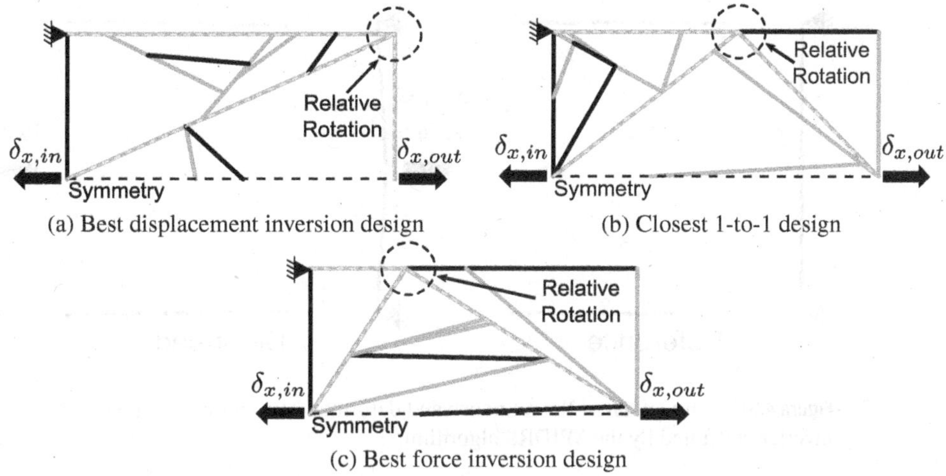

Figure 4.30 Underlying load path geometry for solutions generated by the SPIDRS algorithm. Notice that trends can be identified between the location of the point of relative rotation and mechanism force and displacement inversion behavior.

required for a large $f_{x,out}$ and the reaction force in the $-y$-direction that facilitates the buckling required for a large $\delta_{x,out}$.

While this load path geometry is a prominent feature in SPIDRS-generated topologies, the algorithm does also create some nonintuitive solutions such as the one illustrated in Figure 4.31. This solution, which has $r_\delta = 3.03$ and $r_f = 0.26$, features a similar mechanism to those in Figure 3.10, but here its rotation point is located relatively close to node 4 of the initial graph and it intersects the x-axis nearly halfway along the length of the inverter rather than at the output node. As the input displacement is converted to a translation at this intersection point, a secondary structural member transmits this translation to the right side of the topology, and by extension the output node. Thus, while it can generate solutions with load path geometries that are well-understood and similar to those of traditional topology optimization methodologies, the SPIDRS algorithm is also able to explore the design space and discover nonintuitive solutions to complex multiobjective design problems.

As a demonstration of its flexibility, we now reconsider the problem using the hybrid optimization framework. For this sizing optimization study, only a single material, VeroWhite, is considered (cf. Table 3.2), and a linear structural response is assumed (i.e., effects of local buckling and large displacements/rotations are not captured). Work by Bendsøe and Sigmund has considered the effects of nonlinearity on compliant mechanism design problems (Bendsøe & Sigmund 2003); given these findings, this study is limited in applicability to smaller input displacements and associated linear responses. Using the displacement and force inversion ratios defined in Equation 3.6, the sensitivities of these objective functions with respect to a given structural member thickness t_i can be written as

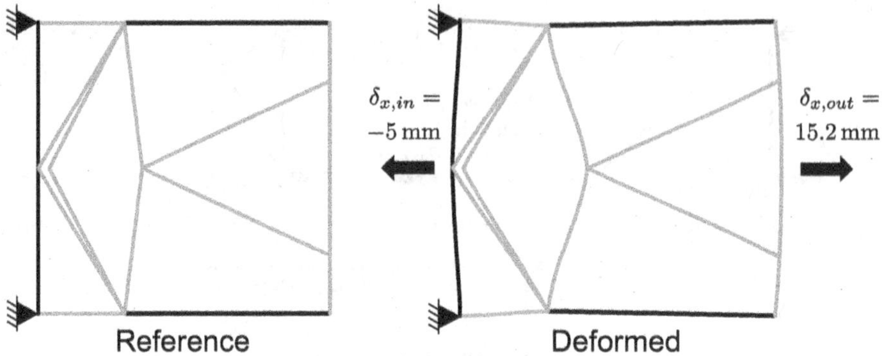

Figure 4.31 Comparison of the reference and deformed configurations of a nonintuitive tensile inverter generated by the SPIDRS algorithm.

$$\frac{\partial r_\delta(t_i)}{\partial t_i} = \frac{\partial}{\partial t_i}\left(\frac{\delta_{x,out}(t_i)}{\delta_{x,in}}\right) = \frac{\partial \delta_{x,out}(t_i)}{\partial t_i}\frac{1}{\delta_{x,in}},$$

and

$$\frac{\partial r_f(t_i)}{\partial t_i} = \frac{\partial}{\partial t_i}\left(\frac{-f_{x,out}(t_i)}{f_{x,in}(t_i)}\right) = -\frac{\partial f_{x,out}(t_i)}{\partial t_i}\frac{1}{f_{x,in}} + \frac{\partial f_{x,in}(t_i)}{\partial t_i}\frac{f_{x,out}}{f_{x,in}^2},$$

respectively, where $\delta_{x,in}$ is a constant input displacement of -1 mm, $\delta_{x,out}$ is the resulting output displacement solved for using Equation 2.36, $\partial\delta_{x,out}(t_i)/\partial t_i$ is obtained by evaluating Equation 2.49, $f_{x,in}$ and $f_{x,out}$ are the resulting input and output reaction forces determined using Equation 2.46, and $\partial f_{x,in}(t_i)/\partial t_i$ and $\partial f_{x,out}(t_i)/\partial t_i$ are the sensitivities of the calculated reaction forces with respect to a given structural member thickness. For this problem, the ε-constraint is applied to $r_f(\boldsymbol{t})$ with $N_\varepsilon = 10$, and the single set of objective function values returned to the $\mathscr{F}(\hat{\boldsymbol{J}}_i)$ for the ith topology (cf. Equation 4.9) are defined as the point along the ε-constraint-generated Pareto frontier closest to the utopia point. The thickness of each structural member is bounded such that $0\,\text{mm} \leq t \leq 4\,\text{mm}$. Additionally, to ensure compatibility between the generated solutions and the linear FEA approach employed, buckling and stress constraints of the form

$$f_x > -\frac{\pi^2 EI}{L^2} \quad \text{and} \quad \sigma_x \leq \sigma_{crit} \tag{4.12}$$

are applied to each structural member, where f_x is the axial internal reaction force in the element coordinate system, $-\pi^2 EI/L^2$ is the Euler critical load for first mode buckling (Timoshenko & Gere 2009), σ_x is the axial stress in the element coordinate system, and $\sigma_{crit} = 33.8$ MPa (Bielefeldt, Akleman, Reich, Beran & Hartl 2019).

Applying the hybrid optimization framework developed in Section 4.4.1, with a specified population size of 40 for 100 generations and two L-system recursions, yields the Pareto frontier shown in Figure 4.32. It is immediately noticeable that this frontier agrees well with the derived ideal solution for the tensile inverter problem given in Equation 3.8. Also shown in Figure 4.32 are several topologies of interest

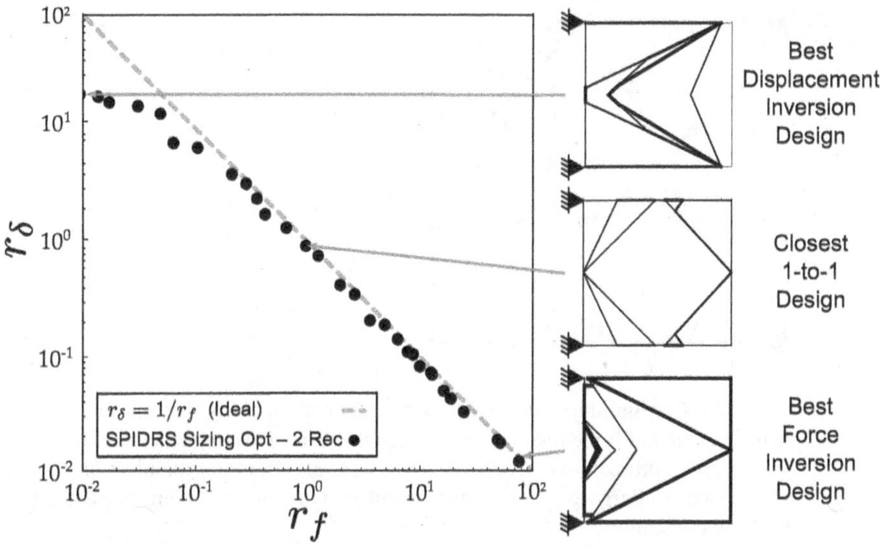

Figure 4.32 Pareto frontier associated with the tensile inverter design problem generated by applying the hybrid optimization framework (cf. Section 4.4.1) to SPIDRS-generated topologies; frontier shows good agreement with the derived ideal solution (Equation 3.8). Line thickness in each topology corresponds to actual structural member thickness.

generated by the hybrid optimization framework, from which one can observe several notable qualities. First, each design features a similar load path geometry to those previously observed in SPIDRS-generated tensile inverters. Second, the effect of the sizing optimization scheme is perhaps more noticeable in this design problem than in the cantilevered truss problem previously described, as each mechanism features structural members of varying thickness rather than the binary "min/max" thicknesses prevalent in the frame topologies. This is attributed to two effects: (i) in the cantilevered frame problems member thickness is more directly related to one objective (minimizing mass), and (ii) the presence of the buckling and stress constraints (cf. Equation 4.12) in the tensile inverter dictate that any structural members carrying axial loads (tensile or compressive) must be sufficiently thick to do so.

Finally, consider the SPIDRS-generated design closest to demonstrating displacement and force inversion ratios of unity (i.e., $r_\delta = r_f = 1$) illustrated in Figure 4.33. In this design, the structural member that connects the input point to the point of relative rotation in the load path (denoted by the dashed red line) is loaded in tension and has an optimized thickness of $t = 0.13$ mm (i.e., acting as a "string"). While intuitive given the kinematics and input conditions, this result would not be possible to achieve via SIMP unless the design domain was very finely discretized (estimated at 1E6 total elements). The structural member that extends from the point of relative rotation to the output point, however, has an optimized thickness of $t = 2.25$ mm to resist compressive buckling (Equation 4.12).

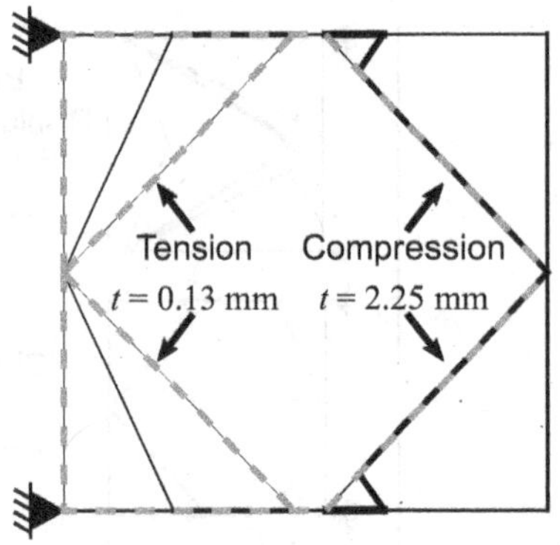

Figure 4.33 Configuration of the SPIDRS-generated topology that most closely demonstrates displacement and force inversion ratios equal to one. This design clearly demonstrates the effect sizing optimization has on solutions, as some structural members in tension can be represented similarly to strings, while members in compression are thickened to satisfy buckling constraints.

Case Study 4.4: 2-D Kinematic Rectifier

Our fourth case study considers a nonintuitive compliant mechanism called a *kinematic rectifier* (or mechanical motion rectifier (Li, Zuo, Kuang & Luhrs 2012)). In this mechanism, an input displacement drives an output displacement such that the direction of the output displacement is independent of that of the input displacement. The initial graph, structural member materials/cross-sections, and boundary conditions illustrated in Figure 4.34 are identical to that of the tensile inverter in Case Study 4.3 with the exception of the output point. To ensure physically meaningful topologies capable of producing motion under load, a spring with a constant $k = 1$ N/m is modeled as attached to node 2; this spring constant is based on *a priori* knowledge of the expected output force and desired output displacement of a compliant mechanism with these dimensions and boundary conditions (cf. Case Study 4.3).

A summary of the multiobjective optimization of the kinematic rectifier is shown in Table 4.4. As in Case Study 4.3, two distinct load cases are considered to determine a structural topology that maximizes the output displacement $\delta_{x,out}$ to the left (i.e., in the $-x$-direction). First, a displacement of $\delta_{x,in} = 5$ mm is applied to node 1, and the resulting output displacement $\delta_{x,out}$ is measured. Given the desired output response of the mechanism, this indicates the "inversion ratio" of the associated structural topology, which is defined as

$$r_{inv} = \left. \frac{-\delta_{x,out}}{\delta_{x,in}} \right|_{f_{x,out}=-k\delta_{x,out}}.$$

In the second load case, an input displacement of equal magnitude is applied in the opposite direction ($\delta_{x,in} = -5$ mm), and the resulting output displacement is again measured. This indicates the "transmission ratio" of the associated structural topology, defined as

Table 4.4 Summary of the multiobjective topology optimization problem for a kinematic rectifier.

Design Problem Statement	
Maximize:	r_{inv}, r_{trans}
by varying:	2 axiom characters,
	5 rule assignments (18 genes each),
subject to:	$\sigma_{Mises}^{max} \leq 33.8\,\text{MPa}$
NSGA-II Parameters (Deb et al. 2000)	
60 members for 400 generations,	
$P_{cross} = 0.9$, $\eta_{cross} = 50$,	
$P_{mut} = 0.1$, $\eta_{mut} = 20$	

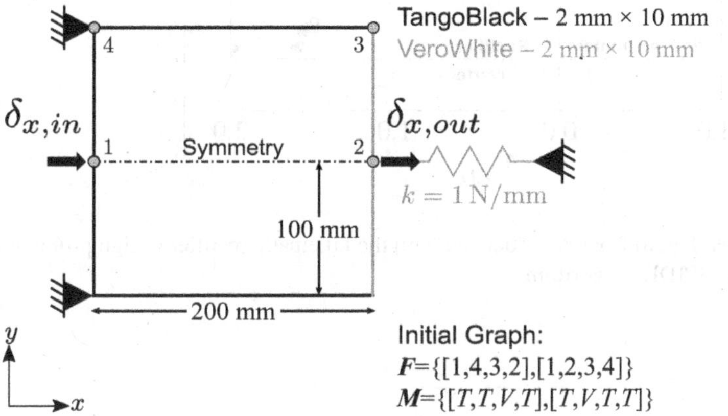

Figure 4.34 Initial graph description and boundary conditions associated with Case Study 4.3 (input displacement $\delta_{x,in}$ will be applied to assess rectification behavior).

$$r_{trans} = \left.\frac{\delta_{x,out}}{\delta_{x,in}}\right|_{f_{x,out}=-k\delta_{x,out}}.$$

The number of design variables, type of structural analysis employed, optimization parameters, and constraints all remain consistent with those of Case Study 4.3.

The Pareto frontier generated by SPIDRS for the kinematic rectifier is shown in Figure 4.35 and, unlike the previous two case studies, contains points that have negative objective values. To clarify, designs found in quadrant I are true kinematic rectifiers and perform as desired (invert motion when $\delta_{x,in} > 0$, transmit motion when $\delta_{x,out} < 0$). Designs found in quadrant II are kinematic transmitters, where $\delta_{x,out}$ is always in the same direction as $\delta_{x,in}$. Designs found in quadrant IV are kinematic inverters, where $\delta_{x,out}$ is always in the opposite direction as $\delta_{x,in}$.

To better understand the behavior of a SPIDRS-generated rectifier, consider the design closest to obtaining displacement ratios equal to one ($r_{inv} = 1.01$, $r_{trans} = 1.09$) in Figure 4.36. Note that the response of this design is governed by the behavior of a select number of structural members denoted in Figures 4.36(c)–4.36(d). Specifically, when $\delta_{x,in} > 0$ (Figure 4.36(c)) a load path similar to those observed in the tensile

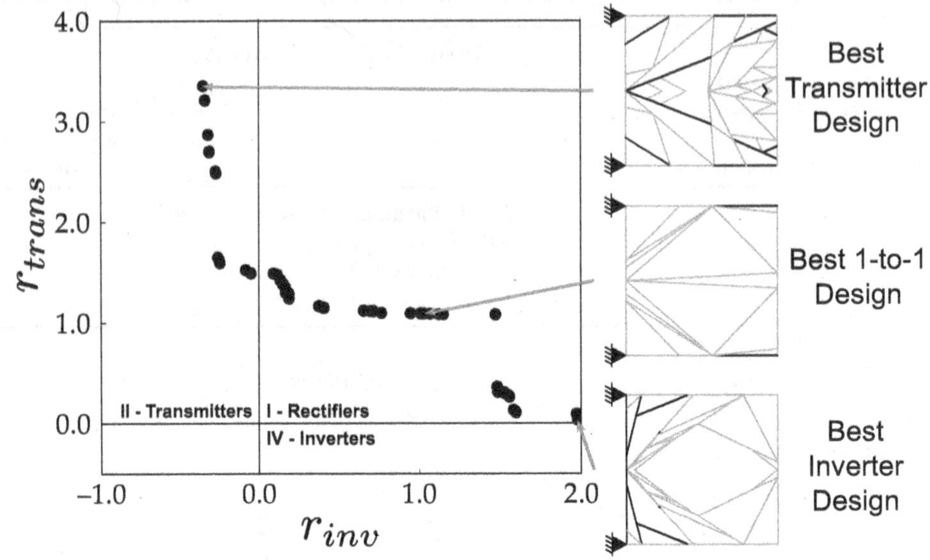

Figure 4.35 Pareto frontier associated with the kinematic rectifier design problem generated using the SPIDRS algorithm.

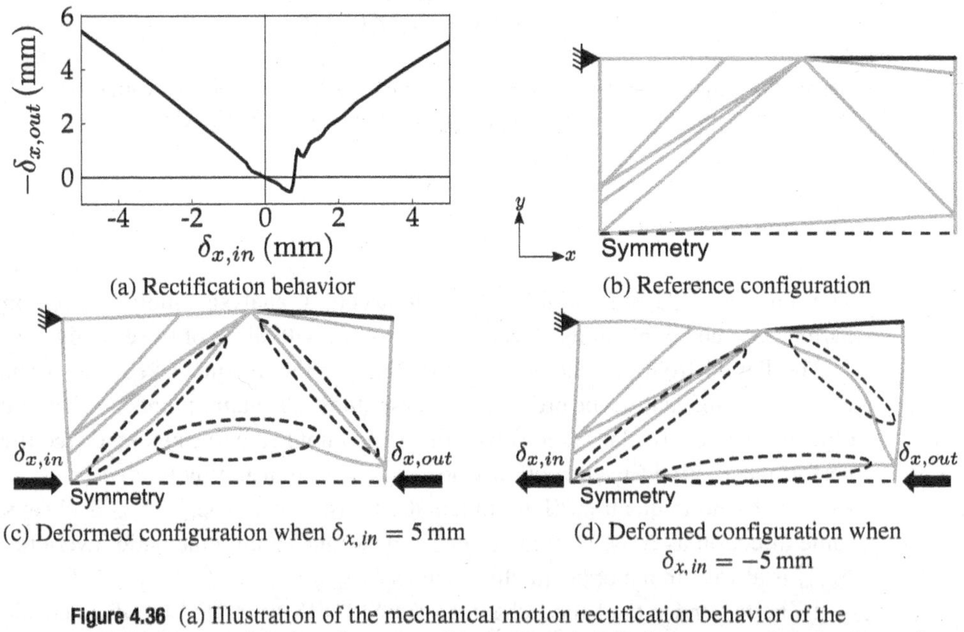

Figure 4.36 (a) Illustration of the mechanical motion rectification behavior of the SPIDRS-generated design closest to achieving a 1-to-1 displacement ratio behavior, and comparison of the (b) undeformed and (c)–(d) deformed configurations of this mechanism (deformations are unscaled). Notice that the functionalities of specific (circled) branches govern the rectification response of the mechanism.

inverters previously discussed is noticeable. The positive input displacement is translated through the structural member on the left, causing the top of the structure to lift up in the positive y-direction. This causes the structural member on the right to be stiff in tension, which in turn pulls the output point in the desired direction. Notice that, during this loading process, the bottom structural member, which is in compression, initially pushes the output point in the opposite direction until it eventually buckles at $\delta_{x,in} \approx 1$ mm and allows the output point to move in the desired direction as seen in Figure 4.36(a). When $\delta_{x,in} < 0$ (Figure 4.36(d)), the structural members on the left and bottom, which are stiff in tension, simply pull the output point in the desired direction while the structural member on the right buckles in compression.

To further explore the ability of the SPIDRS algorithm to develop solutions for nonintuitive problems, a "reverse" kinematic rectifier is considered. This problem has the same boundary conditions, initial graph, and optimization details as discussed previously, but modifies the goal to be maximizing $\delta_{x,out}$ to the right (i.e., positive as in Figure 4.28). The Pareto frontier obtained for this rectifier design problem after considering 60 individuals for 400 generations is shown in Figure 4.37. Given that the desired output direction has reversed, the behavior associated with designs in quadrants II and IV is also reversed such that quadrant II now contains kinematic inverters and quadrant IV now contains kinematic transmitters. The number of solutions along the frontier located in quadrant I, indicating rectification behavior, suggests that reversing the direction of the desired output makes this a much more difficult problem to solve. Considering the potential mechanisms behind each loading case for both rectifier problems, this proves to be an intuitive result. For the original rectifier problem, the kinematic inversion step (i.e., $\delta_{x,in} > 0$) is simply a compressive inverter, which, while not covered in this work, has been briefly explored using the turtle graphics algorithm (Hartl et al. 2017) and has a deformation mechanism that is well-understood. The kinematic transmission step (i.e, $\delta_{x,in} < 0$) simply requires that structural members remain in tension and pull the output point in the same direction as the input point. For the reverse rectifier problem, the kinematic inversion step (i.e., $\delta_{x,in} < 0$) requires the same behavior as a tensile inverter, which has been discussed in great detail in this work. However, the kinematic transmission step (i.e., $\delta_{x,in} > 0$) requires structural members to push the output point in the same direction as the input point. In other words, this loading case requires structural members to stay stiff while in compression, which is contrary to how these members want to behave.

Despite the complexity of the problem, SPIDRS is still able to generate several solutions that demonstrate rectification behavior. Specifically, consider the design closest to obtaining displacement ratios equal to one ($r_{trans} = 0.64$, $r_{inv} = 1.09$) in Figure 4.38. As before, the response of the design is governed by only a select number of structural members denoted in Figures 4.38(c)–4.38(d). When $\delta_{x,in} > 0$ (Figure 4.38(c)), the structural member on the left translates the input displacement and attempts to cause the top of the structure to lift up in the positive y-direction as in Figure 4.38(c). However, the two structural members in the middle, which are oriented almost vertically, resist this movement. Since the displacement of the right side of the structure is constricted in the y-direction by these members, the displacement in the

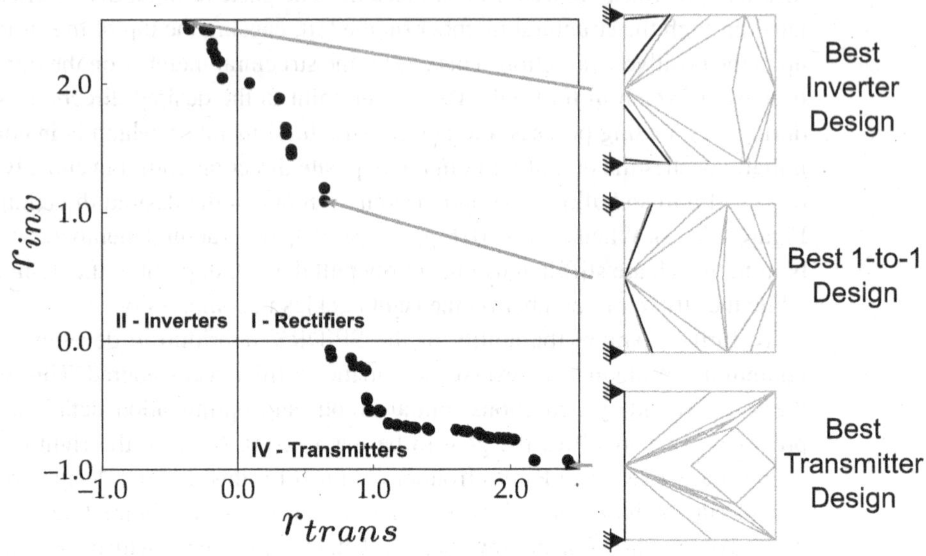

Figure 4.37 Pareto frontier associated with the "reverse" kinematic rectifier design problem generated using the SPIDRS algorithm.

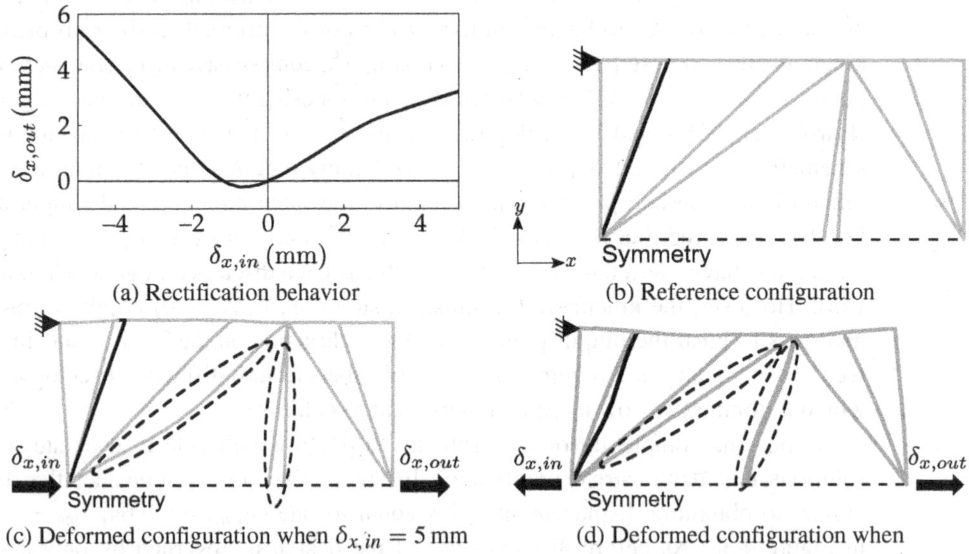

(a) Rectification behavior

(b) Reference configuration

(c) Deformed configuration when $\delta_{x,in} = 5$ mm

(d) Deformed configuration when $\delta_{x,in} = -5$ mm

Figure 4.38 (a) Illustration of the "reverse" mechanical motion rectification behavior of the SPIDRS-generated design closest to achieving a 1-to-1 displacement ratio behavior, and comparison of the (b) undeformed and (c)–(d) deformed configurations of this mechanism (deformations are unscaled). Notice that the functionalities of specific (circled) branches govern the rectification response of the mechanism.

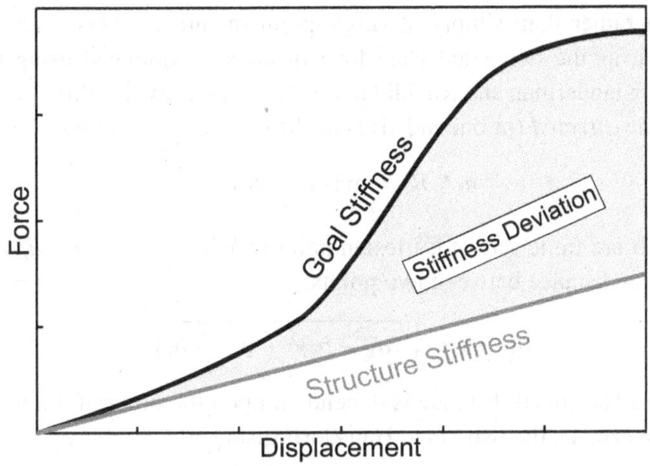

Figure 4.39 Example of variable stiffness structural design problem, where the objective is to minimize the deviation between structure stiffness and a specified goal stiffness.

x-direction that is translated by the now buckling structural member on the left dominates and moves the output point in the desired direction. Thus, the SPIDRS algorithm has developed a transmitting structure capable of remaining stiff while some structural members buckle. When $\delta_{x,in} < 0$ (Figure 4.38(d)), the structure behaves, like the tensile inverters discussed throughout this work, with the structural members in the middle initially resisting deformation (indicated by $\delta_{x,out}$ being negative for $\delta_{x,in} \geq\approx -1$ mm) before eventually buckling.

Case Study 4.5: 2-D Variable Stiffness Structures

To this point, the case studies considered using the SPIDRS algorithm have focused on structural responses for a single set of loading and boundary conditions. For example, in Case Study 4.4, the displacement ratios r_{inv} and r_{trans} were calculated only for $\delta_{x,in} = \pm 5$ mm, and thus the optimization allowed for responses contrary to those desired at lower values of $\delta_{x,in}$ (cf. Figures 4.36(a) and 4.38(a)). However, it can often be desirable within structural design to tailor the response of a structure across a full loading envelope. Existing work in this area of topology optimization is sparse and has been applied primarily to the spatial response of a compliant mechanism (Tai & Prasad 2007, Sharma, Deb & Kishore 2008, Wang & Tai 2008). Here, we will discuss the approach proposed by Ward, Bielefeldt and Hartl (2020)[5] toward the optimization of structures capable of matching specified tensile and compressive force responses across a range of displacements as exemplified in Figure 4.39.

The ability to match a specified stiffness response requires an error metric capable of accurately reflecting deviations between the desired and current response over a

[5] Used with permission of SPIE, from Ward et al. (2020); permission conveyed through Copyright Clearance Center, Inc.

range of data rather than simply at single-point quantities. This study considers a method known as the two-sided Hausdorff distance, formulated using the work of Huttenlocher, Klanderman and Rucklidge (1993) to quantify this difference. To begin, we consider the *directed* (or one-sided) Hausdorff distance given by

$$h(A,B) = \max_{a \in A} \min_{b \in B} d(a,b),$$

where A and B are finite sets of Cartesian point pairs (e.g., $a_i = (a_{i_x}, a_{i_y})$) and $d(a,b)$ is the Euclidean distance between two points:

$$d(a,b) = \sqrt{(a_x - b_x)^2 + (a_y - b_y)^2}.$$

As the directed Hausdorff distance is dependent upon the order of the parameter sets (i.e., $h(A,B) \neq h(B,A)$), the two-sided Hausdorff distance,

$$H(A,B) = \max(h(A,B), h(B,A)), \tag{4.13}$$

is preferred because it lacks this parameter order dependence. Equation 4.13 decreases as A and B become more closely aligned, with a Hausdorff distance of zero indicating that the elements of both sets are identical. The Hausdorff distance has found applications in several fields, such as image recognition operations, but is often avoided due to its sensitivity to outliers within data sets (Huttenlocher et al. 1993, Schutze, Esquivel, Lara & Coello 2012); however, provided the two sets are normalized, here this sensitivity motivates its use. Figure 4.40 provides a comparison of the calculated error between a specified goal curve and two arbitrary test curves using both the two-sided Hausdorff distance and the well-known root mean square error (RMSE). The RMSE metric prefers the test curve in Figure 4.40(a), which is a better match for the middle of the domain but noticeably differs from the goal curve at the endpoints. In contrast, the Hausdorff distance favors the test curve in Figure 4.40(b), which is much closer to the goal curve at the endpoints and captures its general behavior over the entire domain. This example demonstrates the reduced sensitivity of RMSE to outliers (e.g., the endpoints) in comparison to the two-sided Hausdorff distance; when considering stiffness curves, endpoint behavior becomes an important consideration, and thus the test curve in Figure 4.40(b) would be preferred when attempting to match the specified goal curve.

The structural design domain for this case study has a height and length of 76.2 mm as illustrated in Figure 4.41. All structural members created are assumed to consist of Nylon 6,6 with a Young's modulus of $E = 2.5$ GPa and a Poisson's ratio of $\nu = 0.41$ based on experimental characterization. Given the assumption of a single material, the Change Attribute graphical operation (cf. Section 4.3.2) is modified to consider thickness rather than material assignments; specifically, this attribute allows the in-plane thickness of structural members to vary between 1.5 mm and 4.5 mm, with the out-of-plane thickness of all members assumed to be a constant 6.35 mm. Making use of problem symmetry, only the top right quadrant of the design space, measuring 38.1 mm in both length and height, is considered for topological modification; horizontal and vertical symmetry conditions are enforced along the left and bottom edges

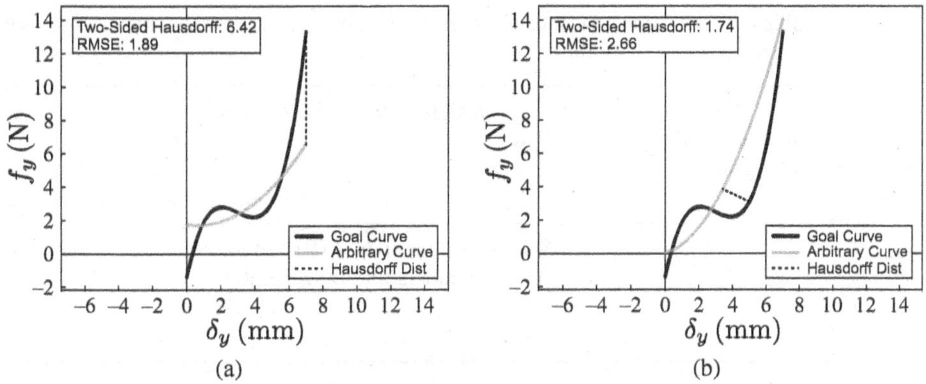

Figure 4.40 Example of the two-sided Hausdorff distance metric compared to RMSE for the specified goal curve (black) given two arbitrary test curves (gray). The Hausdorff distance's sensitivity to outliers at the endpoints make it the preferred option to match a goal curve across an entire domain. Reprinted from Ward, Bielefeldt & Hartl (2020) with permission.

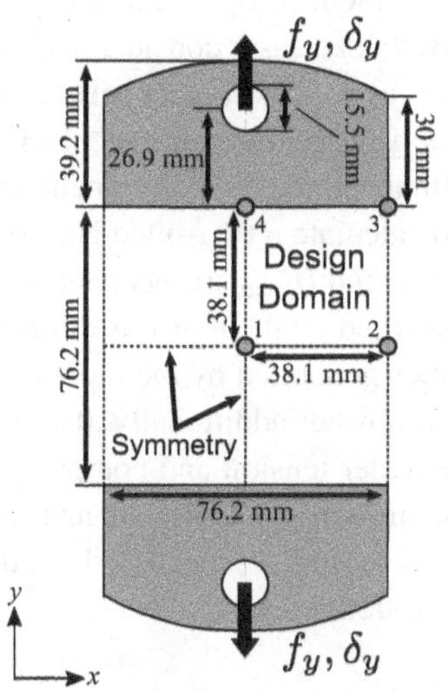

In-Plane Thickness: $1.5\,\text{mm} \leq t \leq 4.5\,\text{mm}$
Out-of-Plane Thickness: 6.35 mm
Initial Graph: $\boldsymbol{F} = \{[1, 4, 3, 2], [1, 2, 3, 4]\}$

Figure 4.41 Initial graph description and boundary conditions associated with Case Study 4.4. Reprinted from Ward, Bielefeldt & Hartl (2020) with permission.

of this region, respectively. Note that while the initial graph is required by the SPIDRS algorithm, the edges comprising it are not considered as part of the final structure and are thus removed.

A summary of the multiobjective optimization of the variable stiffness structure is shown in Table 4.5. To evaluate the ability of a given structural topology to match the variable stiffness response of a specified goal curve, two distinct load cases are

Table 4.5 Summary of the multiobjective topology optimization problem for a variable stiffness structure.

Design Problem Statement	
Minimize:	H_{tens}, H_{comp}
by varying:	2 axiom characters,
	5 rule assignments (18 genes each),
NSGA-II Parameters (Deb et al. 2000)	
60 members for 200 generations,	
$P_{cross} = 0.9$, $\eta_{cross} = 50$,	
$P_{mut} = 0.1$, $\eta_{mut} = 20$	

considered. First, the resulting reaction force f_y, measured as an input displacement of $\delta_y = -2$ mm, is smoothly applied to the structure via a rigid surface (which represents the connector portion of the structure shown in Figure 4.41) at a rate of 0.083 mm/s. Next, f_y is again measured as an input displacement of $\delta_y = 2$ mm is applied via the same rigid surface at the same loading rate. Both load cases consider buckling and large displacements/rotations of structural members as well as contact between the structure and both planes of symmetry to more accurately reflect the behavior of a fully mirrored structure. This reaction force data, along with the applied displacement, is normalized and compared to the specified goal curve to calculate a two-sided Hausdorff distance for both tensile and compressive loadings (H_{tens} and H_{comp}, respectively), with the goal being to minimize these two values. The specified goal curve was defined using an *a priori* knowledge of reasonable reaction forces generated by the input displacements, given both the design domain and the material used; additionally, it was desired that this curve be defined such that (i) response under tension and compression could be independently defined, (ii) it remains monotonic across its domain, and (iii) it remains fully differentiable except at the endpoints, including at piecewise boundaries. Given these specifications, the goal stiffness curve is defined as

$$f_y(\delta_y) = \begin{cases} 70\delta_y & -2 \leq \delta_y \leq 0 \\ 115{\delta_y}^3 - 120{\delta_y}^2 + 70\delta_y & 0 < \delta_y \leq 2 \end{cases} \tag{4.14}$$

and is shown in Case Study 4.5.

The Pareto frontier for this problem after considering 60 individuals for 200 generations is shown in Figure 4.42. The best-performing individual responses for the tension and compression load cases, have Hausdorff distances of $H_{tens} = 0.047$ and $H_{comp} = 0.008$ and are shown in Figure 4.42. These values indicate that the framework was more adept at matching the specified compressive stiffness; this is expected, as matching the linear response in compression represents a simpler problem compared to the desired polynomial tensile response. Also shown is topology associated with the response closest to the utopia point ($H_{tens} = 0.138$, $H_{comp} = 0.076$), which represents the best trade-off between the two objectives.

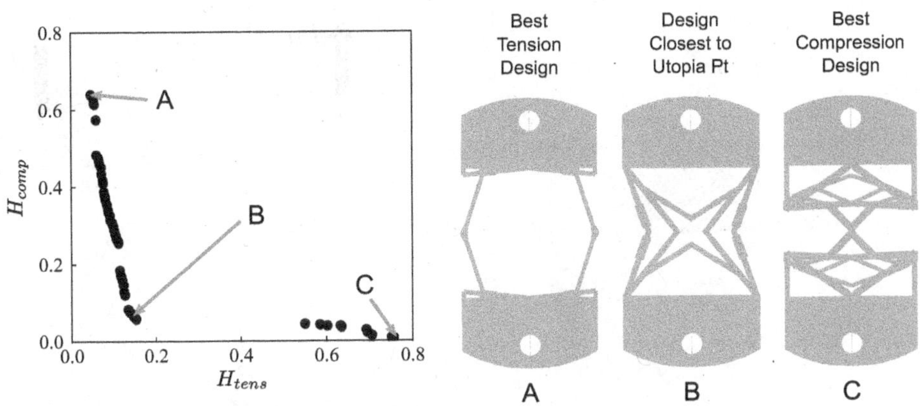

Figure 4.42 Pareto frontier associated with the variable stiffness structure design problem. Reprinted from Ward, Bielefeldt & Hartl (2020) with permission.

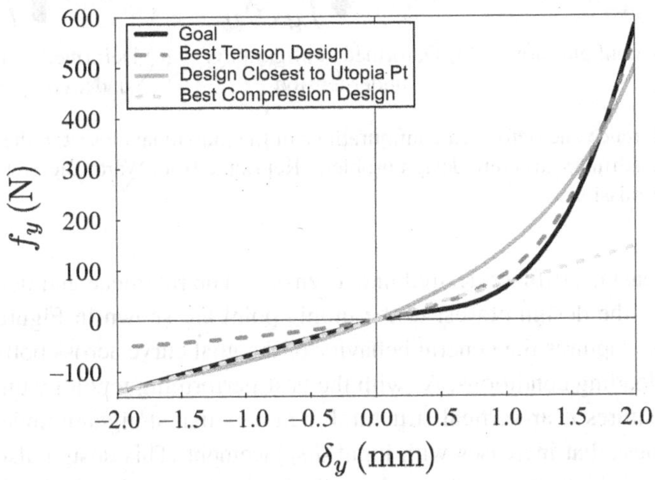

Figure 4.43 Comparison between the goal stiffness curve and the best individual responses with respect to both tension and compression, along with the design closest to the utopia point which represents the best trade-off between the two objectives. Reprinted from Ward, Bielefeldt & Hartl (2020) with permission.

The stiffness curves for the three structural topologies are illustrated in Figure 4.42 and shown in Figure 4.43, along with the specified goal curve defined by Equation 4.14. The best-performing structural topology in tension is characterized by two thin members that straighten under tensile loading to produce increasing stiffness with displacement; however, under compression they offer diminishing stiffness as they begin to buckle outwards. By contrast, the best-performing topology in compression features thick members forming triangular subassemblies that do not undergo large displacements/rotations when subjected to input displacements. Such linearity enables the structure to match the desired stiffness under compression but results in a poor

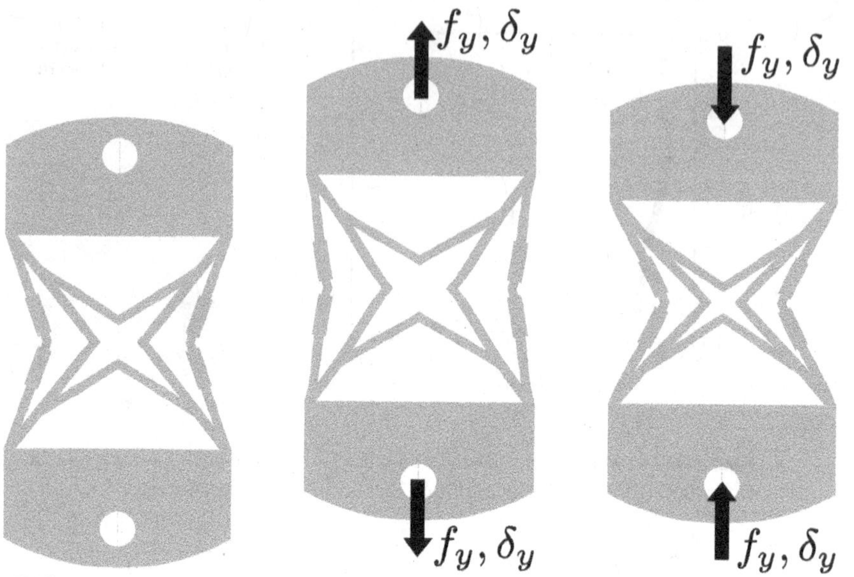

(a) Reference configuration (b) Deformed configuration under tension (c) Deformed configuration under compression

Figure 4.44 Reference and deformed configurations of the individual closest to the utopia point for the variable stiffness structure design problem. Reprinted from Ward, Bielefeldt & Hartl (2020) with permission.

fit to the increasing stiffness desired under tension. The reference and deformed configurations for the design closest to the utopia point are shown in Figure 4.44. This stiffness curve captures the general behavior of the goal curve across both tensile and compressive loading conditions. As with the best-performing topology under tension, this design features near-vertical structural members that straighten under tension to produce stiffness that increases with input displacement. This design also contains a configuration of thick structural members similar to that seen in the best-performing topology under compression, which provides a near-linear stiffness response under compressive loading.

Case Study 4.6: 3-D Cantilevered Frame

The first 3-D topology optimization case study considers a cantilevered frame subjected to transverse loading. As illustrated in Figure 4.45, the structural design domain is defined by the initial graph as

$$\boldsymbol{F} = \{[1,2,3,4],[6,5,8,7],[1,4,8,5],[3,2,6,7],[1,5,6,2],[4,3,7,8]\},$$

where these nodes are assigned spatial coordinates such that the domain has a length of 120 mm in the x-direction, a height of 40 mm in the y-direction, and a width of 30 mm in the z-direction. This initial graph and subsequently created structural members are assumed to consist of VeroWhite (cf. Table 3.2) with a square cross-section of

Table 4.6 Summary of the multiobjective topology optimization problem for a 3-D cantilevered frame subjected to transverse loading.

Design Problem Statement	
Minimize:	normalized mass $\hat{m}$
and Maximize:	normalized transverse stiffness $\hat{k_{trans}}$
by varying:	2 axiom characters, 4 rule assignments (18 genes each)
NSGA-II Parameters (Deb et al. 2000)	
100 members for 1,000 generations, $P_{cross} = 0.9$, $\eta_{cross} = 50$, $P_{mut} = 0.1$, $\eta_{mut} = 20$	

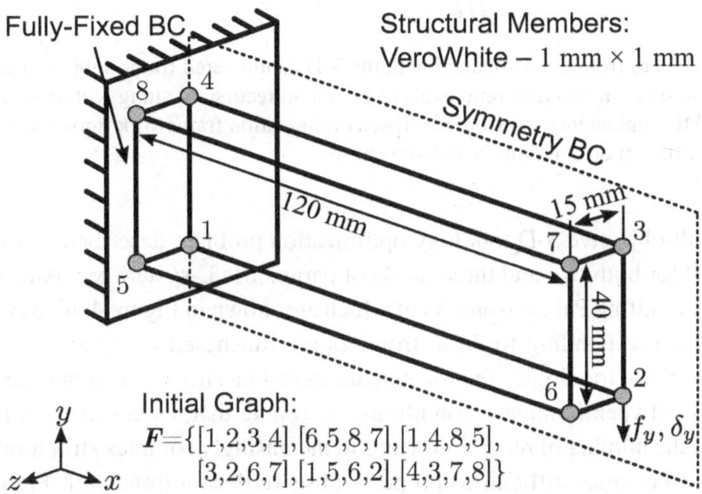

Figure 4.45 Initial graph description and structural boundary conditions associated with the case study of a 3-D cantilevered frame subjected to transverse loading.

1 mm × 1 mm. As in Case Study 4.1, geometric symmetry is assumed about the x–y plane associated with the face [1, 2, 3, 4] such that only half of the topology is generated by the SPIDRS algorithm. A fully fixed boundary condition (i.e., $u_x = u_y = u_z = \theta_x = \theta_y = \theta_z = 0$) is placed on all nodes in the plane associated with the face [1, 4, 8, 5], and a concentrated transverse force of $f_y = -0.05$ N is applied at node 2.

Details regarding the multiobjective optimization of the 3-D cantilevered frame subjected to transverse loading are provided in Table 4.6. The objectives of this problem are the minimization of normalized mass $\hat{m} = m/m_{null}$ (where $m_{null} = 9 \times 10^{-4}$ kg) and maximization of normalized transverse stiffness $\hat{k}_{trans} = k_{trans}/k_{trans,null}$ (where $k_{trans} = f_y/\delta_y$ and $k_{trans,null} = 4.48$ N/mm). As mentioned in Section 4.4.2, the Change Attribute operation in the SPIDRS algorithm is neglected, and thus the problem requires only 74 independent variables. A fully linear structural analysis assuming plane strain conditions allows for $100 \times 1{,}000 = 100{,}000$ individual designs to be considered with relative computational efficiency.

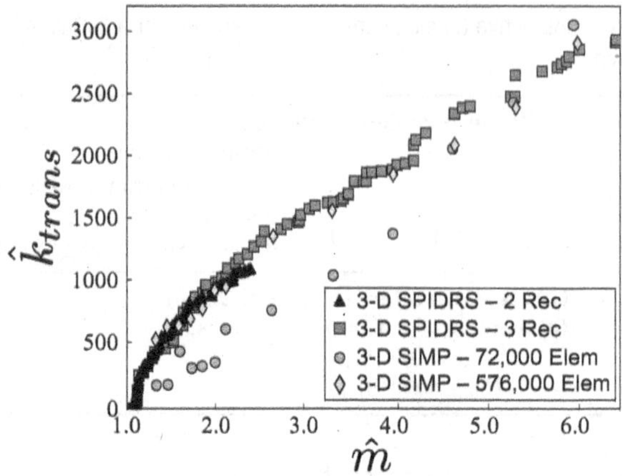

Figure 4.46 Pareto frontiers associated with the 3-D cantlievered frame subjected to transverse loading case study at two different levels of L-system recursion, along with frontiers generated using a SIMP implementation from the Tosca optimization framework implemented in Abaqus using two different design domain discretizations.

The multiobjective 3-D topology optimization problem described in Table 4.6 was considered for both two and three levels of parametric L-system recursion (cf. Section 4.4.2), the resulting Pareto frontiers of which are shown in Figure 4.46. Several designs of interest corresponding to these frontiers are illustrated in Figures 4.47 and 4.48. Notice that these topologies feature elegant truss-like structural layouts and deliberate load paths between boundary conditions. Also note that, as observed in Figure 4.20, increasing the number of recursions leads to increasingly complex structural topologies such that it becomes difficult to properly observe design trends (cf. Figure 4.48(c)). Due to this result, subsequent 3-D design problems will consider only two levels of recursion.

Also shown in Figure 4.46 are frontiers obtained using a SIMP formulation from the Tosca optimization suite available in Abaqus (Dassault Systèmes of America Corp. 2014). This implementation considers a design problem identical to that shown in Figure 4.45 and material properties consistent with those given in Table 3.2. Two different discretizations of the initial structural domain are considered: (i) a mesh made up of elements measuring 1 mm × 1 mm × 1 mm, corresponding to 72,000 design variables, and (ii) a mesh made up of elements measuring 0.5 mm × 0.5 mm × 0.5 mm, corresponding to 576,000 design variables. Selected resulting structural topologies for both discretizations considered are shown in Figures 4.49 and 4.50.

Comparing the resulting topologies and associated frontiers to those generated using the SPIDRS algorithm, several trends emerge. First, the SPIDRS-generated structural topologies illustrated in Figures 4.47 and 4.48 and SIMP generated topologies illustrated in Figures 4.49 and 4.50 share many geometric similarities, suggesting that SPIDRS is effective in exploring a 3-D structural design space and constructing efficient load paths. Second, it is noticeable that the SIMP-generated frontier associated with the coarser discretization made up of 72,000 elements is outperformed by

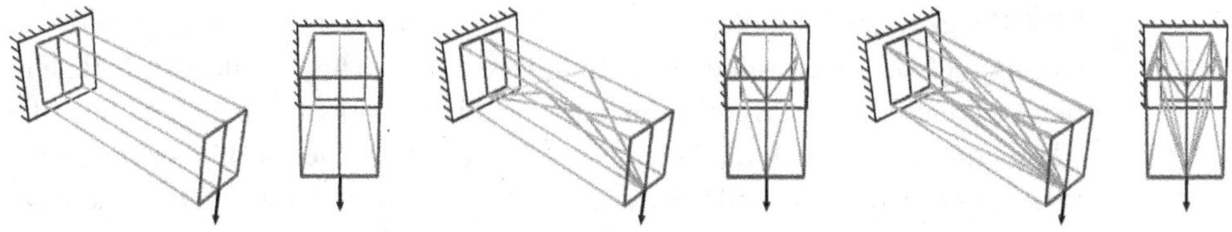

(a) Lightest design ($\hat{m} = 1.0$, $\hat{k}_{trans} = 1.0$)

(b) Design closest to the utopia point ($\hat{m} = 1.51$, $\hat{k}_{trans} = 667.9$)

(c) Stiffest design ($\hat{m} = 2.40$, $\hat{k}_{trans} = 1,127$)

Figure 4.47 3-D cantilevered frame topologies of interest generated by SPIDRS using two recursions. Structural members at each end are shaded to better illustrate the topologies.

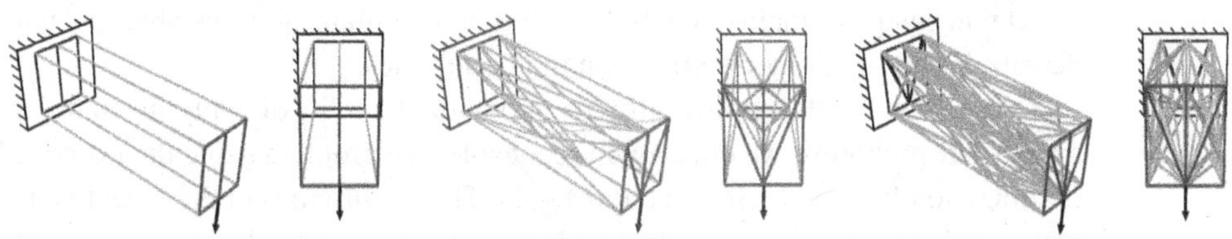

(a) Lightest design ($\hat{m} = 1.0$, $\hat{k}_{trans} = 1.0$)

(b) Design closest to the utopia point ($\hat{m} = 2.55$, $\hat{k}_{trans} = 1,425$)

(c) Stiffest design ($\hat{m} = 6.46$, $\hat{k}_{trans} = 2,964$)

Figure 4.48 3-D cantilevered frame topologies of interest generated by SPIDRS using three recursions. Structural members at each end are shaded to better illustrate the topologies.

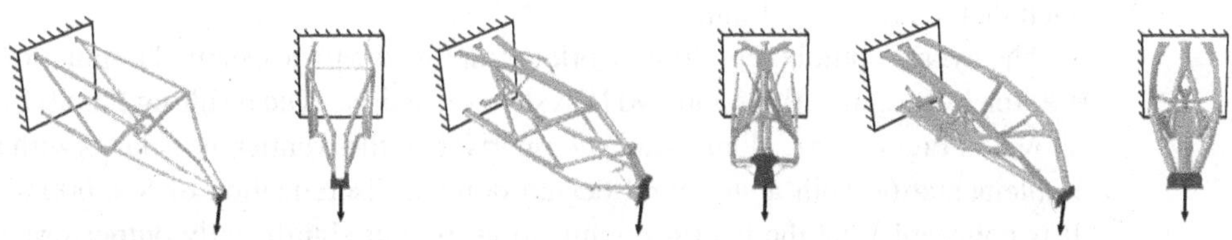

(a) Design with a volume fraction of 0.012 ($\hat{m} = 1.60$, $\hat{k}_{trans} = 459.7$)

(b) Design with a volume fraction of 0.030 ($\hat{m} = 3.99$, $\hat{k}_{trans} = 1,405$)

(c) Design with a volume fraction of 0.045 ($\hat{m} = 5.98$, $\hat{k}_{trans} = 3,080$)

Figure 4.49 3-D cantilevered frame topologies of interest generated using a 72,000 element-SIMP implementation. Structural members at each end are shaded to better illustrate the topologies.

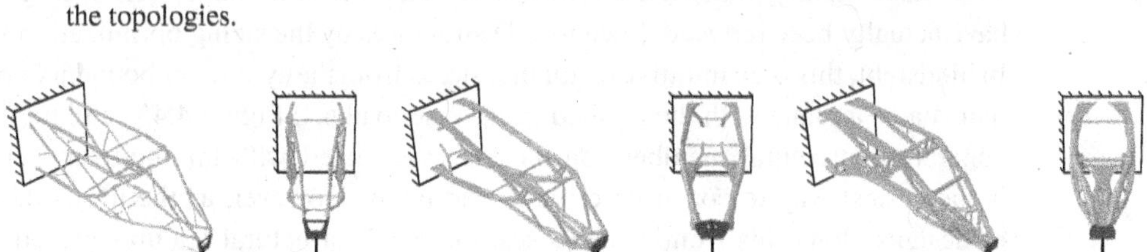

(a) Design with a volume fraction of 0.012 ($\hat{m} = 1.59$, $\hat{k}_{trans} = 667.4$)

(b) Design with a volume fraction of 0.030 ($\hat{m} = 3.99$, $\hat{k}_{trans} = 1,880$)

(c) Design with a volume fraction of 0.045 ($\hat{m} = 6.03$, $\hat{k}_{trans} = 2,936$)

Figure 4.50 3-D cantilevered frame topologies of interest generated using a 576,000 element-SIMP implementation. Structural members at each end are shaded to better illustrate the topologies.

SPIDRS-generated topologies (regardless of the number of L-system recursions considered) at lower values of normalized mass. This is due to SIMP having difficulty creating meaningful load paths with a relatively small amount of material when considering such a coarse discretization. However, at higher normalized mass values, performances of SIMP-generated structures with this course discretization are comparable to those created by SPIDRS because they both have more material with which to create load paths and are not limited to a constant cross-sectional area. It is also interesting to note that there is good agreement between the frontiers associated with the SIMP implementation considering 576,000 design elements and SPIDRS algorithm considering 3 L-system recursions. In particular, the performance of these SIMP-generated structural topologies at lower normalized mass values highlights the effect that design domain discretization has on potential solutions obtained when using density-based topology optimization methodologies.

To demonstrate the applicability and potential benefits of applying sizing optimization to 3-D problems, we once again reconsider the problem using the proposed hybrid optimization framework (cf. Section 4.4.1). The sensitivities of the objective functions, mass and stiffness, are equivalent to Equations 4.10 and 4.11. As in Case Study 4.1, the ε-constraint is applied to the mass objective with $N_\varepsilon = 15$, and the single set of objective function values returned for each individual structural topology (cf. Equation 4.9) is defined by calculating the last value of $\varepsilon^{(j)}$ for which there is a significant (i.e., $\geq 1\%$) increase in stiffness. For a direct comparison to the results obtained above, the thickness (and therefore out-of-plane width) for each structural member is bounded such that $0\,\text{mm} \leq t \leq 1\,\text{mm}$.

The hybrid optimization framework is employed with a specified population size of 100 for 1,000 generations and two L-system recursions. The resulting Pareto frontier is shown in Figure 4.51, along with a comparison to the frontier associated with a SIMP implementation with a structural design domain discretization of 576,000 elements. It is noticeable that the hybrid optimization frontier significantly outperforms SIMP-generated designs despite the extremely fine discretization of the design domain. Additionally, even with a large number of elements, SIMP is unable to truly converge on a solution at normalized mass below approximately 1.3.[6] Several resulting topologies from the frontier generated by the hybrid optimization approach are shown in Figure 4.52. Interestingly, topologies with low normalized mass values (cf. Figure 4.52(a)) have actually been reduced down to 2-D structures by the sizing optimization scheme. In hindsight, this is an intuitive result that stems from the symmetry boundary condition that shares a plane with the applied transverse load (cf. Figure 4.45) and the fact that generating structural members on the face associated with this boundary condition is the easiest way to construct critical load paths. However, as the normalized mass of designs along this frontier increases, more 3-D structural features are utilized, as observed in Figures 4.52(b)–4.52(c).

[6] The design shown with $\hat{m} \approx 0.65$ does not feature a load path, and its normalized stiffness value is believed to represent some numerical artifact from structural analysis.

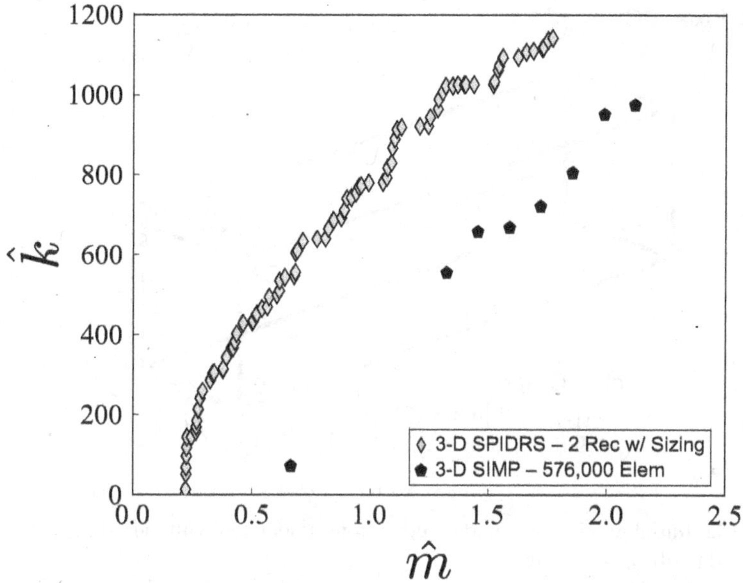

Figure 4.51 Pareto frontier of the design of a 3-D cantilevered frame subjected to transverse loading generated using the hybrid topology/sizing optimization framework with member thickness bounds of $0\,\text{mm} \leq t \leq 1\,\text{mm}$, along with comparisons to the SPIDRS topology optimization with sizing optimization applied to the frontier generated using SIMP implementation with 576,000 elements.

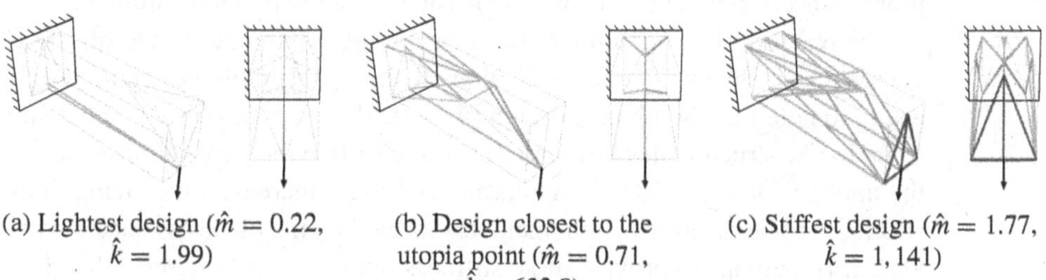

(a) Lightest design ($\hat{m} = 0.22$, $\hat{k} = 1.99$)

(b) Design closest to the utopia point ($\hat{m} = 0.71$, $\hat{k} = 632.8$)

(c) Stiffest design ($\hat{m} = 1.77$, $\hat{k} = 1,141$)

Figure 4.52 Several 3-D cantilevered frame topologies of interest generated by SPIDRS using a hybrid optimization approach, where members removed from the structure are represented by thin lines. At low normalized mass values, topologies are reduced to 2-D structures, while topologies with higher normalized mass values feature more 3-D structural features.

Given the demonstrated effect of symmetry boundary conditions on resulting designs generated using a hybrid topology/sizing optimization approach, an additional 3-D cantilevered frame case study is considered as illustrated in Figure 4.53. The 3-D structural design domain is defined to have a length of 120 mm, a height of 40 mm, and a width of 30 mm. The initial state of the graph is defined as

$$F = \{[1,2,3,4],[6,5,8,7],[1,4,8,5],[3,2,9,6,7],[1,5,6,9,2],[4,3,7,8]\}.$$

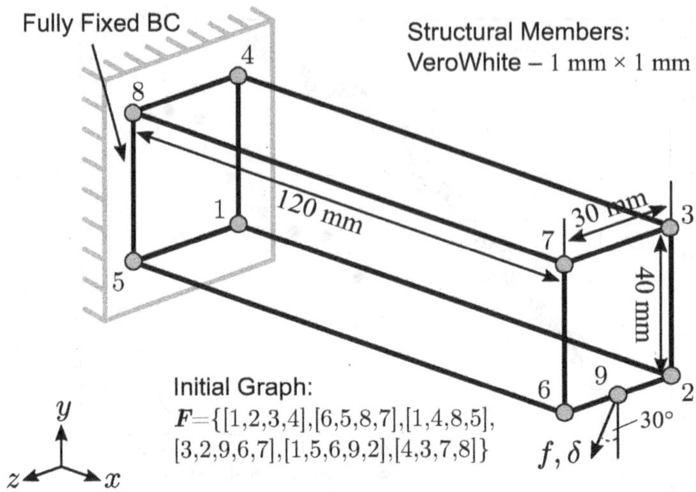

Figure 4.53 Initial graph and boundary conditions associated with the 3-D cantilevered frame subjected to off-axis loading.

A fully fixed boundary condition (i.e., $u_x = u_y = u_z = 0$) is placed on all nodes in the plane associated with the face $[1,4,8,5]$. To provide an indication of the stiffness of a given structural topology, a concentrated force of $f = 0.05$ N is applied to node 9 in a direction 30° off the y-axis (cf. Figure 4.53), while the displacement δ *in the same direction* is measured. The multiobjective hybrid topology/sizing optimization problem is considered in a manner consistent with the previous problem.

The resulting Pareto frontier after considering a population size of 100 for 1,000 generations is shown in Figure 4.54, along with several topologies of interest. It is interesting to note that the lightest design ($\hat{m} = 0.30$, $\hat{k} = 186.0$) is a simple two-member triangle-like structure that is oriented at an angle off of the y-axis almost equivalent with the applied force ($\approx 29.8°$). As normalized mass increases, the sizing optimization scheme tends to keep structural members located on the plane associated with the face $[1,2,3,4]$, perhaps to provide more stiffness in the y-direction given that the majority of the applied force acts in this direction, with additional members extending from nodes 5 and 8 to the edge between nodes 2 and 3, providing additional stiffness in the z-direction. The design with the largest normalized mass and highest stiffness retains all of its structural members after sizing optimization with the exception of members in the same plane as the fully-fixed boundary condition, which is consistent with the results of the 3-D sizing optimized truss previously discussed. Thus, without the ability to rely on a symmetry boundary condition to easily produce critical load paths, the hybrid optimization framework is still capable of producing elegant solutions to 3-D structural design problems using SPIDRS-generated topologies.

Case Study 4.7: 3-D Cantilevered Frame with Torsional Loading

The second 3-D topology optimization case study considers a cantilevered frame subjected to torsional loading as shown in Figure 4.55. The initial graph is defined

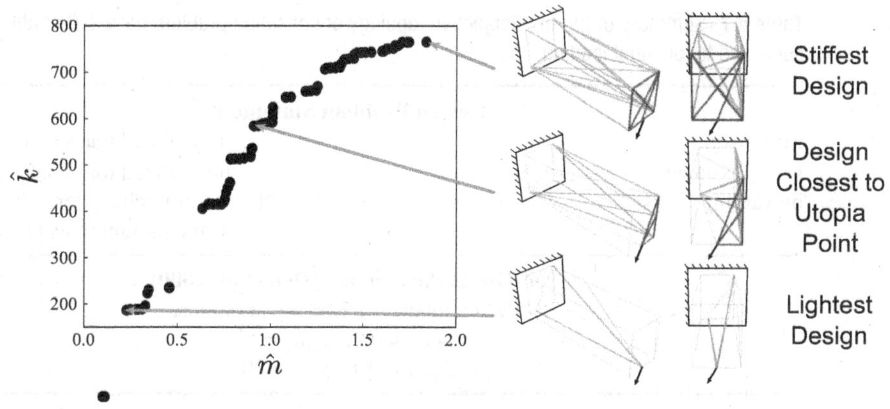

Figure 4.54 Pareto frontier and several topologies of interest after considering the case study illustrated in Figure 4.53 for 1,000 generations with a population size of 100 using the proposed hybrid optimization framework. Structural members removed from each design by the sizing optimization scheme are denoted by thin lines.

to remain consistent with that of the transverse load case, with the nodes in this graph given spatial coordinates such that the frame has a length of 120 mm in the x-direction, a height of 40 mm in the y-direction, and a width of 40 mm in the z-direction; note that unlike the transverse load case, this problem does not consider geometric symmetry. A fully fixed boundary condition is specified for all nodes in the plane associated with the face [1, 4, 8, 5], and a moment of $M_x = 5\,\text{N}\cdot\text{mm}$ is applied to all nodes in the plane associated with the face [3, 2, 6, 7]. Details of the multiobjective optimization problem are given in Table 4.7; the objectives of this problem are the minimization of normalized mass $\hat{m} = m/m_{null}$ (where $m_{null} = 9.4 \times 10^{-4}$ kg) and maximization of normalized torsional stiffness $\hat{k}_{tors} = k_{tors}/k_{tors,null}$ (where $k_{tors} = M_x/\theta_x$ and $k_{tors,null} = 5.29\,\text{N} \cdot \text{mm}$). The structural analysis approach and optimization parameters remain consistent with those considered in the transverse load case.

The Pareto frontier for the multiobjective 3-D topology optimization problem illustrated in Figure 4.55 is shown in Figure 4.56, along with several topologies of interest. As in the transverse loading case, the lightest and least stiff design is simply the topology defined by the initial graph. The stiffest and heaviest design features a series of triangular structures that alternate in tension and compression to oppose the rotation generated by the applied moment. Perhaps the most interesting topology generated by the SPIDRS algorithm is the design closest to the utopia point, which features a similar series of triangular structures. However, this topology demonstrates symmetry across the plane illustrated in Figure 4.57. This is a fascinating result when considering that no symmetry boundary conditions were specified in the design problem, and further demonstrates the ability of the SPIDRS algorithm to effectively explore a 3-D structural design domain.

Table 4.7 Summary of the multiobjective topology optimization problem for a 3-D cantilevered frame subjected to torsional loading.

Design Problem Statement	
Minimize:	normalized mass $\hat{m}$
and Maximize:	normalized torsional stiffness $\hat{k}_{tors}$
by varying:	2 axiom characters,
	4 rule assignments (18 genes each)
NSGA-II Parameters (Deb et al. 2000)	
100 members for 1,000 generations,	
$P_{cross} = 0.9$, $\eta_{cross} = 50$,	
$P_{mut} = 0.1$, $\eta_{mut} = 20$	

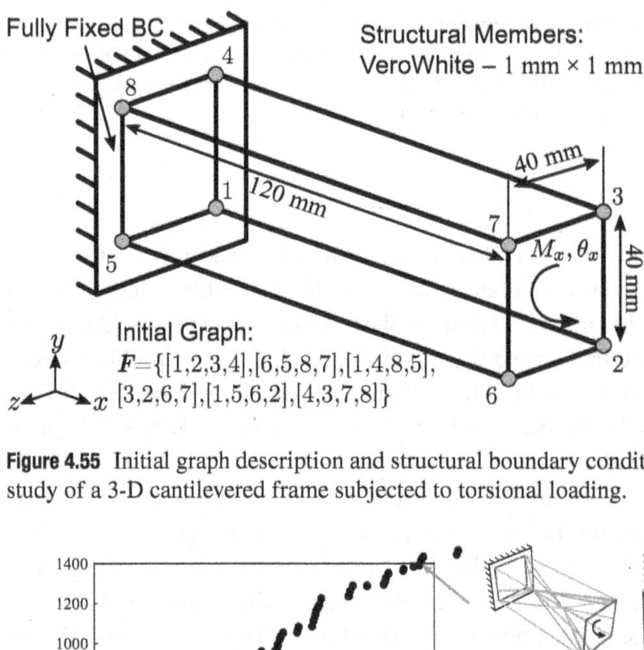

Figure 4.55 Initial graph description and structural boundary conditions associated with the case study of a 3-D cantilevered frame subjected to torsional loading.

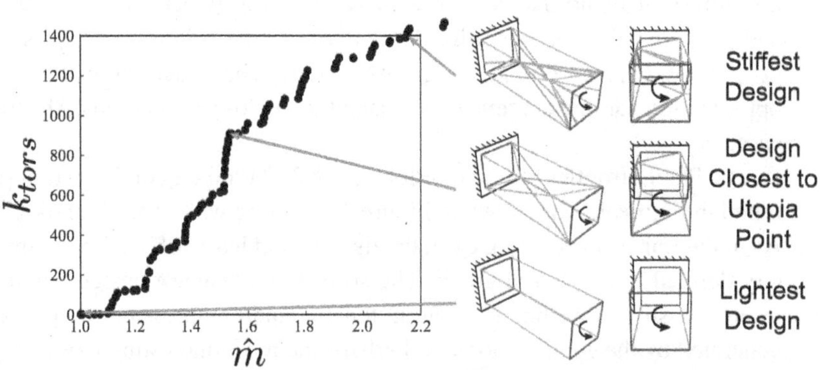

Figure 4.56 Pareto frontiers associated with the 3-D cantilevered frame subjected to torsional loading case study along with several SPIDRS-generated topologies of interest. Structural members at each end are shaded to better illustrate the topologies.

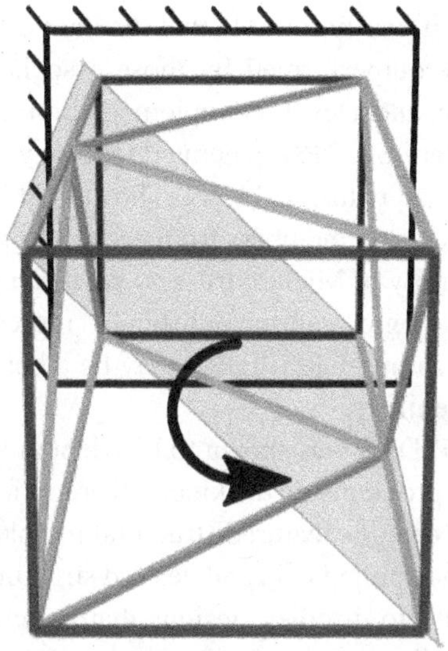

Figure 4.57 Despite the lack of symmetry boundary conditions defined in the problem (cf. Figure 4.55), the design closest to the utopia point generated by the SPIDRS algorithm demonstrates symmetry about the highlighted plane.

4.6 Conclusions

In this chapter, we discussed the development and implementation of a parametric L-system interpretation algorithm known as SPIDRS. Using the principle of connectivity found in graph theory, SPIDRS represents the topology of a structure using the connectivity of a planar graph. After interpreting a character from a parametric L-system string, which provides for greater design freedom than the traditional L-system utilized in Chapter 3, SPIDRS induces topological modifications by manipulating the connectivity of the graph based on the operation associated with that character. These operations were developed to ensure that movement, structure creation, and material assignments remain deliberate and natural.

Several case studies and extensions were then explored to demonstrate the effectiveness and potential of the SPIDRS algorithm. Based on these discussions, the following conclusions can be drawn:

- In both canonical problems considered, topologies generated by the SPIDRS algorithm outperformed those associated with the turtle graphics interpretation approach detailed in Chapter 3. Whereas the creation and intersection of structural members/load paths occur in a randomized fashion using turtle graphics, the subdividing behavior of the SPIDRS algorithm allows for these to be developed in an efficient and deliberate manner. This is especially noticeable in the cantilevered frame case study, where SPIDRS-generated designs along the Pareto frontier quickly develop efficient load paths, build critical load-carrying members off of preexisting members, and then adjust the positions of these members to achieve varying performances.

- In the cantilevered frame case study, it was noticeable that the Pareto frontier generated by SPIDRS was significantly outperformed by those associated with the Michell truss, a SIMP implementation, and a level set implementation. This is due to the presence of the initial graph in every SPIDRS-generated topology regardless of the existence of more advantageous load paths, and it was shown that by removing the mass and stiffness contributions of those members from each design the frontier compares much more favorably to family of Michell trusses. We also observed that both the SIMP and level set methodologies can have difficulty representing high aspect ratio structural members, indicating that SPIDRS may be a better approach when desiring solutions with such members.
- In addition to canonical problems, SPIDRS was shown to be adept at finding solutions to nonintuitive structural design problems. In the kinematic rectifier case study, we observed that SPIDRS-generated designs featured structural members with different functionalities depending on the applied load and desired structural response. SPIDRS also demonstrated the ability to discover designs that match a specified stiffness response across an entire loading envelope rather than at a single specified point.
- Leveraging its foundation in graph theory, the SPIDRS algorithm enables the consideration of 3-D structural design domains without an increase in the number of design variables required – an impressive feat when compared to the increase in design space dimensionality that occurs when considering appropriate discretizations of 3-D domains using density-based methods. The Pareto frontiers generated by SPIDRS for a 3-D cantilevered frame subjected to transverse loading compares favorably against those generated by a SIMP implementation, which once again had difficulties representing high aspect ratio structural members.
- Finally, integrating a sizing optimization scheme within the SPIDRS topology optimization framework was shown to further improve the structural performance of generated designs. This integration was complicated by the use of a gradient-based optimizer, which requires a single-scalar-objective value; thus, multiobjective problems had to be reduced to a series of constrained single-objective problems using the ε-constraint approach. Using this hybrid optimization framework, the SPIDRS-generated Pareto frontier for a 2-D cantilevered frame was shown to outperform topologies obtained using a SIMP formulation and to converge to the frontier associated with the Michell truss. Additionally, the addition of the geometry optimization step discussed in Section 2.3.3 to the hybrid optimization framework generated an optimal design that outperforms each topology optimization approach introduced in Chapter 2.

As a means of summarizing the most promising methods considered in this book through the end of this chapter, Figure 5.44 is provided. It repeats most of the key results of Chapter 2 associated with solving the example topology optimization problem of Figure 2.3. The solution obtained by the SPIDRS approach, considering both geometric and sizing optimizations, is also included, and its effectiveness is clear. In Chapter 5, we will explore another highly capable approach for generating likewise effective structures and networks while utilizing L-systems in a very different way.

5 Arrangement L-system Interpretation

Chapter 4 introduced an interpretation algorithm for parametric L-systems based on the idea of subdividing faces on a graph. Such an approach is similar to a different class of L-systems known as the map L-system, which was first proposed by Lindenmayer and collaborators (Nakamura, Lindenmayer & Aizawa 1986) as a model for the cellular division processes in organisms. Like SPIDRS, the map L-system introduces topological modifications via subdivisions of a domain, though the map L-system more closely follows the evolutionary developmental principles of biology and encodes these processes directly into the rewriting of the L-system rather than relying on an intermediate interpretation step.

This chapter focuses on arrangement L-systems (ALS), which generalizes the map L-system to include multiply connected regions, cell merging, and curved edges to be created or deleted in the developmental stages of the layout. The chapter starts with a detailed review of the original map L-system, where the effects of its various parameters on the developed layout are examined. This review is followed by the presentation of several extensions of the original map L-system leading to the ALS, and the introduction of directed cellular dynamics and the cellular division via two-point topological derivative (TPTD) for accelerated convergence of the method. The chapter closes with several case studies, illustrating the application of the method to engineering problems.

5.1 Map L-system

Formally, a map is defined as a finite set of regions, with each region bounded by a sequence of edges that intersect at vertices. Every edge is part of the boundary of a region, and the regions are simply connected as illustrated in Figure 5.1. From a biological perspective, these maps are analogous to cellular layers, where the regions represent the cells and the edges their walls. There are numerous approaches to developing maps using an L-system as discussed in Chapters 3 and 4 and in books such as Prusinkiewicz and Lindenmayer (2004); herein, we discuss the binary propagating map 0L-systems with markers (mBPM0L-systems) proposed by Nakamura, Lindenmayer & Aizawa (1986) as these systems are more powerful than similar context-free map L-systems (Prusinkiewicz & Lindenmayer 2004). Such systems are called *binary* because the cells divide into two, and they are called *propagating* because cells cannot

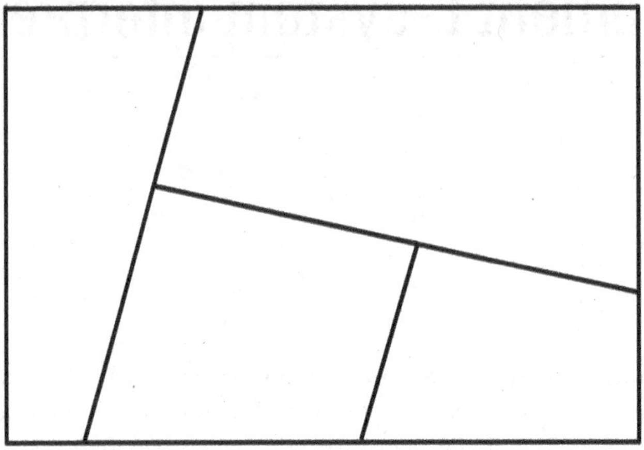

Figure 5.1 A map is a class of planar graphs with cycles defined by a finite set of regions. Each region is bounded by a finite, closed sequence of edges that meet at vertices.

fuse or vanish. The designation 0L-systems stems from context-free parallel rewriting systems that do not allow for the interaction among the cells. Finally, markers define points along the edges where cells can divide. The inspiration for markers. is the attachment sites in cells, where walls may divide during mitosis. For simplicity, hereafter, we refer to mBPMOL-systems as MLS.

Like the L-systems defined in Chapters 3 and 4, MLS consists of an alphabet $\boldsymbol{\Sigma}$, an axiom ω^0, and a finite set of rewriting rules $\boldsymbol{P}$. However, because the interpretation of words and rewriting processes differ markedly from the methods previously discussed, these now-familiar components are briefly reviewed in the context of MLS.

- An *alphabet* is a finite, non-empty set, $\boldsymbol{\Sigma}$, whose elements are called characters or letters. Characters can be one of three types: nonterminal, terminal, or a special symbol. A nonterminal character is so called because it can be rewritten (i.e., a variable character as defined in Section 3.1), while a terminal character and a special symbol are constants in the rewriting process (i.e., constant characters as defined in Section 3.1). For example,

$$\boldsymbol{\Sigma} = \{\mathtt{A}, \mathtt{B}, \mathtt{C}, \mathtt{D}, \mathtt{E}, \mathtt{F}, \mathtt{x},\ \mathtt{[}, \mathtt{]}, \mathtt{+}, \mathtt{-},\ \rightarrow,\ \leftarrow\}$$

  is an acceptable alphabet; herein, we adopt the convention that uppercase Latin letters (e.g., A, B) are nonterminal characters, lowercase Latin letters (e.g., x) are terminal characters, and the remaining characters (e.g., [,]) are special symbols. These characters in the alphabet are used in the definition of the axiom and production rules and to label all edges in a map.
- As in English, *words* are groups of letters, and an *axiom* ω^0 is a non-empty word. For example,

$$\omega^0 = \overrightarrow{\mathtt{A}}\overleftarrow{\mathtt{B}}\overleftarrow{\mathtt{A}}\overrightarrow{\mathtt{B}}$$

is an axiom. Each terminal/nonterminal character of the axiom has an arrow set over it, and, together, the character and the arrow, is called a *token*. Thus, $\overrightarrow{\mathtt{A}}$ is the first token in the axiom above, $\overleftarrow{\mathtt{B}}$ is the second token, etc.

- As the name suggests, the set of *production rules* $\boldsymbol{P}$ consists of rewriting rules to be applied to nonterminal characters of the alphabet. Each rule of $\boldsymbol{P}$ is of the form $X \rightarrow \alpha$, where X, called the *predecessor*, denotes a nonterminal character in $\boldsymbol{\Sigma}$ with a right arrow set on top of it, and α, called the *successor*, is a non-empty word. As an example, consider two different production rules for the nonterminal character A:

$$\overrightarrow{\mathtt{A}} \rightarrow \overleftarrow{\mathtt{X}}, \tag{5.1}$$

$$\overrightarrow{\mathtt{A}} \rightarrow \overleftarrow{\mathtt{B}}\,\overleftarrow{\mathtt{C}}[+\overleftarrow{\mathtt{A}}]\overrightarrow{\mathtt{B}}. \tag{5.2}$$

  These production rules demonstrate the two possible types of tokens in a successor: bracketed tokens and non-bracketed tokens. A bracketed token in a successor is called a *marker*, while a non-bracketed token is called a *non-marker*. In Equation 5.2, the tokens $\overleftarrow{\mathtt{B}}$, $\overleftarrow{\mathtt{C}}$, and $\overrightarrow{\mathtt{B}}$ are non-marker tokens, while the token $[+\overleftarrow{\mathtt{A}}]$ is a marker token. For reasons that will be discussed later, all successors must have at least one non-marker. These two types of tokens perform different roles in the cellular division process, as will be discussed in Section 5.1.2.

Using the alphabet, axiom, and production rules, the layout is developed in a series of developmental cycles, which is analogous to the process of recursion discussed in Sections 3.1 and 4.2. However, this requires the generation of a labeled and oriented initial map, which is discussed in Section 5.1.1.

5.1.1 Initial Map Labeling and Reorienting

Consider a starting unlabeled map as shown in Figure 5.2(a) and the axiom $\omega^0 = \overrightarrow{\mathtt{A}}\,\overleftarrow{\mathtt{B}}\,\overleftarrow{\mathtt{A}}\,\overrightarrow{\mathtt{B}}$. The starting map, M_0, must be provided with an enumeration and initial orientation of its edges; in Figure 5.2(a), we proceed counterclockwise starting from the bottom. However, any other enumeration or orientation of the edges of M_0 would be valid. This enumeration is used to pair the edges of M_0 with the tokens of the axiom ω^0. To begin, the first token of the axiom, $\overrightarrow{\mathtt{A}}$, is paired with the first edge, numbered "0," of M_0, the second token of the axiom, $\overleftarrow{\mathtt{B}}$, is paired with the second edge, numbered "1," of M_0, and so on. The letter in the token is then used to label the corresponding edge such that edges 0, 1, 2, and 3 are labeled A, B, A, and B, respectively. Finally, each labeled edge in M_0 is assigned an orientation using the following convention: the $\rightarrow$ means the original orientation of the edge is preserved, while the $\leftarrow$ means the original orientation of the edge is reversed. Thus, the arrow in the first token of the axiom, $\overrightarrow{\mathtt{A}}$, is to the right, and original orientation of edge 0 is preserved; the arrow on top of the second token $\overleftarrow{\mathtt{B}}$ is to the left and therefore reverses the original orientation of edge 1; the third token $\overleftarrow{\mathtt{A}}$ reverses the original orientation of edge 2; and the fourth token $\overrightarrow{\mathtt{B}}$

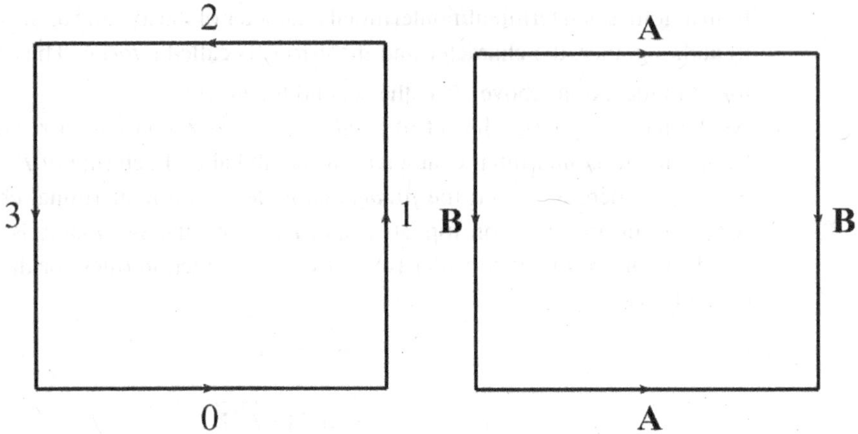

(a) Map provided with a starting edge enumeration and orientation

(b) Initial map after edge labeling and reorientation using the axiom

Figure 5.2 Geometrical representation of the initial map. The orientation of each edge is indicated by the corresponding arrow.

keeps the orientation of edge 3. The result of the labeling and reorientation of the edges is the initialized map shown in Figure 5.2(b).

With the labeled and oriented initial map, a cellular development algorithm is applied recursively to the initial map until stopping criteria are met. The developmental algorithm itself is composed of three stages: edge rewriting, cellular division, and cellular dynamics, each of which are discussed in detail below.

5.1.2 Edge Rewriting Stage

At the edge rewriting stage, all edges in the map are traversed and a production rule is applied to each based on their corresponding label. Specifically, edges labeled `A` are rewritten to reflect the production rule associated with the predecessor `A`, edges labeled `B` are rewritten to reflect the production rule associated with the predecessor `B`, and so on. Rewriting of an edge entails edge division, labeling, and orientation of the new sub-edges, and marker placement. To begin, the edge being rewritten, or *predecessor edge*, is divided into N_{NM} equal segments, where N_{NM} is the number of non-marker tokens in the rule. Next, each of the N_{NM} new segments is paired with the corresponding N_{NM} non-markers in the rewriting rule, and each segment is labeled and oriented according to the rule associated with the non-marker token; the labeling and orientation of the new segments is performed in an analogous fashion to the labeling and orientation of the initial map. Finally, the sets of markers between the non-markers are paired with the corresponding $N_{NM} + 1$ joints of the newly rewritten edge, with the placement of each based on the following interpretation of the characters within the marker. The first symbol sets the placement of the marker with respect to the predecessor edge (+ indicates that the marker is placed in the face to the right of the predecessor,

Table 5.1 Pairings for the production rule $\overrightarrow{\mathtt{A}} \rightarrow \overleftarrow{\mathtt{B}}\,\overleftarrow{\mathtt{C}}[+\overleftarrow{\mathtt{A}}]\overrightarrow{\mathtt{B}}$ applied to the predecessor edge shown in Figure 5.3(a).

Segment Number	Token	Joint Number	Markers
1	$\overleftarrow{\mathtt{B}}$	1	$\emptyset$
2	$\overleftarrow{\mathtt{C}}$	2	$\emptyset$
3	$\overrightarrow{\mathtt{B}}$	3	$\{[+\overleftarrow{\mathtt{A}}]\}$
–	–	4	$\emptyset$

while - indicates that the marker is placed in the face to the left of the predecessor edge; note that left and right directions are defined with respect to the orientation of the predecessor edge). The second token labels the marker and uses the arrow to set its orientation (a left arrow indicates that the marker should point toward the joint, while a right arrow indicates the marker should point away from the joint).

As an example, consider the rewriting of the predecessor edge in Figure 5.3(a) using the production rules in Equations 5.1–5.2. With the orientation of the predecessor edge, we define the left vertex as the source (i.e., starting vertex) and the right vertex as the target (i.e., ending vertex). The production rule in Equation 5.1 has only one non-marker token, $\overleftarrow{\mathtt{x}}$. As such, no edge division is performed, and the predecessor edge is simply relabeled x and its orientation reversed since the arrow at the top of the token is opposite to the direction of the predecessor, resulting in the new edge shown in Figure 5.3(b). Conversely, the production rule in Equation 5.2 possesses multiple non-markers and a marker. As $N_{NM} = 3$, the predecessor edge is partitioned into three equal segments, each segment paired with a corresponding non-marker (the first segment is paired with the first token, the second segment with the second token, and so on). Thus, according to Equation 5.2, the first segment from the source is labeled B and its orientation reversed, the second segment is labeled C and its orientation reversed, and the third segment is labeled B and its orientation is left unchanged. Finally, we consider the placement of the single marker $[+\overleftarrow{\mathtt{A}}]$. As this marker is located between the second and third non-markers, it will correspond to the joint between the second and third segments on the newly rewritten edge. The + symbol indicates that the marker is placed to the right of the predecessor, the letter A denotes the marker label, and the left arrow signifies that the marker points toward the joint. The vertex where this marker is placed is called the *support vertex*. The resulting edge is shown in Figure 5.3(c); herein, markers are represented by arrows that reference their orientation as being either toward or away from the predecessor edge. In summary, the full rewriting process consists of two pairings: (i) non-markers with newly created segments and (ii) markers and the joints between these segments. For illustration purposes, these pairings for the production rule in Equation 5.2 are shown in Table 5.1. Note that joints without corresponding markers are assigned an empty set $\emptyset$ for typesetting purposes only.

Equations 5.1–5.2 represent simple examples of production rules. More generally, a rule may contain markers of several labels that are unevenly distributed between non-

Table 5.2 Pairings for the production rule defined in Equation 5.3.

Segment Number	Token	Joint Number	Set of Markers
1	$\overrightarrow{B}$	1	$\{[+\overrightarrow{A}][-\overleftarrow{B}]\}$
2	$\overleftarrow{D}$	2	$\emptyset$
3	$\overrightarrow{C}$	3	$\emptyset$
4	$\overrightarrow{B}$	4	$\{[+\overrightarrow{A}]\}$
–	–	5	$\{[-\overleftarrow{C}][+\overrightarrow{A}][-\overleftarrow{B}]\}$

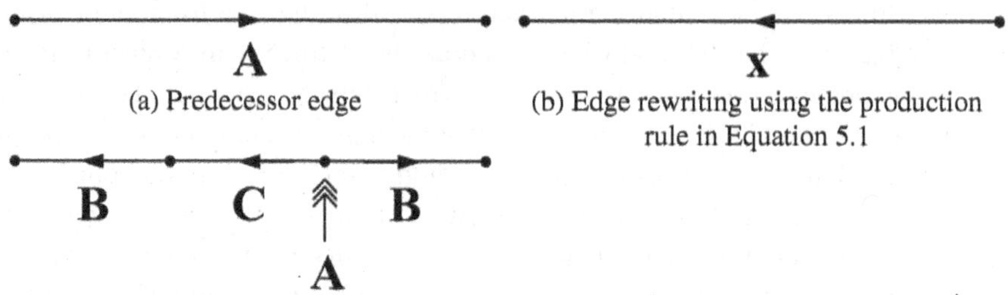

(a) Predecessor edge

(b) Edge rewriting using the production rule in Equation 5.1

(c) Edge rewriting using the production rule in Equation 5.2

Figure 5.3 Examples of the edge rewriting stage. The predecessor edge is labeled A and oriented according to the arrow. The orientation in an edge defines the source and the target vertices: the former is the starting vertex and the latter the end vertex.

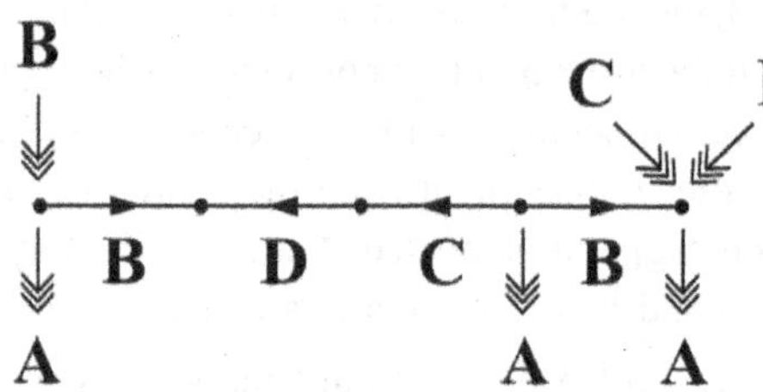

Figure 5.4 Edge rewriting using the production rule in Equation 5.3.

markers. To illustrate the variety of rewriting, consider the following production rule for the same predecessor edge (cf. Figure 5.3(a)):

$$\overrightarrow{A} \rightarrow [+\overrightarrow{A}][-\overleftarrow{B}]\overrightarrow{B}\overleftarrow{D}\overleftarrow{C}[+\overrightarrow{A}]\overrightarrow{B}[-\overleftarrow{C}][+\overrightarrow{A}][-\overleftarrow{B}]. \quad (5.3)$$

This rule has four non-markers ($N_{NM} = 4$), leading to the pairings listed in Table 5.2. With these pairings, the rewriting of the edge follows the protocol defined above: (i) the edge is partitioned into N_{NM} equal segments, (ii) the segments are labeled and oriented according to the non-marker tokens in the pairings, and (iii) the markers are placed on the joints according to the marker tokens in the pairings. The result is shown in Figure 5.4.

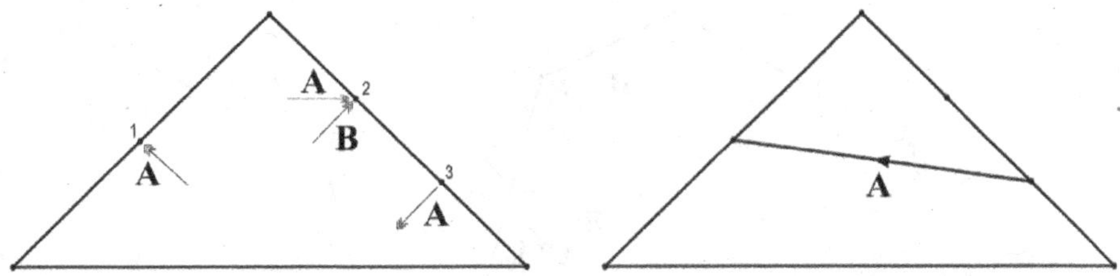

(a) Face at the start of the cellular division stage. Note the markers at vertices 1, 2, and 3

(b) New faces and edge created after the cellular division

Figure 5.5 Example of cellular division.

5.1.3 Cellular Division Stage

After the edge rewriting stage, new edges are produced and markers are placed at their support vertices in the new map. In the cellular division stage, the cells of the new map are traversed and each cell is screened for matching markers. Matches are defined such that the pair of markers has the same label, compatible orientations, and does not intersect existing edges. A compatible orientation is defined such that one of the markers in the pair points away from its support vertex while the second marker points toward its support vertex. If these criteria are met, a new edge in the face is created with the label and orientation determined by the matching markers. Note that other desirable attributes, such as manufacturability constraints on the length of the edges or the angle between them, can be used as criteria for the cellular division and will be demonstrated in the case studies below.

To illustrate the cellular division process, we consider a cell in a map as shown in Figure 5.5(a). The labels and orientations of the edges in this cell are omitted, as these attributes play no role in the cellular division stage. In this example we begin at vertex 1, which supports only one marker labeled A oriented toward the vertex; we call this marker a *trial marker*. For each trial marker, we seek a match from the remaining support vertices in the cell. In Figure 5.5(a), vertex 2 supports two *test markers*: a marker labeled A and oriented toward the vertex and a marker labeled B also oriented toward the vertex. The first test marker at vertex 2 has a compatible label, A, but an incompatible orientation, since both are oriented toward their respective support vertices. The second test marker has an incompatible label, B. Thus, neither test marker meets the criteria for cellular division and the screening proceeds to vertex 3. Vertex 3 supports a single test marker labeled A and oriented away from the vertex, which meets the criteria for labeling and orientation. Therefore, a new directed edge oriented from vertex 3 to vertex 1 and labeled A is created. This process divides the cell in two and adds the new edge to the map, as shown in Figure 5.5(b).

When checking for matching markers in a given cell, the possibility of finding no matches or multiple matches exists. If there are no marker matches, then no cellular division occurs in the cell and the screening moves to the next cell. However, if multiple matches occur in a cell, an edge is created for the first match discovered and the remaining matches are discarded. We consider the latter case in the example shown

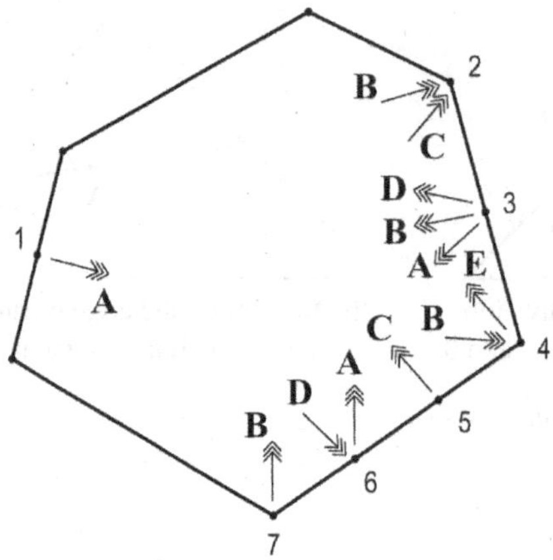

(a) Face at the start of the cellular division stage

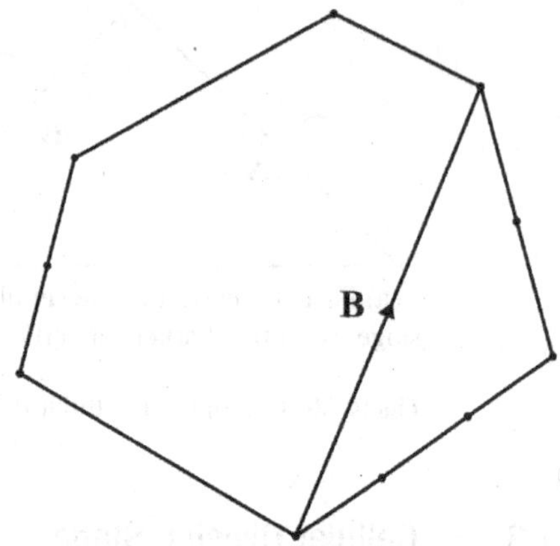

(b) New faces and edge created after the cellular division

Figure 5.6 Example of cellular division.

in Figure 5.6(a), where as before only the labels and number relevant to the process are kept. To determine the first pair of matching markers, an enumeration is needed for the markers supported by each vertex. This enumeration is generated during the edge rewriting stage, where markers and their supporting vertices in a given cell are appended to a list as they are placed. For this example, we consider an enumeration given by the following ordering: 1 : (`A`), 2 : (`B`, `C`), 3 : (`D`, `B`, `A`), 4 : (`E`, `B`), 5 : (`C`), 6 : (`A`, `D`), 7 : (`B`). Starting from vertex 1, there is no match for the trial marker `A`, as the remaining test markers are incompatible in either label or orientation. Moving to vertex 2, the first test marker is labeled `B` and there exist two matching markers in the cell: the first test marker on vertex 3 and the second test marker on vertex 7. The first match on vertex 3 is located on the same edge as vertex 2; as a new edge between these vertices would intersect an existing edge, this match is discarded. This leaves the second match on vertex 7 as the first feasible match discovered, and therefore a new edge oriented from vertex 7 to vertex 2 and labeled `B` is created as shown in Figure 5.6(b). After this cellular division, all markers in the cell are discarded.

This example demonstrates the role that the enumeration of markers plays in the cellular division process. Had the order in vertex 2 been defined as 2 : (`C`, `B`), a new edge oriented from vertex 5 to vertex 2 and labeled `C` would have been created instead.

5.1.4 Cellular Dynamics Stage

The third and final step in a developmental cycle is the cellular dynamics stage. In MLS, this stage mimics the dynamics of cellular growth and movement. Specifically, the shape and size of the cells develop according to the mechanics of an elastic wall under osmotic pressure perpendicular to the cell walls. In this model, vertices are assumed to be connected by linear springs, while the osmotic pressure is assumed to be inversely proportional to the area of the cell and is represented by a pair of forces of equal magnitude directed outward from the cell wall and applied to each wall's vertices. Having defined these forces, the equilibrium condition is applied to each vertex

$$\sum_{w \in W} \mathbf{F}_w = 0, \tag{5.4}$$

where W is the set of walls w incident to a vertex and $\mathbf{F}_w$ are the forces corresponding to a wall $w \in W$.

The forces acting on a vertex are illustrated in Figure 5.7, where a generic vertex i has incident cells 1, 2, and 3 and is connected to vertices j, k and l. For readability, the forces due to the osmotic pressure are placed at the midpoint of each wall, though in practice each force contributes half of its magnitude to each vertex in the edge. For example, $\mathbf{F}_{p_{12}}$ is split into two equal forces, one placed on vertex i and the other on vertex j, both with magnitude $\frac{1}{2}\left|\mathbf{F}_{p_{12}}\right|$ and same direction as $\mathbf{F}_{p_{12}}$. Thus, we see in Figure 5.7 that each wall contributes three forces to each vertex: the linear spring force and two forces caused by the outward pressures from the wall's adjacent cells. As an example, for wall ij, these forces are given as

$$\mathbf{F}_{ij} = \mathbf{F}_{s_{ij}} + \frac{1}{2}\left(\mathbf{F}_{p_{12}} + \mathbf{F}_{p_{21}}\right).$$

The linear spring force $\mathbf{F}_{s_{ij}}$ for edge ij is defined as

$$\mathbf{F}_{s_{ij}} = -k(l_{ij} - l_{ij,0}),$$

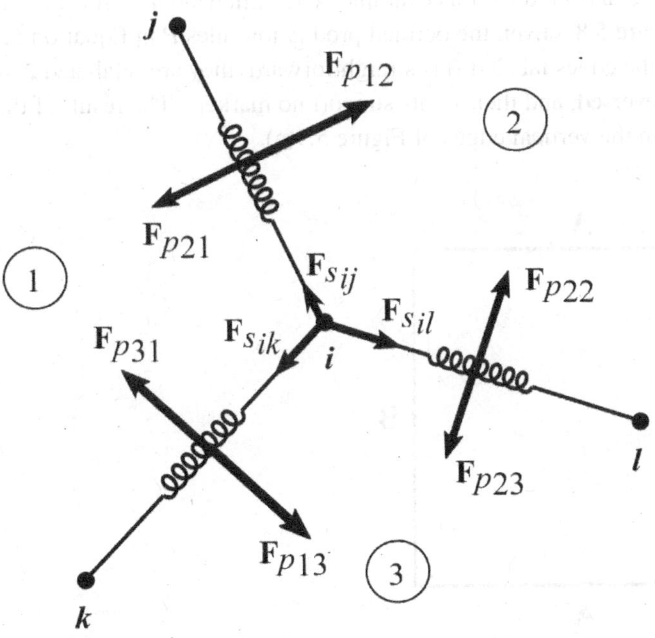

Figure 5.7 Forces acting on the map vertices: osmotic pressure $\mathbf{F}_p$, elastic springs $\mathbf{F}_s$, as forces applied on vertex i.

where $l_{ij,0}$ and l_{ij} are the undeformed and deformed edge lengths of wall ij, respectively. The force due to the osmotic pressure $\mathbf{F}_{p_{12}}$ is computed as

$$\mathbf{F}_{p_{12}} = \frac{P_0 \cdot l_{ij,0}}{A_1},$$

where A_1 is the area of cell 1. Writing Equation 5.4 for every vertex yields a nonlinear system of equations for the vertex coordinates (x_i, y_i), which is solved iteratively. For efficiency, this equilibrium calculation can be used for a limited number of iterations or applied only to the final layout, if not bypassed entirely.

5.1.5 Examples

Having introduced the different components of MLS, this section presents a series of examples to elucidate the stages in the developmental cycles of an MLS.

Example 5.1 Consider a single developmental cycle of the following map L-system:

$$\boldsymbol{\Sigma} = \{\mathtt{A}, \mathtt{B}, \mathtt{x}\}, \tag{5.5}$$

$$\omega = \mathtt{ABAB}, \tag{5.6}$$

$$P = \{\overrightarrow{\mathtt{A}} \rightarrow \overrightarrow{\mathtt{B}}[\mathtt{-}\overrightarrow{\mathtt{A}}][\mathtt{+}\overleftarrow{\mathtt{A}}]\overleftarrow{\mathtt{C}}, \overrightarrow{\mathtt{B}} \rightarrow \overleftarrow{\mathtt{A}}, \overrightarrow{\mathtt{C}} \rightarrow \mathtt{x}\}. \tag{5.7}$$

For simplicity, special symbols from the alphabet are omitted and no cellular dynamics are considered for this example. The initialized map for this example is shown in Figure 5.8. Given the defined production rules P in Equation 5.7, the rewriting of the edges labeled B is straightforward: they are relabeled A with their orientation reversed, and their joints support no markers. The result of this rewriting can be seen in the vertical edges of Figure 5.9(a).

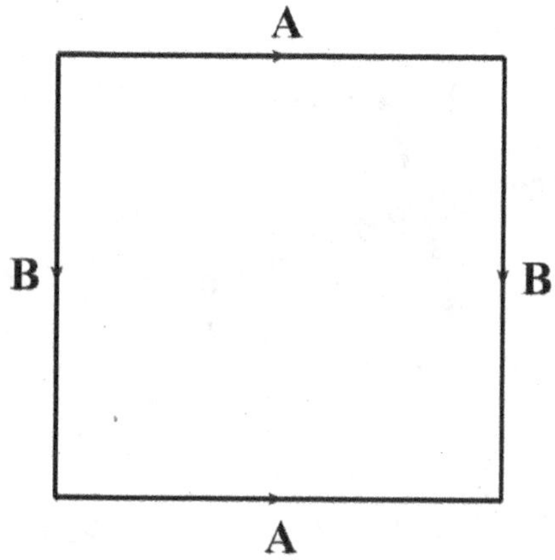

Figure 5.8 Initialized map for Example 5.1. The orientation of each edge is indicated by the corresponding arrow.

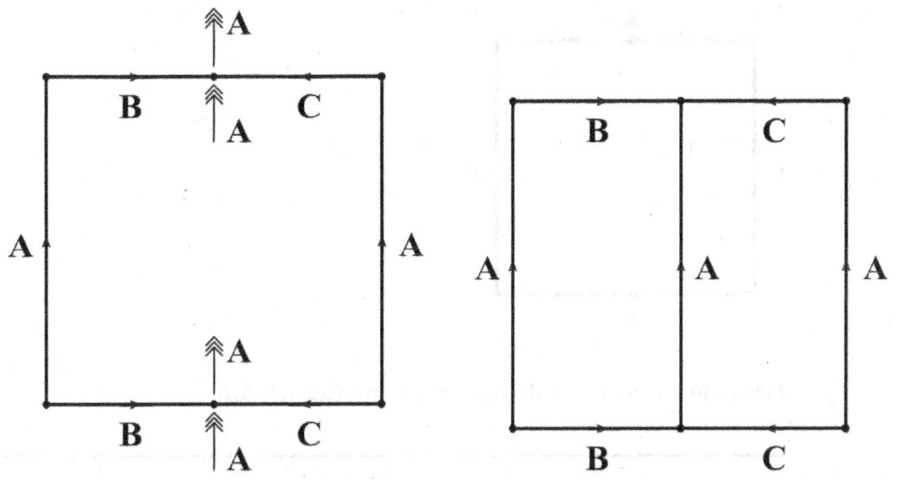

(a) Edge rewriting stage, where each edge is relabeled and subdivided according to the rewriting rules

(b) Cellular division stage, where each cell is traversed and divided according to division criteria and the existence of markers in the cell

Figure 5.9 First developmental cycle of the MLS in Example 5.1.

For edges labeled A, the first step is to divide the edge into two segments $N_{NM} = 2$, where N_{NM} is the number of non-markers in the rule (cf. Equation 5.7). After this division, each segment is labeled according to the non-marker paired with it. Starting from the source vertex, the first segment is labeled B with the same orientation and the second is labeled C with its orientation reversed as defined by the production rules P. The rewriting rule for A has two markers between the two non-markers that will be supported by the second joint from the source vertex; the first and third joints are assigned no markers. To determine the placement of the two markers, we consider first the bottom edge in Figure 5.9(a). The first marker $[-\overrightarrow{\mathtt{A}}]$ is placed to the left of the predecessor edge (i.e., on the interior of the cell due to the orientation of the predecessor) and oriented away from the joint, while the second marker $[+\overleftarrow{\mathtt{A}}]$ is placed to the right of the predecessor (i.e., on the exterior of the cell) and oriented toward the joint. The markers on the top edge are placed in a similar manner, resulting in the map shown in Figure 5.9(a).

At the start of the cellular division stage, the map consists of a single cell that contains two markers. These markers have both compatible labels and orientations, and thus a new edge labeled A and oriented from bottom to top divides the cell. The result is a map with two cells as shown in Figure 5.9(b). As no cellular dynamics step is considered for this example, this concludes the first developmental cycle of the MLS.

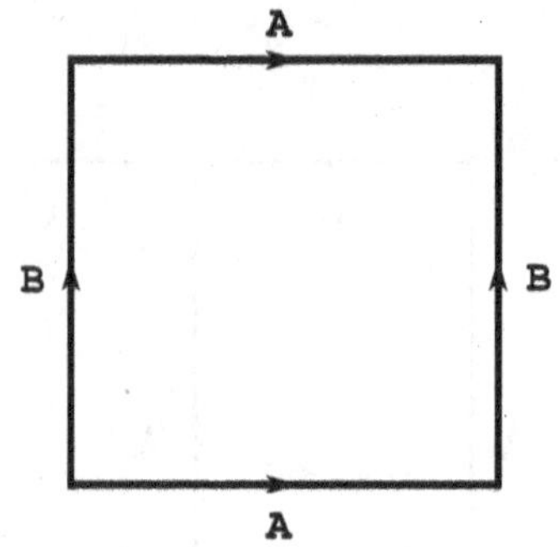

Figure 5.10 Initial map with edge labeling for Example 5.2.

Example 5.2 In this example, we consider an MLS that creates a Cartesian grid at any desired level of refinement; to focus on the edge rewriting stage, we again bypass the cellular dynamics stage for this example. Specifically, we consider the following map L-system:

$$\Sigma = \{\mathtt{A}, \mathtt{B}\},$$
$$\omega = \mathtt{ABAB},$$

and rewriting rules

$$\overrightarrow{\mathtt{A}} \rightarrow \overleftarrow{\mathtt{B}}[+\overleftarrow{\mathtt{A}}][-\overrightarrow{\mathtt{A}}]\overleftarrow{\mathtt{B}}, \tag{5.8}$$
$$\overrightarrow{\mathtt{B}} \rightarrow \overrightarrow{\mathtt{A}}. \tag{5.9}$$

The initial map with the edges labeled according to the axiom is shown in Figure 5.10. As the production rules in this example are similar to the one in Example 5.1, it is left as an exercise for the reader to show that the recurrent application of the edge rewriting and cellular division stages results in the sequence of maps shown in Figure 5.11 for the first four developmental cycles of the MLS.

Example 5.3 This example shows the effect of the cellular dynamics on the layout of the intermediate cycles of the developmental process. The MLS is given as

$$\omega = \mathtt{ABAB}, \tag{5.10}$$
$$\Sigma = \{\mathtt{A}, \mathtt{B}, \mathtt{x}\}, \tag{5.11}$$
$$\omega = \mathtt{ABAB}, \tag{5.12}$$

and rewriting rules

$$\overrightarrow{\mathtt{A}} \rightarrow \overrightarrow{\mathtt{B}}[+\overrightarrow{\mathtt{A}}]\overrightarrow{\mathtt{x}}[-\overleftarrow{\mathtt{A}}]\overrightarrow{\mathtt{B}},$$
$$\overrightarrow{\mathtt{B}} \rightarrow \overleftarrow{\mathtt{A}}.$$

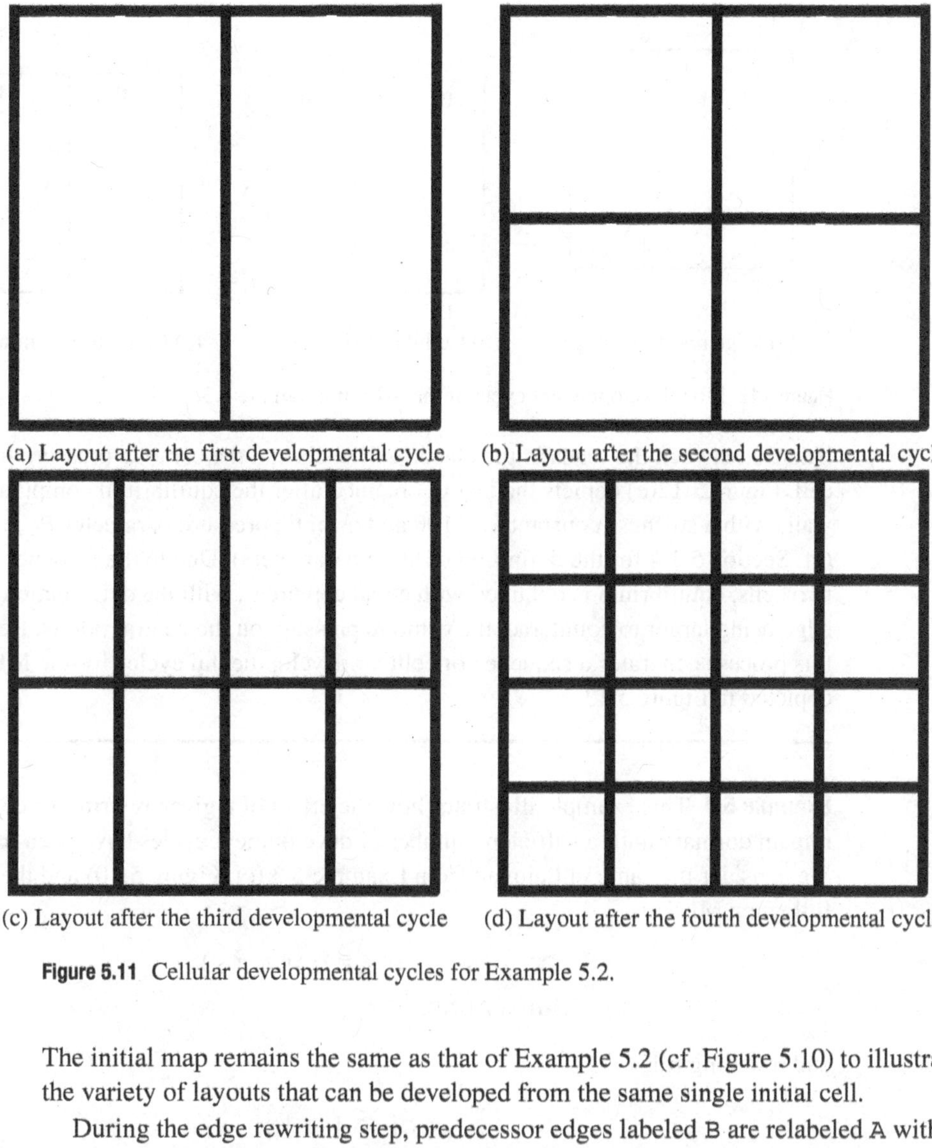

(a) Layout after the first developmental cycle (b) Layout after the second developmental cycle

(c) Layout after the third developmental cycle (d) Layout after the fourth developmental cycle

Figure 5.11 Cellular developmental cycles for Example 5.2.

The initial map remains the same as that of Example 5.2 (cf. Figure 5.10) to illustrate the variety of layouts that can be developed from the same single initial cell.

During the edge rewriting step, predecessor edges labeled `B` are relabeled `A` with no changes to orientation and no subdivision. The edges labeled `A` are partitioned into three new and equal segments; starting from the source vertex, these segments are labeled `B`, `x`, and `B`, respectively, all with an orientation consistent with that of the predecessor. After the first segment, a marker `A` is placed to the right and oriented away the joint (in accordance with the + and right arrow symbols in the marker). A second marker `A` is placed to the left of the predecessor after the second segment and oriented toward the joint (in accordance with the - and left arrow symbols in the marker). After this rewriting process, only two of the markers exist within the cell of the initial map. These markers are compatible in both label and orientation, resulting in a cellular division along the edge labeled `A` connecting these markers (cf. Figure 5.12(b)). The final step is the cellular dynamics stage, where we seek a

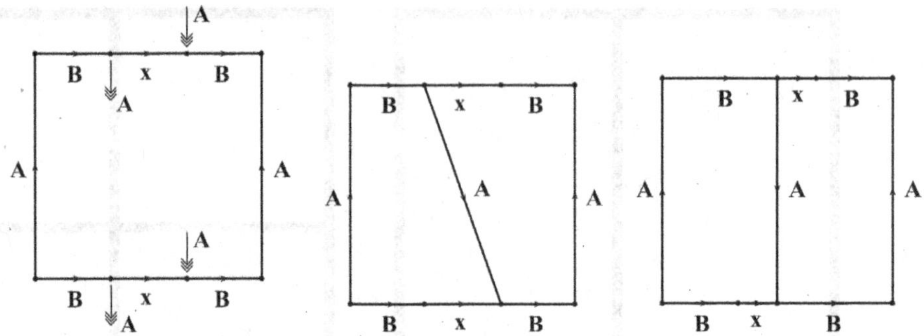

(a) Edge rewriting stage (b) Cellular division stage (c) Cellular dynamics stage

Figure 5.12 First developmental cycle of the MLS in Example 5.3.

balance between the loads in the elastic cell walls and the osmotic pressure in the cell. Figure 5.12(c) depicts the layout obtained after the equilibrium computation for walls with a stiffness constant $k = 100$ and osmotic pressure parameter $P_0 = 1$ (cf. Section 5.1.4 for the definition of these parameters). Due to the presence of only two cells, equilibrium is obtained with equal cell areas, with the deformation of the x edge being larger to counteract the osmotic pressure on the central points. Repeating this process generates a sequence of cellular developmental cycles for the MLS as depicted in Figure 5.13.

Example 5.4 This example illustrates how the effect of a given rewriting rule can remain dormant until a sufficient number of development cycles have been reached. We consider the same initial map from Example 5.3 (cf. Figure 5.10) and the following MLS:

$$\Sigma = \{\mathtt{A}, \mathtt{B}, \mathtt{C}, \mathtt{D}, \mathtt{E}, \mathtt{F}, \mathtt{G}, \mathtt{H}, \mathtt{I}, \mathtt{J}, \mathtt{x}\},$$
$$\omega = \mathtt{ABAB},$$

and rewriting rules

$$\overrightarrow{\mathtt{A}} \rightarrow \overrightarrow{\mathtt{C}}[+\overleftarrow{\mathtt{D}}][-\overrightarrow{\mathtt{D}}]\overrightarrow{\mathtt{C}}, \tag{5.13}$$
$$\overrightarrow{\mathtt{B}} \rightarrow \overrightarrow{\mathtt{D}}, \tag{5.14}$$
$$\overrightarrow{\mathtt{C}} \rightarrow \overrightarrow{\mathtt{E}}, \tag{5.15}$$
$$\overrightarrow{\mathtt{D}} \rightarrow \overrightarrow{\mathtt{F}}[+\overrightarrow{\mathtt{E}}][-\overleftarrow{\mathtt{E}}]\overrightarrow{\mathtt{F}}, \tag{5.16}$$
$$\overrightarrow{\mathtt{E}} \rightarrow \overrightarrow{\mathtt{G}}[+\overleftarrow{\mathtt{H}}][-\overrightarrow{\mathtt{H}}]\overrightarrow{\mathtt{G}}, \tag{5.17}$$
$$\overrightarrow{\mathtt{F}} \rightarrow \overrightarrow{\mathtt{H}}, \tag{5.18}$$
$$\overrightarrow{\mathtt{G}} \rightarrow \overrightarrow{\mathtt{I}}, \tag{5.19}$$
$$\overrightarrow{\mathtt{H}} \rightarrow \overrightarrow{\mathtt{J}}[+\overrightarrow{\mathtt{I}}][-\overleftarrow{\mathtt{I}}]\overrightarrow{\mathtt{J}}, \tag{5.20}$$
$$\overrightarrow{\mathtt{I}} \rightarrow [-\overrightarrow{\mathtt{x}}]\mathtt{x}[+\overleftarrow{\mathtt{x}}], \tag{5.21}$$
$$\overrightarrow{\mathtt{J}} \rightarrow \overrightarrow{\mathtt{x}}. \tag{5.22}$$

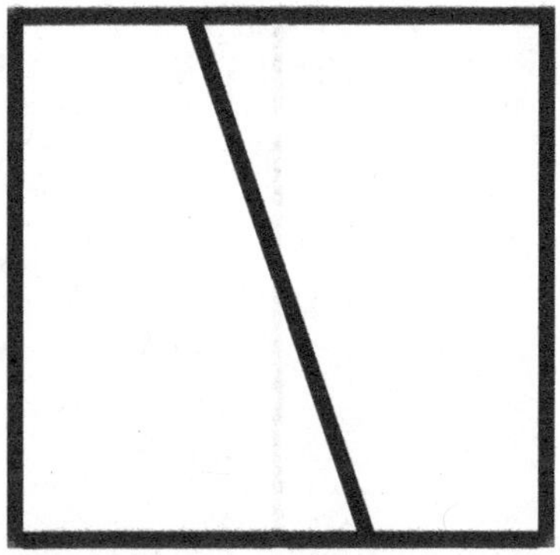

(a) First developmental cycle – cellular division

(b) First developmental cycle – cellular dynamics

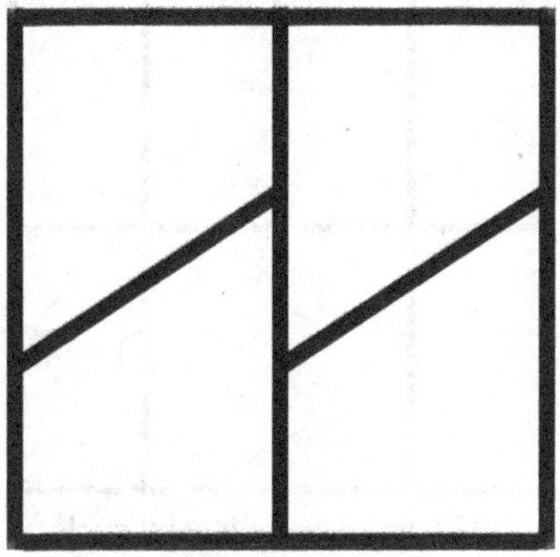

(c) Second developmental cycle – cellular division

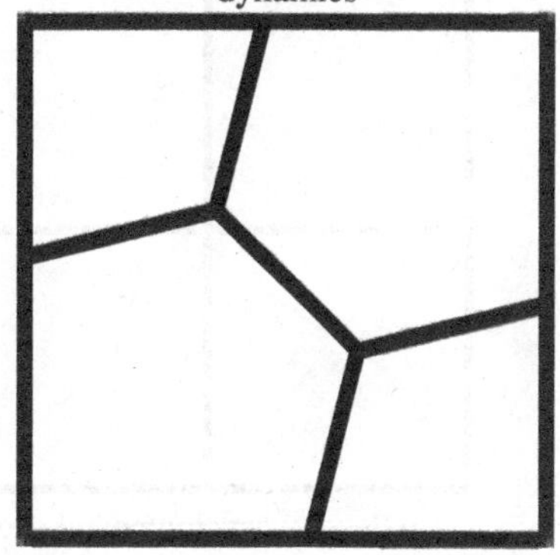

(d) Second developmental cycle – cellular dynamics

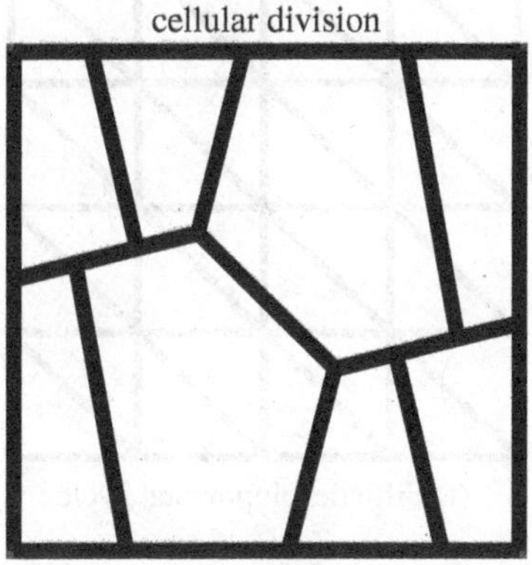

(e) Third developmental cycle – cellular division

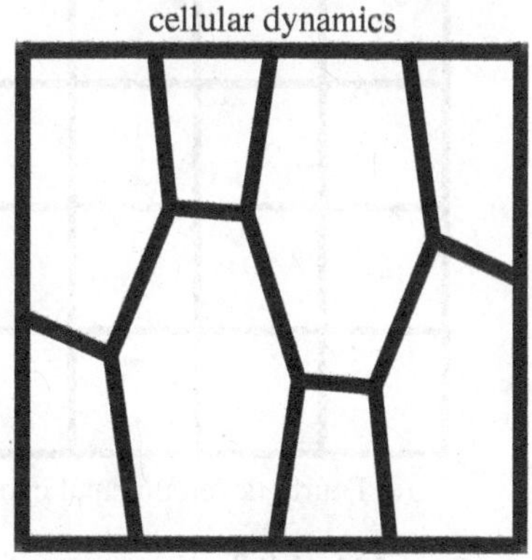

(f) Third developmental cycle – cellular dynamics

Figure 5.13 Cellular developmental cycles for Example 5.3.

Here, the delayed effect is the creation of diagonal edges that will divide square cells at the end of the fifth developmental cycle. To obtain this delayed effect, the rules in Equations 5.13, 5.16, 5.17, and 5.20, when combined with the rules in Equations 5.14, 5.15, 5.18, and 5.19, perform the task of generating a Cartesian grid of equal squares (cf. Equation 5.2) with different labelings. These rules only differ in their labelling, as new labels for the same actions must be created to allow for the delay;

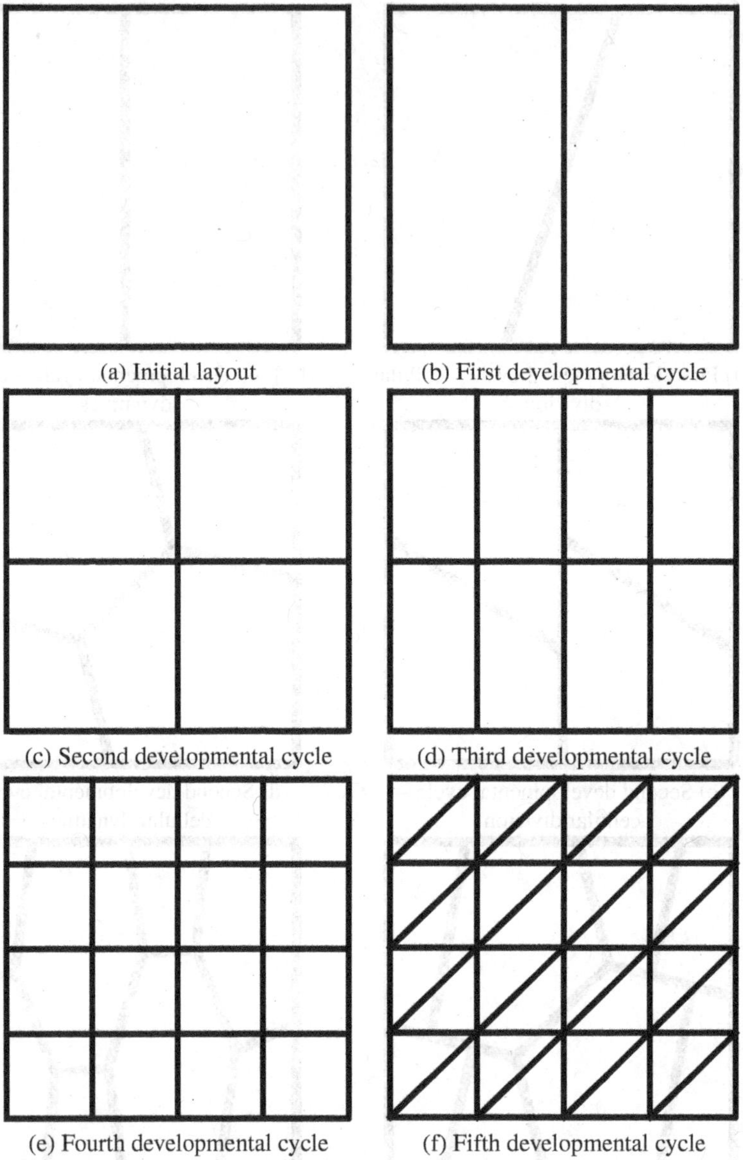

(a) Initial layout (b) First developmental cycle

(c) Second developmental cycle (d) Third developmental cycle

(e) Fourth developmental cycle (f) Fifth developmental cycle

Figure 5.14 Cellular developmental cycles for Example 5.4.

the need for relabeling follows from the fact that a subset of rules that are self-referencing cannot “call” rules outside of their subset. The last two rules, Equations 5.21–5.22, apply the delayed effect: the rule in Equation 5.21 starts and ends with markers and has only one non-marker. Such a rule does not divide the edge, but places markers at the beginning and end of each edge labeled `I`. The result is the cellular development shown in Figure 5.14.

5.2 Variations of the Theme

In Section 5.1, we presented the MLS proposed by Nakamura, Lindenmayer & Aizawa (1986). This section presents variations of the MLS that broaden the scope of possible layouts that can be generated by the MLS while condensing the number and/or length of the rewriting rules needed to attain these layouts.

5.2.1 Mutable/Eraser Rules

Example 5.4 required ten rewriting rules to place a detail in the layout at the end of its fifth developmental cycle. However, for each of the first four cycles the underlying task of the rules were the same: to produce a Cartesian grid. As Example 5.2 illustrated, generating such Cartesian grid can be accomplished with only two rewriting rules (cf. Equations 5.8–5.9). The need for additional rules results from the absence of a "function call" mechanism in the MLS, which would allow for the execution of independent subsets of rewriting instructions. This section introduces the concept of mutable rules that address this issue while also making the set of rewriting rules more concise.

A *mutable rule* is a production rule of the form $\mathtt{A} \xrightarrow{s} (\alpha_0, \alpha_1, \ldots, \alpha_p)$, where the edge $\mathtt{A} \in \Sigma$ is the predecessor, as before, $s\colon \mathbb{N} \to \{0, \ldots, p\}$ is a switch function, and the finite sequence of strings $(\alpha_i)_{i=1,\ldots,p}$ is the novel successor. The switch function takes the value of the cycle and returns an integer, $i \in \{0, \ldots, p\}$. The integer i is used to call the correspondent component of the successor α_i, and each component of the successor α_j, $i \in \{0, \ldots, p\}$ is a rewriting rule as defined for the MLS. When the switch function is a predicate (i.e., a function that evaluates to true or false), the mutable rule is written more concisely as follows:

$$A \to s \mathbin{?} \alpha_t : \alpha_f,$$

where s is a predicate, and α_t and α_f represent the "true" and "false" rules. Parallels can be drawn between the switch function described here and the condition statement within production rules for parametric L-systems introduced in Section 4.2, though the condition statement is a function of parameter(s) rather than the cycle number.

To illustrate the effect of using mutable rules, we revisit Example 5.4. Using the mutable rules, the new MLS can be written concisely as

$$\Sigma = \{\mathtt{A}, \mathtt{B}\},$$
$$\omega = \mathtt{ABAB},$$

and rewriting rules

$$\overrightarrow{\mathtt{A}} \to (n < n_0) \mathbin{?} \overleftarrow{\mathtt{B}}[+\overleftarrow{\mathtt{A}}][-\overrightarrow{\mathtt{A}}]\overleftarrow{\mathtt{B}} : [-\overrightarrow{\mathtt{x}}]\mathtt{x}[+\overleftarrow{\mathtt{x}}], \tag{5.23}$$

$$\mathtt{B} \to (n < n_0) \mathbin{?} \overrightarrow{\mathtt{A}} : \mathtt{x}, \tag{5.24}$$

where $n \in \mathbb{N}$ is the cycle number, and $n_0 \in \mathbb{N}$ is the cycle number when the rules switch (e.g., to reproduce Example 5.4 $n_0 = 5$). For the first $n_0 - 1$ developmental cycles, the

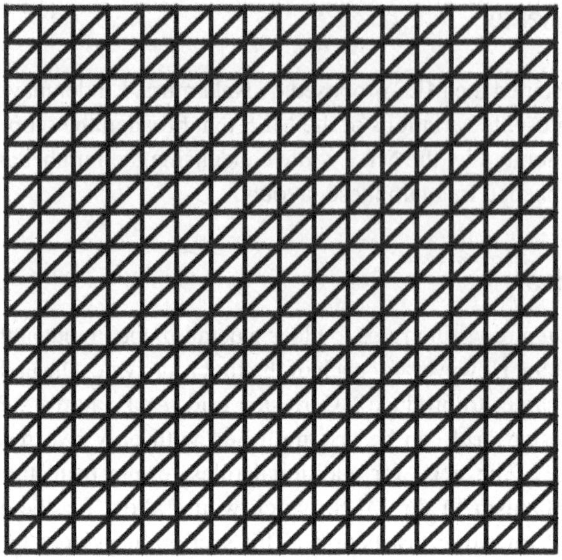

Figure 5.15 Final layout generated using mutable rules in Equations 5.23–5.24 with $n_0 = 9$.

MLS builds a Cartesian grid according to the rules for the "true" predicate (i.e., the first rule before the colon). Then, at the n_0 developmental cycle, both rewriting rules change to the rules for the detail, and the diagonal edges are generated. Thus, mutable rules provide the desired function call mechanism that executes a set of instructions for a number of cycles and then shifts to alternative instructions. With the mutable rules, a longer delay is now easily obtained by modifying the predicate. For instance, $n_0 = 9$ delays the application of the detail until after the eighth cycle. The result is the layout shown in Figure 5.15.

In contrast to living organisms, whose cells fuse during their development (Oren-Suissa & Podbilewicz 2007), the original MLS cannot fuse cells and, by extension, cannot erase edges. To mimic that aspect of cellular development, we introduce an *eraser rule* $\mathtt{e}$ that deletes an edge when the rule is applied. For illustration, we consider the following MLS:

$$\Sigma = \{\mathtt{A}, \mathtt{B}, \mathtt{C}, \mathtt{D}, \mathtt{E}, \mathtt{x}\},$$
$$\omega = \mathtt{ABAB},$$

and rewriting rules

$$\overrightarrow{\mathtt{A}} \rightarrow (n < 5)\ ?\ [-\overleftarrow{\mathtt{C}}]\mathtt{x}[+\overrightarrow{\mathtt{C}}] : \mathtt{x},$$
$$\overrightarrow{\mathtt{B}} \rightarrow (n < 2)\ ?\ \overrightarrow{\mathtt{B}} : [-\overleftarrow{\mathtt{D}}]\mathtt{x}[+\overleftarrow{\mathtt{D}}],$$
$$\overrightarrow{\mathtt{C}} \rightarrow (n < 5)\ ?\ \overleftarrow{\mathtt{D}}[-\overrightarrow{\mathtt{D}}][+\overrightarrow{\mathtt{D}}]\overrightarrow{\mathtt{D}} : \mathtt{x},$$

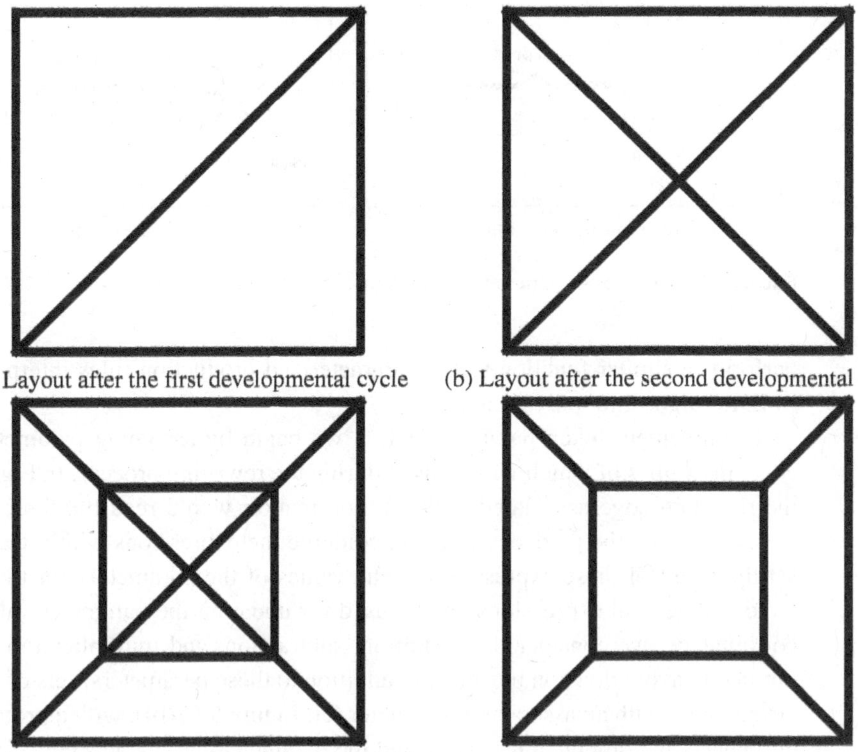

(a) Layout after the first developmental cycle (b) Layout after the second developmental cycle

(c) Layout after the third developmental cycle (d) Layout after the fourth developmental cycle

Figure 5.16 Cellular developmental cycles for a map L-system with the eraser rule.

$$\vec{\mathtt{D}} \rightarrow (n < 5)\ ?\ \overleftarrow{\mathtt{E}}[-\vec{\mathtt{x}}][+\overleftarrow{\mathtt{x}}]\mathtt{x} : \mathtt{x},$$
$$\vec{\mathtt{E}} \rightarrow (n < 5)\ ?\,\mathtt{x} : \mathtt{e}.$$

The resulting developmental cycles are shown in Figure 5.16, which shows the deletion of the central edges in the fourth developmental cycle.

5.2.2 Parameterized Tokens

A map contains two mathematical structures: a combinatorial structure and a geometrical structure. The combinatorial structure is expressed by a graph, which defines the nodes and their connectivity. As discussed in Section 4.1, the vertices or edges of the graph possess no attributes, geometrical or otherwise. In the MLS, geometrical attributes are simply and conveniently assigned to the vertices in the form of Cartesian coordinates in the plane and to edges in the form of line segments. However, in engineering applications, other attributes such as the thickness of an edge or parameters defining an edge may be desirable. To accommodate such attributes, the tokens of a rewriting rule can be parameterized as illustrated in Figure 5.17. Note that this concept

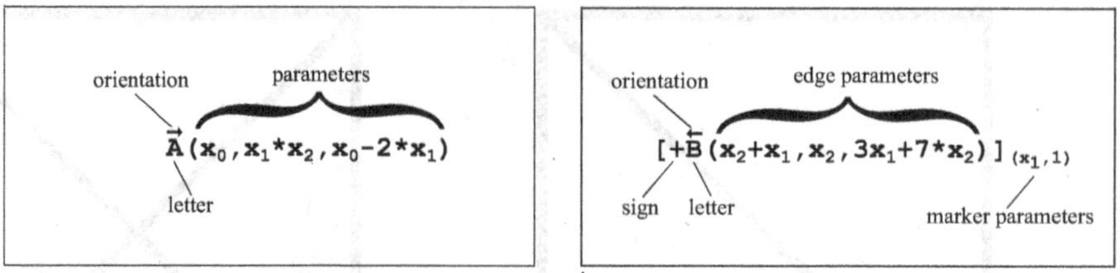

(a) Non-marker token (b) Marker token

Figure 5.17 Examples of parameterized tokens.

performs a similar function to the parameterized production rules interpreted by the SPIDRS algorithm in Section 4.3.3.

To implement token parameterization we begin by assigning parameters to each edge, the values of which will be used during the rewriting process. In Figure 5.17(a), the rewritten edge associated with the non-marker would read the three parameters (x_0, x_1, x_2) from the predecessor edge, compute the expressions within the token, and set the result of these expressions as the values of the parameters for the new edge. Note that general expressions can be used for updating the parameter values, but for convenience, we consider only addition, subtraction, and multiplication (division is omitted to avoid division by zero). In addition to these parameters, sets of markers are also defined with an associated parameter (cf. Figure 5.17(b)), with a protocol used to define how to compute a new edge and the parameters of any new edge formed using the parameters associated with the two markers involved. For concreteness, the parameters of a new edge are taken as the arithmetic average of the parameters in the two markers generating the edge. Note that this process propagates the parameters through the cellular development process, but the specific interpretation of these attributes for the new edge must be defined by the user.

To illustrate the use of the parameterized tokens, we consider the following MLS with one parameter, x_0:

$$\Sigma = \{\mathtt{A}, \mathtt{B}, \mathtt{C}, \mathtt{D}, \mathtt{E}, \mathtt{x}\},$$
$$\omega = \mathrm{A}(0)\mathrm{B}(0)\mathrm{A}(1)\mathrm{B}(1),$$

and rewriting rules

$$\vec{\mathtt{A}}(x_0) \rightarrow \vec{\mathtt{D}}(x_0)[-\overleftarrow{\mathtt{C}}(x_0)][+\vec{\mathtt{C}}(x_0)]\vec{\mathtt{D}}(2 * x_0),$$
$$\vec{\mathtt{B}}(x_0) \rightarrow \vec{\mathtt{C}}(x_0),$$
$$\vec{\mathtt{C}}(x_0) \rightarrow \vec{\mathtt{B}}(x_0)[-\vec{\mathtt{A}}(x_0)][+\overleftarrow{\mathtt{A}}(x_0)]\vec{\mathtt{B}}(2 * x_0),$$
$$\vec{\mathtt{D}}(x_0) \rightarrow \vec{\mathtt{B}}(x_0),$$

where x_0 is defined to represent the thickness of the edge. To illustrate the flexibility in the method, for this example, we add a third criterion for cellular division: a new edge must have a minimum length of 0.1. Such a criterion can arise, for instance, from manufacturing considerations.

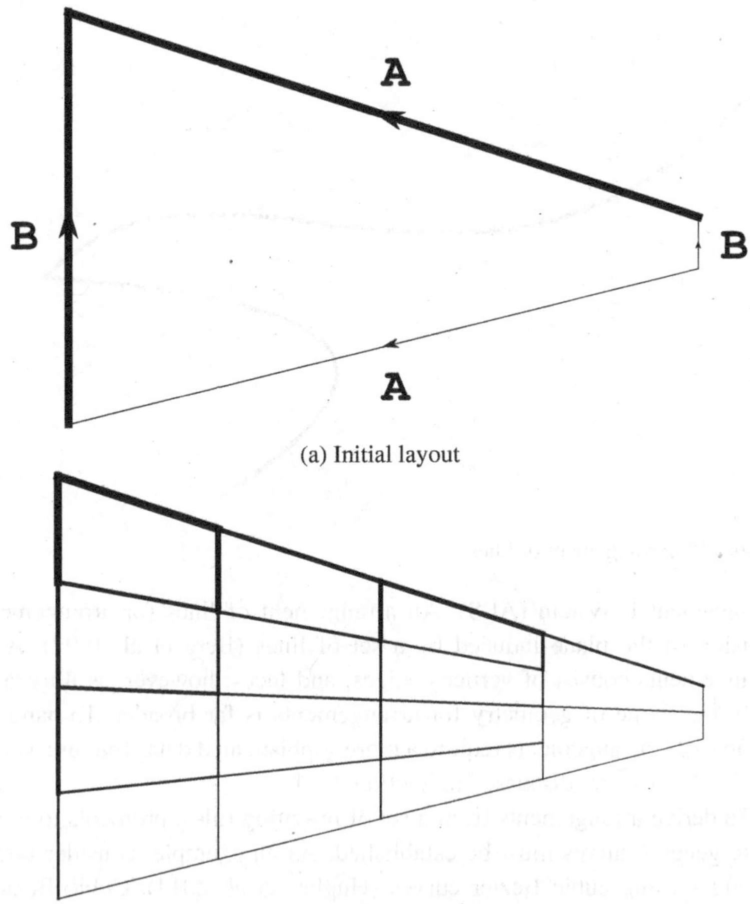

(a) Initial layout

(b) Layout after five developmental cycles

Figure 5.18 Cellular developmental cycles using parameterized tokens.

Figure 5.18 shows the initial map and the map after five developmental cycles. The thickness variations within the resulting developed structure are noticeable and are a useful attribute in actual engineering design. Other desirable attributes (e.g., multiple materials) may be assigned by adding a parameter to the list providing associated interpretations. Note also the effect of the minimum length criterion for the creation of new edges; the two cells at the right of the trapezoid are not subdivided, as they would lead to edges that violate the prescribed constraint at the wing tip.

5.2.3 Arrangement L-system

Map L-systems, such as the mBPM0L-system and extensions introduced above, are edge-control map-rewriting systems (Prusinkiewicz & Lindenmayer 2004). In this section, we introduce a more general edge-controlled rewriting system, which we dub

Figure 5.19 Arrangement of lines.

arrangement L-system (ALS). An arrangement of lines (or arrangement) is a subdivision of the plane induced by a set of lines (Berg et al. 1997). As with maps, arrangements consist of vertices, edges, and faces; however, as illustrated in Figure 5.19, the scope of geometry for arrangements is far broader. To handle the greater complexity, arrangements require a more sophisticated data structure, such as the half-edge data structure discussed in Section 4.3.1.

To derive arrangements from a set of rewriting rules, protocols to model and generate general curves must be established. As an example, consider a representation of edges using cubic Bézier curves (Hughes et al. 2014). Cubic Bézier curves are parametric curves defined by four control points (p_0, p_1, p_2, and p_3) such that

$$B(t) = (X(t), Y(t)) = \sum_{i=0}^{3} p_i \frac{3!}{i!\,(3-i)!} t^i (1-t)^{3-i},$$

where $t \in [0, 1]$. The cubic Bézier curves defining the edges of the initial map are specified in the initial arrangement. New edges, however, require a protocol and additional parameters, which originate from parameterized tokens (cf. Section 5.2.2), to define their associated curves. When two markers are selected for cellular division, each one contributes its marker parameters at the marker support point. For cubic Bézier curves, two marker parameters are required: a scale, ρ, and a rotation angle, θ. With the scale and rotation, the velocity at the marker can be computed by applying the scale and rotation to the vector, $r = p_3 - p_0$. Specifically, given two markers with support vertices at p_0 and p_3, the former being the source and the latter the target vertex of the edge, and parameters ρ_0, θ_0 and ρ_3, θ_3, respectively, we compute the two additional control points p_1 and p_2 in two steps as follows:

1. *Compute the velocity at each end of the curve.* Define $r = p_3 - p_0 = x_3 - x_0 + i(y_3 - y_0)$, and $s_k = \rho_k(\cos\theta_k + i\sin\theta_k) = a_k + ib_k$, $k = 0, 3$. The velocity at p_k can then

be computed as $v_k = s_k r = (x_3 - x_0)a_k - (y_3 - y_0)b_k + i[(x_3 - x_0)b_k + (y_3 - y_0)a_k]$, for $k = 0, 3$.

2. *Use the velocity computed in item 1 to determine the control points p_1 and p_2.* For point p_1, we obtain: $v_0 = 3(p_1 - p_0) \Rightarrow p_1 = p_0 + v_0/3$; and, similarly, $p_2 = p_3 + v_3/3$.

Higher order Bézier curves (as well as other curve families such as conics, spirals, or polylines) can be treated using a similar protocol, where the parameters of the curve are carried as an attribute at the markers and interpreted as appropriate curve parameters when constructing the curve representing the new edge. Finally, in general, in the ALS, any predefined number of nonintersecting curves can be added to a cell, whereas in MLS, only one cellular division is allowed.

Example 5.5 This example shows how arrangements can be generated with the protocol above. Consider the following ALS:

$$\Sigma = \{\mathrm{A, B, C, D, E, F, G, x}\},$$
$$\omega = \mathrm{ABC},$$

and rewriting rules

$$\vec{\mathrm{A}} \to (n < 2)\ ?\ \vec{\mathrm{F}}[+\vec{\mathrm{D}}]_{(3/4,-3/14)}[-\vec{\mathrm{E}}]_{(3/4,3/14)}\vec{\mathrm{F}} : \mathrm{x},$$
$$\vec{\mathrm{B}} \to (n < 2)\ ?\ \vec{\mathrm{G}}[+\overleftarrow{\mathrm{D}}]_{(3/4,3/14)}\vec{\mathrm{E}} : \mathrm{x},$$
$$\vec{\mathrm{C}} \to (n < 2)\ ?\ \vec{\mathrm{G}}[+\overleftarrow{\mathrm{E}}]_{(3/4,-3/14)}\vec{\mathrm{D}} : \mathrm{x},$$
$$\vec{\mathrm{D}} \to (n < 4)\ ?\ \vec{\mathrm{E}}[+\vec{\mathrm{D}}]_{(3/4,3/14)}[-\overleftarrow{\mathrm{D}}]_{(3/4,-3/14)}\vec{\mathrm{E}} : \mathrm{x},$$
$$\vec{\mathrm{E}} \to (n < 4)\ ?\ \vec{\mathrm{D}} : \mathrm{x}[+\overleftarrow{\mathrm{D}}]_{(3/4,3/14)},$$
$$\vec{\mathrm{F}} \to (n < 4)\ ?\ \vec{\mathrm{F}} : \vec{\mathrm{A}}[+\vec{\mathrm{A}}]_{(3/4,-3/14)}[+\vec{\mathrm{B}}]_{(3/4,3/14)}[+\vec{\mathrm{C}}]_{(3/4,3/14)}[+\vec{\mathrm{D}}]_{(3/4,3/14)}\vec{\mathrm{A}},$$
$$\vec{\mathrm{G}} \to (n < 4)\ ?\ \vec{\mathrm{G}} : \vec{\mathrm{A}}[+\vec{\mathrm{A}}]_{(3/4,3/14)}[-\overleftarrow{\mathrm{B}}]_{(3/4,-3/14)}\vec{\mathrm{A}}.$$

Figure 5.20 depicts the initial arrangement with the corresponding labeling and orientation of the edges, as well as the control points and control polylines formed by the line segments p_0p_1, p_1p_2, p_2p_3. The first four developmental cycles are shown in Figure 5.21. Note that two nonintersecting curves are allowed per face in this example, while the MLS only allows one division per cell.

Example 5.6 The ALS can generate patterns that are not possible to obtain using the MLS. As an example, here we combine Example 5.4 with eraser rules (cf. Section 5.2.1) to create a center detail of Bézier curves and add it to a Cartesian grid. The ALS is defined as follows:

$$\Sigma = \{\mathrm{A, B, C, D, E, F, G, x}\},$$
$$\omega = \mathrm{ABAB},$$

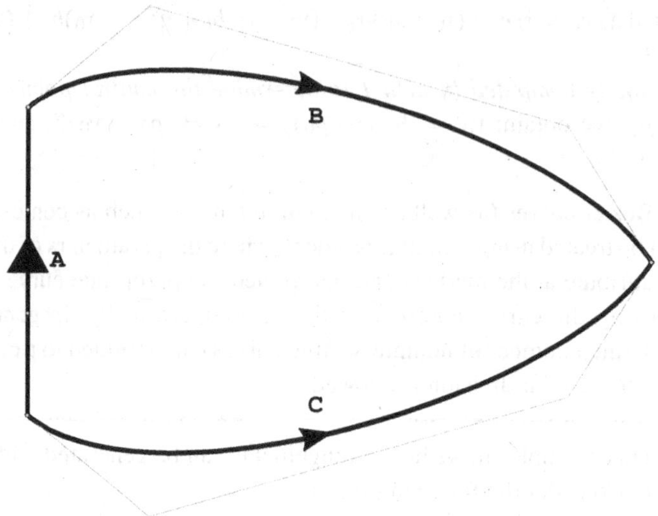

Figure 5.20 Initial arrangement for Example 5.5. Gray lines are the control polylines formed by the line segments p_0p_1, p_1p_2, p_2p_3.

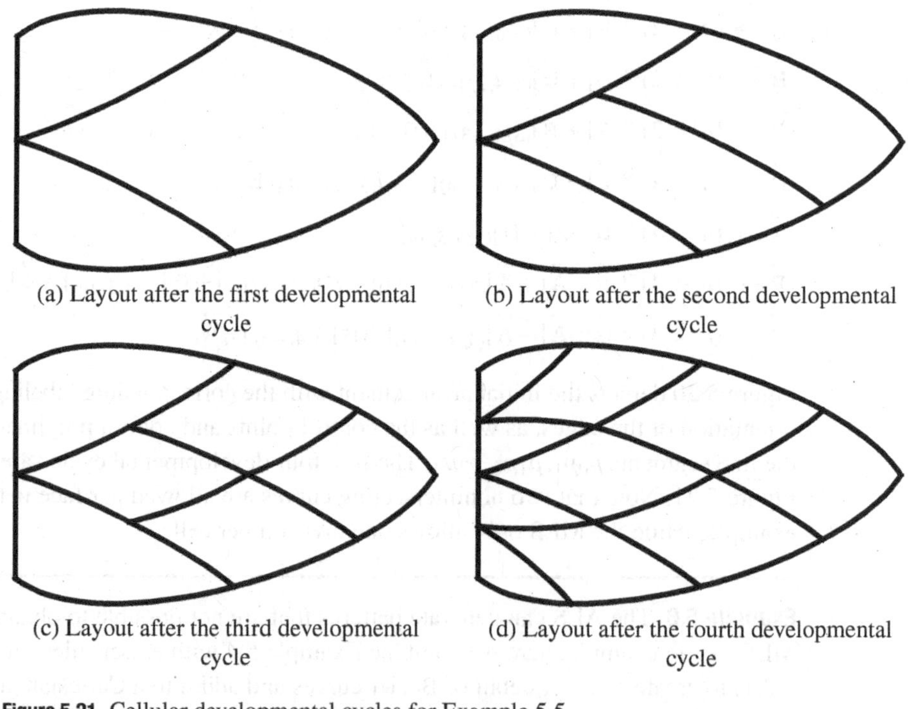

(a) Layout after the first developmental cycle

(b) Layout after the second developmental cycle

(c) Layout after the third developmental cycle

(d) Layout after the fourth developmental cycle

Figure 5.21 Cellular developmental cycles for Example 5.5.

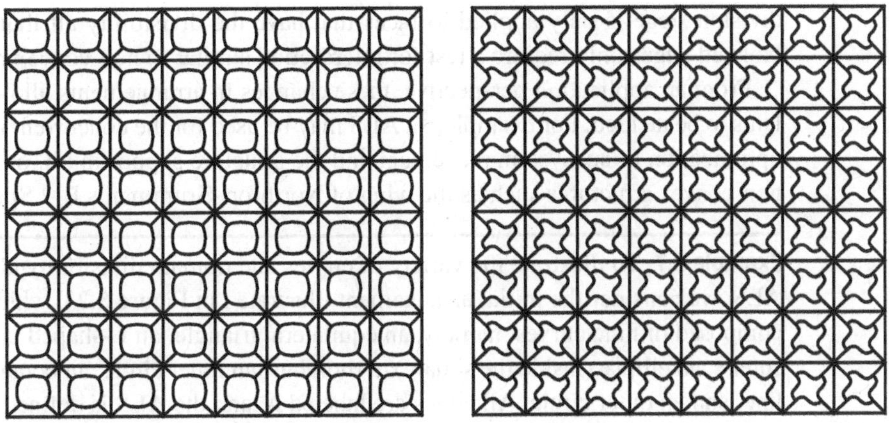

(a) Arrangement L-system with parameters: $\rho = 3/4$, $\theta = 3/4$ and $\bar{\theta} = -3/4$

(b) Arrangement L-system with parameters: $\rho = 2$, $\theta = 4/5$ and $\bar{\theta} = 4/5$

Figure 5.22 Final layout after 11 developmental cycles for Example 5.6 using two different sets of parameters.

and rewriting rules,

$$
\begin{aligned}
&\overrightarrow{\mathtt{A}} \rightarrow (n < 9)\ ?\ \overleftarrow{\mathtt{B}}[+\overleftarrow{\mathtt{A}}][-\overrightarrow{\mathtt{A}}]\overleftarrow{\mathtt{B}} : \overrightarrow{\mathtt{C}},\\
&\overrightarrow{\mathtt{B}} \rightarrow (n < 9)\ ?\ \overrightarrow{\mathtt{A}} : \overrightarrow{\mathtt{D}},\\
&\overrightarrow{\mathtt{C}} \rightarrow (n < 14)\ ?\ [-\overleftarrow{\mathtt{E}}]\mathtt{x}[+\overrightarrow{\mathtt{E}}] : \mathtt{x},\\
&\overrightarrow{\mathtt{D}} \rightarrow (n < 11)\ ?\ \overrightarrow{\mathtt{D}} : [-\overleftarrow{\mathtt{F}}]\mathtt{x}[+\overrightarrow{\mathtt{F}}],\\
&\overrightarrow{\mathtt{E}} \rightarrow (n < 14)\ ?\ \overleftarrow{\mathtt{F}}[-\overrightarrow{\mathtt{F}}][+\overrightarrow{\mathtt{F}}]\overrightarrow{\mathtt{F}} : \mathtt{x},\\
&\overrightarrow{\mathtt{F}} \rightarrow (n < 14)\ ?\ \overleftarrow{\mathtt{G}}[-\overrightarrow{\mathtt{x}}]_{(\rho,\theta)}[+\overleftarrow{\mathtt{x}}]_{(\rho,\bar{\theta})}\mathtt{x} : \mathtt{x},\\
&\overrightarrow{\mathtt{G}} \rightarrow (n < 14)\ ?\ \mathtt{e} : \mathtt{x},
\end{aligned}
$$

where ρ, θ, and $\bar{\theta}$ are marker parameters defining the Bézier curves of new edges. In this example, two different sets of parameter definitions are considered: the first with $\rho = 3/4$, $\theta = 3/4$, and $\bar{\theta} = -3/4$, and the second with $\rho = 2$, $\theta = 4/5$, and $\bar{\theta} = 4/5$. The resulting layouts after 11 developmental cycles are shown in Figure 5.22.

In addition to general curves, an ALS can consider non-convex regions, multiply connected regions, and isolated edges. Non-convex regions are handled by adding the criterion for cellular division that new edges cannot intersect existing edges, which only requires an intersection test for new candidate edges. Multiply connected regions require similar intersection tests between new edges and all connected components of the face, as well as assigning a Boolean attribute to each face defining its divisibility: if this attribute is true, then cellular division is applied to the cell. With this attribute, each face is screened for divisibility before the cellular division step, and the cellular division stage is only applied to faces that have the divisibility attribute set to true. Isolated edges only require a test for intersection similar to non-convex regions.

From an application perspective, these changes in arrangements allow novel problems to be tackled. For example, an ALS may be used for the concurrent modeling and optimization of subsystem placement and the structure supporting it, or the design of non-convex structures such as the wing of Northrop-Grumman's B-2 Spirit aircraft.

Example 5.7 To illustrate the variety of curves and domain divisibility allowed by an ALS, we consider the initial arrangement illustrated in Figure 5.23, which is composed of nine curves, namely, an equilateral triangle, an L-shaped polygon, a square, an ellipse (inside the square), a rhombus, an inner circle, an octagon, a straight line, and an outer circle. All edges are labeled A, and the ALS is defined as follows:

$$\Sigma = \{\mathtt{A}\},$$

$$\omega = \mathtt{A}\ldots\mathtt{A},$$

$$P = \{[+\overrightarrow{\mathtt{A}}][-\overrightarrow{\mathtt{A}}][+\overleftarrow{\mathtt{A}}][-\overleftarrow{\mathtt{A}}]\overrightarrow{\mathtt{A}}[+\overrightarrow{\mathtt{A}}][-\overrightarrow{\mathtt{A}}][+\overleftarrow{\mathtt{A}}][-\overleftarrow{\mathtt{A}}]\overrightarrow{\mathtt{A}}[+\overrightarrow{\mathtt{A}}][-\overrightarrow{\mathtt{A}}][+\overleftarrow{\mathtt{A}}][-\overleftarrow{\mathtt{A}}]\}.$$

This ALS has a single rule that divides all edges in half and creates candidate edges for cellular division at all vertices in a face. To accelerate the subdividing process, ten nonintersecting curves are allowed per cell.

The inner circle, rhombus, ellipses, and octagon are specified as indivisible, while the faces associated with all other closed curves can be divided. In addition to the regular criteria for cellular division (compatibility of labels and orientation of markers and nonintersecting of curves), for this example, a minimum angle constraint is also implemented. Figure 5.24 shows the layouts obtained at the end of the eight cycles with minimum angle constraints of 10° and 61°. As expected, with a smaller angle constraint, the faces tagged for division are finely partitioned by small cells. By contrast, with a larger angle constraint the equilateral triangle remains undivided (since any inner sub-triangle would necessarily have an angle smaller than 61°) and

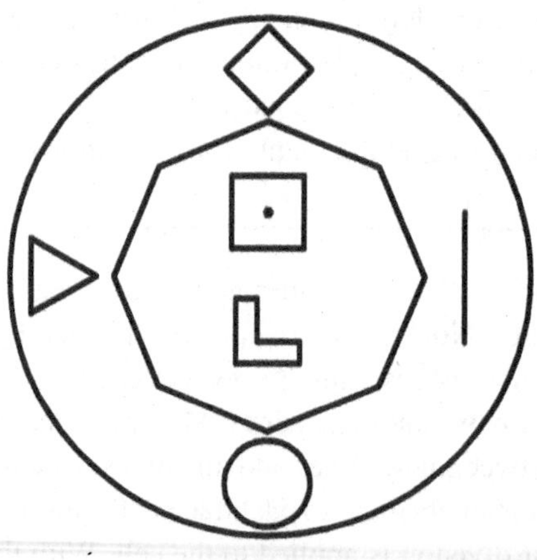

Figure 5.23 Initial arrangement for Example 5.7.

(a) Minimum angle equal to 10° (b) Minimum angle equal to 61°

Figure 5.24 Final layouts after eight development cycles for Example 5.7 using two different minimum angle constraints.

the layout is composed of more regular quadrilaterals. This example is ideal to test the implementation of an ALS, since it allows for the testing of cellular division criterion using qualitative analysis of the complexity of the expected layout. For example, the correctness of the implementation of intersection constraints, and angle constraints can easily be assessed by visual inspection of Figure 5.24.

5.2.4 Directed Cellular Dynamics

The cellular dynamics in the MLS seek to emulate the primary forces acting on an organism cell as it develops. However, for optimization purposes, a better approach would be to direct the cellular dynamics toward optimizing the desired objective function. In this subsection, we illustrate the use of nonlinear programming to direct the cellular dynamics toward an optimal shape for truss structures within the ALS formulation. Concretely, in the cellular dynamics stage of the developmental cycle of an ALS, we apply a limited number of iterations of the shape and sizing optimization presented in Section 2.3.3 for geometrical and topological truss optimization. Each of these iterations modifies both the shape of the layout and the size of its members, but, importantly for the optimization process, does not change the topology of the layout. The latter is only changed during the cellular division stage, when new cells can be created or destroyed. Therefore, with directed dynamics, each cellular developmental cycle of an ALS can be understood as an optimization process where the geometry/member sizes of a new "ground structure" layout (cf. Section 2.3) are adjusted to enhance its performance.

To illustrate the benefits of employing a directed search at the cellular dynamics stage, we consider the optimized solutions obtained with sizing and combined shape/sizing optimization of an ALS-generated ground structure shown in

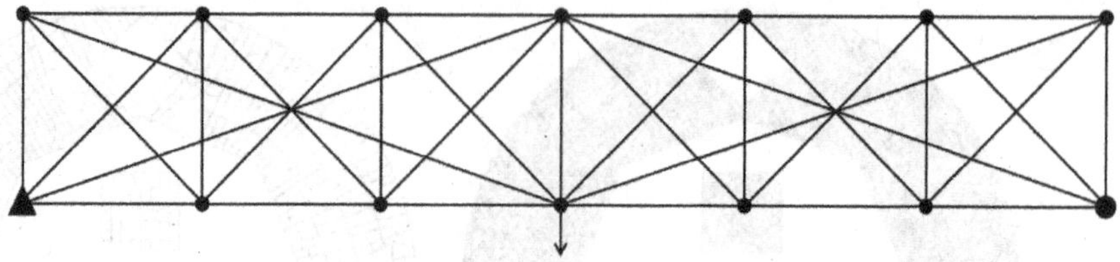

(a) Layout of the ground structure

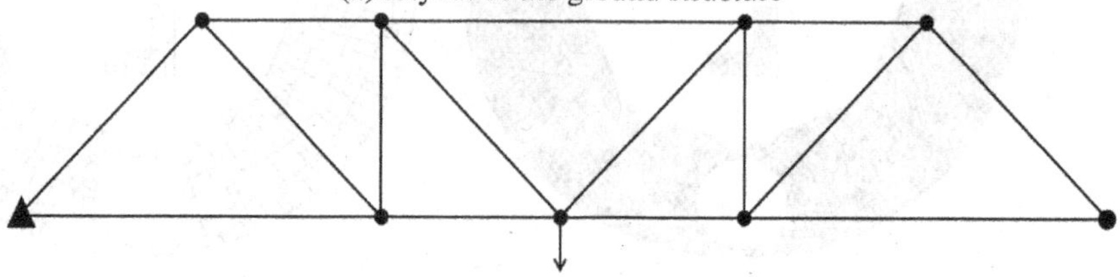

(b) Layout obtained by performing sizing optimization

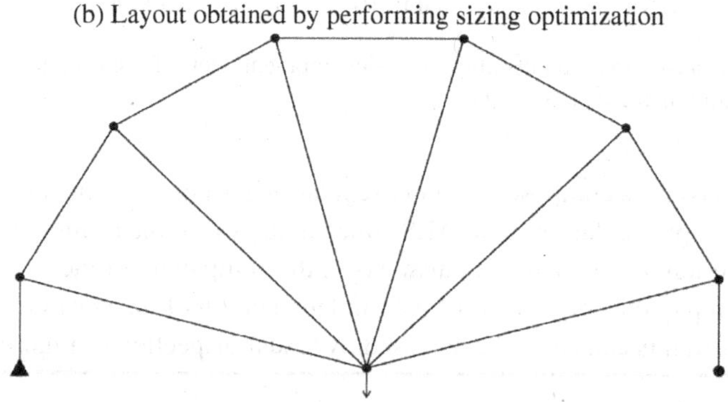

(c) Layout obtained by performing sizing and shape optimization

Figure 5.25 Effect of implementing a directed cellular dynamics approach.

Figure 5.25(a). The optimized truss obtained with sizing is shown in Figure 5.25(b) and has a volume $v_S = 16$. By contrast, the optimized truss resulting from the sizing and shape optimization has the shape shown in Figure 5.25(c) with a volume $V_{DCD} = 9.6462$, where the theoretical minimum derived in Michell (1904*a*) for this problem is $V_{Mi} = 3\pi \approx 9.4248$. For comparison, in the original approach in the MLS, the balance of an osmotic pressure and cell wall elasticity would be used; the equilibrium for such combination would be a shape between Figures 5.25(b) and 5.25(c), with small probability of matching the shape in Figure 5.25(c). The effect of the directed search on the optimization process is further investigated in Section 5.3.

5.2.5 Cellular Division via TPTD

In the ALS, the first set of markers that satisfies a set of criteria is selected for cellular division, and the remaining markers in the cell are ignored. This ad hoc approach creates a dependency of the layout developed on the implementation of the ALS, since the first pair of markers that meets a set of criteria depends on the ordering and traversal of the markers in a cell. For optimization, it is desirable to select the pair of markers for division that generates the greatest reduction in the objective function among those that satisfy the set of division criteria. In this subsection, we introduce the concept of the TPTD (Kobayashi, Canfield & Kolonay 2021) to quantify and identify such a pair of markers.

Let Ω be a bounded set of $\mathbb{R}^N$ $(N = 2, 3)$ and assume a shape functional $\mathcal{J}$ is given on a working domain D such that $\mathbb{R}^N \supset D \supset \Omega$. The TPTD measures the change in $\mathcal{J}$ due to the addition of a thin ligament connecting two points at the boundary $\partial\Omega$. Not all points at $\partial\Omega$ are suitable for that purpose, and so we define the set of admissible pairs as

$$\mathcal{A} = \{(p,q) \in \partial\Omega \times \partial\Omega\},$$

where the line $\overline{pq}$ lies outside of Ω in a region prescribed to be traction free. More general conditions of admissibility are conceivable (e.g., general prescribed traction), but this definition suffices for the introduction of the TPTD.

Given its importance in structural optimization, we consider the two-point derivative of the compliance functional of a body $\hat{J}_c(\Omega)$ (Allaire et al. 2004):

$$\hat{J}_c(\Omega) = \int_\Omega Ae(u) : e(u),$$

where A is the elasticity tensor and $e(u)$ is the small displacement strain tensor, $e(u) = \nabla u + \nabla^T u$. Then, given an admissible pair $p, q \in \partial\Omega$, the TPTD of the compliance can be written as

$$\mathcal{T}\hat{J}_c(p,q) = \frac{u_0^\perp(p) + u_0^\perp(q)}{2} \int_0^l f^\perp(x)dx +$$
$$\frac{u_0^\perp(p) - u_0^\perp(q)}{2l} \int_0^l (l-2x)f^\perp(x)dx +$$
$$\int_0^l \frac{f^\parallel(x)\left\{x\int_0^l\int_0^\eta \frac{f^\parallel(\xi)}{\hat{E}}d\xi d\eta - l\int_0^x\int_0^\eta \frac{f^\parallel(\xi)}{\hat{E}}d\xi d\eta + [(l-x)u(p) + xu(q)].\frac{q-p}{l}\right\}}{l} dx -$$
$$\frac{(\hat{E})}{2l^2}\int_0^l \left[\frac{(u(p)-u(q)).(q-p)}{l} - l\int_0^x \frac{f^\parallel(\xi)}{(\hat{E})}d\xi + \int_0^l\int_0^\eta \frac{f^\parallel(\xi)}{(\hat{E})}d\xi d\eta\right]^2 dx,$$

where f denotes an applied body force field, $\hat{E} = E/(1-\nu^2)$, for plane strain or $\hat{E} = E$ for plane stress/3-D problems, $l = |p-q|$, x is the arc length coordinate along the line $\overline{pq}$, and the superscripts $^\parallel$ and $^\perp$ denote the component parallel or orthogonal to $\overline{pq}$ oriented from p to q, respectively. More details of the TPTD, including its derivation, can be found in the cited reference (Kobayashi et al. 2021).

To illustrate the computation of the TPTD, we consider a long cantilever beam subject to a load at the tip as illustrated in Figure 5.26; the Young's modulus of the material is $E = 10^5$, and Poisson's ratio is $\nu = 0.3$. The significance of the TPTD to such structures stems from its ability to quantify the performance of potential candidates for division of the frame. For this example, the quantification is based on the amount of reduction of the compliance caused by the insertion of a thin ligament in

Figure 5.26 TPTD demonstration problem featuring a cantilevered frame clamped on the left and subject to a constant traction at the tip with a resultant force $F = 10$ in the vertical direction. This figure also defines the coordinates s and t along the bars of the cantilever.

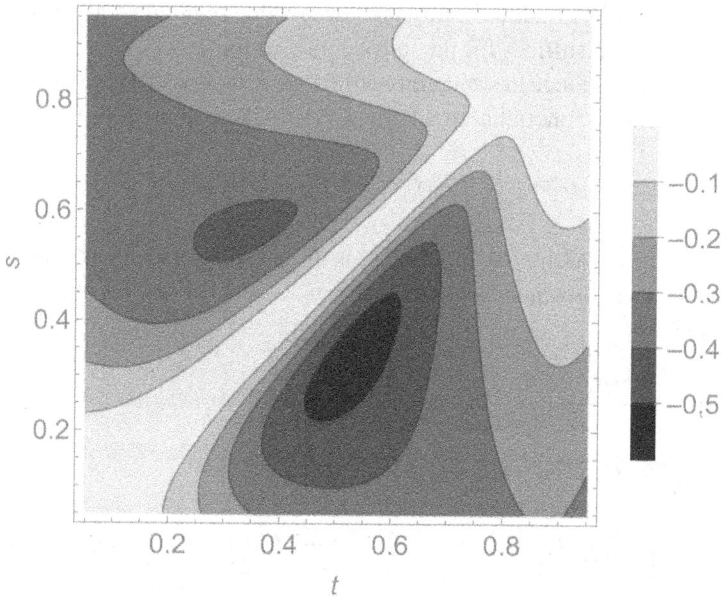

Figure 5.27 Contour map with the distribution of the two-point derivative for candidate points along the lower and upper members in Figure 5.26. The coordinates s and t are defined in Figure 5.26.

the frame. Figure 5.27 shows the contour map of the TPTD for all candidate pairs of points connecting points on the lower and upper bars of the initial frame. Here the candidate pairs are identified with a pair of fraction coordinates (t, s) along the lower and upper edges of the frame, respectively (cf. Figure 5.26). It is clear from this figure that connecting points that are approximately at equal, horizontal distance from their left supports are ineffective in stiffening the frame. Better designs are obtained with diagonally oriented frames close to the center of the bars, with the minimum of the TPTD of about $\hat{\mathcal{TJ}}_c \approx -0.54$ at $t = 0.53$ and $s = 0.34$.

5.3 Case Studies

This section presents a series of case studies that demonstrate the ability of the ALS to generate optimized topologies for structural design problems. Case Study 5.1 considers a benchmark topology optimization problem of a short cantilever structure, while Case Study 5.2 revisits the long cantilever problem studied extensively in Chapter 2. Case Study 5.3 demonstrates the effectiveness of using the directed cellular dynamics approach (cf. Section 5.2.4) on a structure with multiple applied loads. Case Study 5.4 discusses the application of the ALS to the benchmark L-bracket problem, which features a non-convex design domain. Finally, Case Study 5.5 considers the optimized design of the structural layout concurrent with the sub-system placement in a satellite.

Case Study 5.1: Short Cantilever Structure

The first test case is an archetypal benchmark for structural topology optimization (Hamda et al. 2002) and features a rectangular design domain measuring 2 units × 1 unit. The structure is assumed to be in a state of plane stress with a cantilever boundary condition applied to the left side and a load applied at the middle vertex of the right side as illustrated in Figure 5.28. The Young's modulus E, point load P, and density ρ are all set to 1, while the Poisson ratio ν is set to 0.3. The goal is to obtain the lightest cantilever structure for a prescribed maximum deformation. Mathematically, the problem can be expressed as

$$\min \quad W(\mathbf{x}), \tag{5.25}$$

$$\text{s.t.} \quad D(\mathbf{x}) - D_{lim} \leq 0, \tag{5.26}$$

where $\mathbf{x}$ is the vector that encodes the genome, $W(\mathbf{x})$ is the cantilever's weight nondimensionalized by the weight of a solid rectangular cantilever, $D(\mathbf{x})$ is the measured displacement, and D_{lim} is the maximum allowable displacement, assumed here to be 220.

Topologies generated by the ALS hold no canonical physical meaning, and therefore a relationship between the elements of the topology (e.g., cells, edges, and vertices) and relevant physical components must be established. For this problem, edges of the topology are assumed to represent structural members of the frame and the interior regions of the cells represent voids. The ALS used in the generation of the cantilever geometry is defined as follows:

- The initial map is bounded by the rectangle $[0,2] \times [0,1]$. The initial map has five edges to ensure that one of the vertices is located at the coordinates $(2, 0.5)$ where the load is applied. Figure 5.29 depicts the design box and the initial map edges.
- The number of developmental stages is an optimization parameter in the range $[3,6]$.
- The alphabet Σ contains eight letters excluding the terminal symbol x.
- The production rules have eight tokens.
- The equilibrium is computed after every developmental stage. The maximum number of fixed point iterations is set to 10. The osmotic pressure and the spring

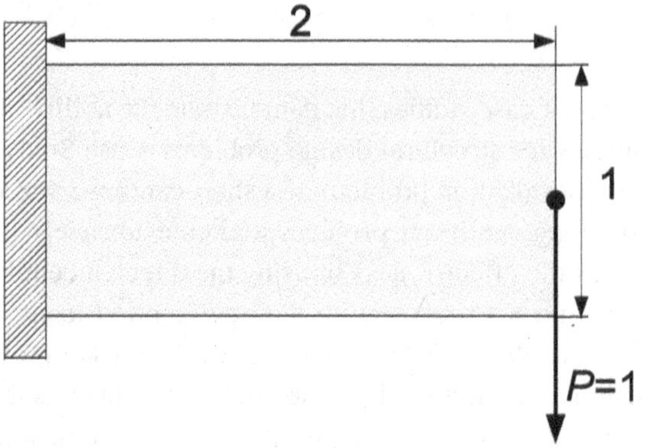

Figure 5.28 The 2 × 1 cantilever benchmark. Reprinted from (Pedro and Kobayashi 2011) with permission.

Figure 5.29 Design domain and initial map for MLS development. Vertices marked in gray are fixed. Reprinted from (Pedro and Kobayashi 2011) with permission.

coefficient are the same for all cells and edges and are equal to 2 and 50, respectively. All vertices in the topology are free to move inside the design domain, except the two attachment points at the clamped side of the cantilever, points $(0,0)$ and $(0,1)$, and the point $(2,0.5)$, where the load P is applied (these points are highlighted in gray in Figure 5.29).

Note that the parameters above were selected to agree with published parameters and are not meant to be optimal. Each edge (i.e., structural member) of the map is thickened using a thickness attribute as introduced in Section 5.2.1, allowing the ALS to assign larger thicknesses to members that experience higher stress levels.

The computation of the fitness and constraints for this problem were performed using the finite element software COMSOL Multiphysics ®. Constraints are enforced using a penalization P_w defined as

$$P_w = C\left(e^{\max(D(\mathbf{x})-D_{lim},0)} - 1\right),$$

where $C = 1$, which added to the fitness value in Equation 5.25. The genetic algorithm (GA) used a population of 50 individuals evolved until no improvement is observed

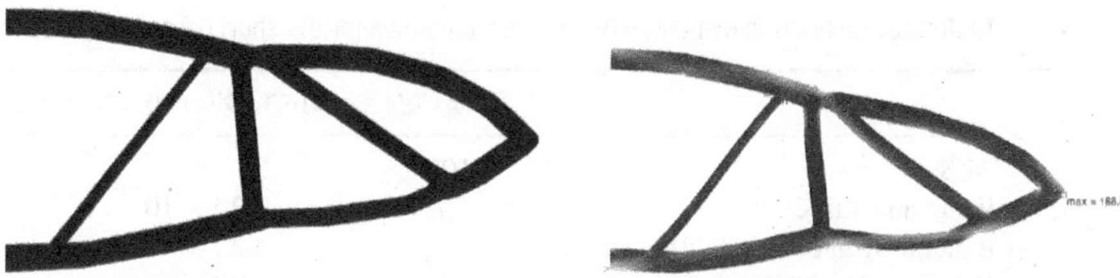

(a) Topological layout with $W(\mathbf{x}) = 0.260$ (b) Von Mises stress distribution in the deformed configuration

Figure 5.30 Best individual solution obtained after evolving a population size of 50 for 100 generations using the ALS. Arrows in the stress map represent principal stresses, with converging pairs representing compression stresses and diverging pairs representing tensile stresses. Reprinted from (Pedro and Kobayashi 2011) with permission.

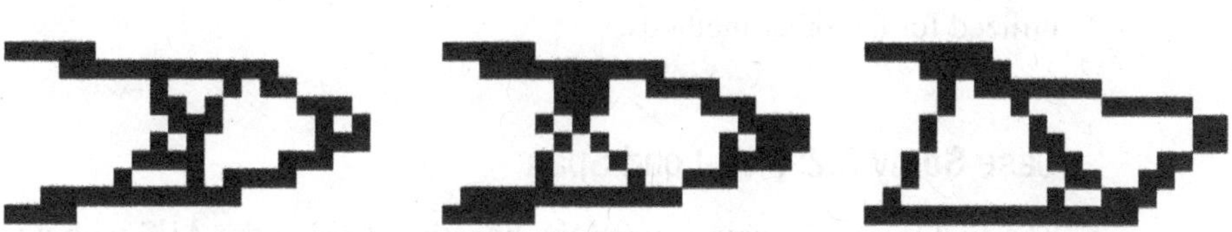

(a) (b) (c)

Figure 5.31 Best designs for the minimum weight problem from (a) Wang and Tai (2005), (b) Balamurugan et al. (2008), and (c) Hamda et al. (2002). Reprinted from (Pedro and Kobayashi 2011) with permission.

for more than 50 generations (with a maximum of 100 generations), with new generations formed by 5% elite individuals, 80% crossover individuals, and 15% mutated individuals.

Figure 5.30 shows the topology and the Von Mises stress distribution for the best design, which achieves $W(\mathbf{x}) = 0.260$. Table 5.3 compares this result against those obtained using evolutionary programming reported in the literature for the same problem. The latter include results using two bit-array representations (Chapman et al. 1994, Wang & Tai 2005*b*) and a Voronoi-based representation (Hamda et al. 2002) for the cantilever. The three optimal structures reported in these works are depicted in Figure 5.31. It should be noted that none of the structures in Figure 5.31 correspond to practical designs since all have one-point connections between several elements, whereas all solutions obtained using the ALS are manufacturable. Observing the minimum weight achieved using the various methods, it is clear that the optimal structure obtained with ALS represents a significant improvement in design performance compared to the reported results using other methodologies (i.e., a 20% weight reduction compared to the best reported weight (Hamda et al. 2002)).

To aid in explaining the reasons for this performance increase, Table 5.3 lists the percentage of feasible individuals for all generations: The efficiency of the GA is drastically reduced if most of the generated topologies are not viable. For bit-array representations, the fraction of viable designs is extremely low (17%) compared to the

Table 5.3 Comparative analysis of several techniques for the short cantilever problem.

	W_{min}	*Feasible*	Mesh Discretization	FE evaluations
ALS	0.260	100%	–	$\sim 5,000$[1]
Wang and Tai (2005)	0.325	17%	20×10	$\sim 20,000$
Balamurugan et al. (2008)	0.340	N/A[2]	20×10	$\sim 100,000$
Hamda et al. (2002)	0.330	N/A	20×10	$\sim 16,000$

[1] 100 generations. [2] Not Available.

ALS, where all generated individuals are feasible. Table 5.3 also provides the number of Finite Element (FE) evaluations necessary to obtain the designs reported; here, the number of evaluations necessary for the ALS is a fraction of the number of evaluations utilized for the other methods.

Case Study 5.2: Five-Load Span

The first test case demonstrated the benefits of using the ALS with a GA, especially when compared to the bitmap GA approach. This second test case builds upon this finding to consider the gains of using the directed search approach in the cellular dynamics stage (Clark 2019). Specifically, we consider the layout optimization of a truss defined over a span of five unit loads evenly spaced on an interval between a pinned support on the left and a roller at right as illustrated in Figure 5.32(a). The alphabet and initial map (cf. Figure 5.32(g)) of the ALS are kept constant for all individuals. The genes developed by the GA are used to encode the axiom, rewriting rules, and the number of developmental cycles. Thus, a trial candidate for the structure is represented by the following triplet:

$$a_{\mathrm{I}} = [\eta_{DC}, \omega, P], \tag{5.27}$$

where $\eta_{DC} \in \{2, \ldots, 17\}$ denotes the number of developmental cycles, ω is the axiom, and P is the set of rewriting rules. The axiom, $\omega \in \Sigma^*$, is generated by assigning a marked label to each of the initial map edges. The production rules $P = [P_0, ..., P_{NT}]$ are those described for the ALS and are applied to the axiom η_{DC} times. The final layout generated by the ALS is then passed to the cellular dynamics stage, which is represented by the nonlinear programming (NP) approach explained in Section 2.3.3 for the simultaneous optimization of truss geometry and topology. The full procedure yields an optimum volume V_I for the individual, which is then set as the fitness of the individual. We note that this process is non-Lamarckian, meaning that the traits acquired in the NP cellular dynamics step are not re-encoded in the genes of the individual. The genetic attributes in Equation 5.27 are encoded as a binary string with chromosomes assigned to each of the three attributes, which compartmentalizes the

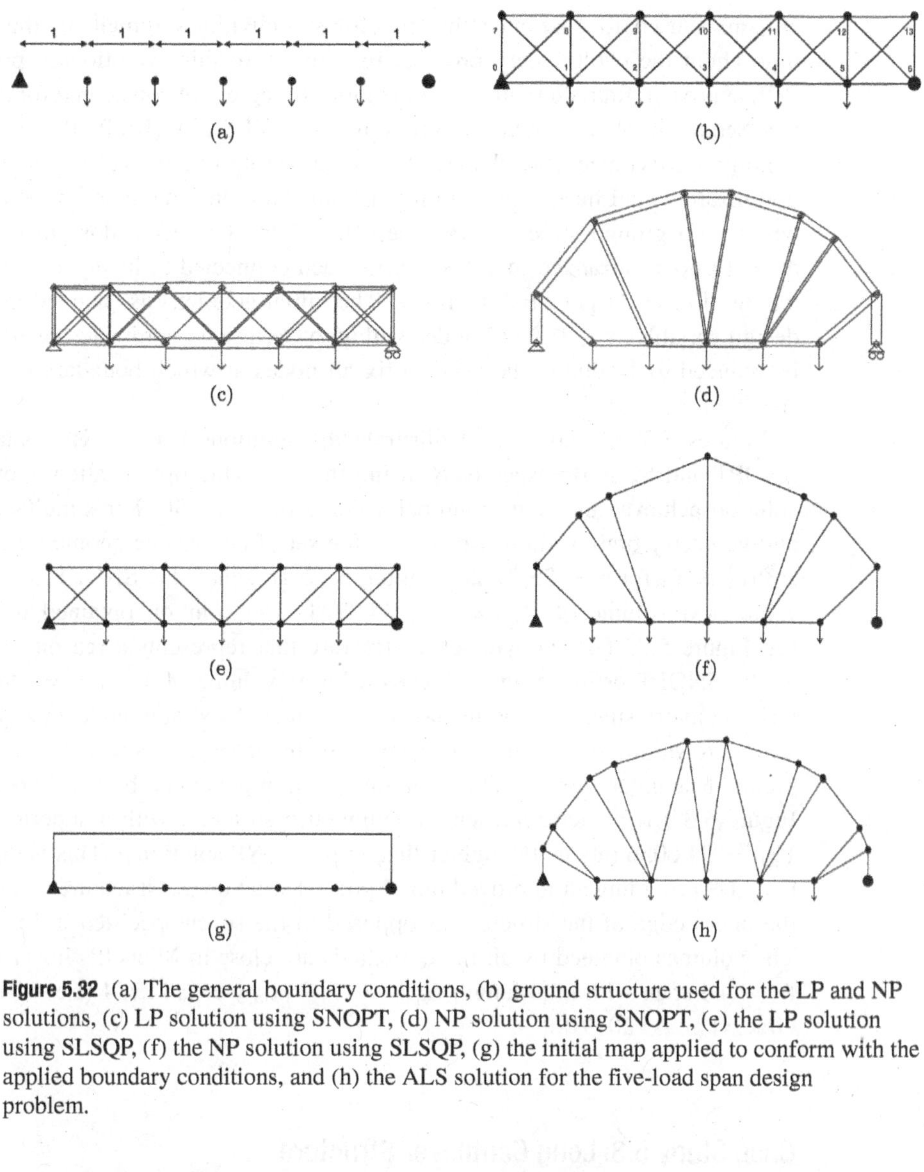

Figure 5.32 (a) The general boundary conditions, (b) ground structure used for the LP and NP solutions, (c) LP solution using SNOPT, (d) NP solution using SNOPT, (e) the LP solution using SLSQP, (f) the NP solution using SLSQP, (g) the initial map applied to conform with the applied boundary conditions, and (h) the ALS solution for the five-load span design problem.

genetic transfer in the crossover to each specific trait. To guarantee that each map generated by its rewriting rules P results in a truss (i.e., the map does not possess unconstrained rigid modes), a constrained Delaunay triangulation is applied to the layout generated by the ALS prior to the cellular dynamics stage.[1]

A population of 50 individuals is evolved for 60 generations using selection, crossover, and mutation. Selection is performed using the tournament method, and

[1] An alternative approach is to constrain the cellular division process of the ALS (Khetan, Lohan & Allison 2015).

elitism is used to guarantee that the fittest individuals remain in the population throughout the evolutionary process. In addition to this evolutionary programming (EP) approach, four alternative optimization strategies are considered for comparison: the Sequential Least SQuares Programming (SLSQP) (Kraft 1994) and Sparse Nonlinear OPTimizer (SNOPT) (Achtziger 2007) algorithms each coupled with sizing optimization via Linear Programming (LP) or shape and sizing optimization using NP based on a ground structure. For these methods, the design domain is defined by $N = 14$ nodes arranged in a 7×2 grid, each connected to its nearest neighbors for a total of $m = 31$ potential members. The initial geometry is encoded in a vector of design variables $\bar{\mathbf{y}} \in \mathbb{R}^{28}$ (12 nodes with two Cartesian coordinates each), which can be reduced to 14 due to the need to fix all nodes at which boundary conditions are specified.

Figures 5.32(c) and 5.32(e) illustrate the solutions $\mathbf{t}', \mathbf{t}'' \in \mathbb{R}^{33}$ obtained using SNOPT and SLSQP, respectively, using the LP sizing optimization approach; each solution achieves the same minimal volume of $V = 56$. For simultaneous sizing and geometry optimization, we specify the set of admissible geometries, $Y = \{\mathbf{y} \in \mathbb{R}^{28} | y_p = \bar{y}_p, \text{for } p = 26, 27, 28\}$. Figure 5.32(d) depicts the SNOPT solution, which achieves a volume of $V_{SNOPT} = 34.9924$. The solution obtained using SLSQP (cf. Figure 5.32(f)) is a symmetric structure that represents a reasonable extension of the SNOPT estimate and achieves a lower volume of $V_{SLSQP} = 34.9881$. The base geometry suggests that this layout is obtained by sliding node 9 away from node 10 and toward node 8, which disengages member l_{39} and activates l_{29}, forming the second triangular support. The optimum solution generated by the ALS is shown in Figure 5.32(h), which represents a symmetric structure with a superior volume of $V_{EP} = 34.6056$ (about 1% lighter than either of NP solutions). This is due in part to the ALS providing an improved initial ground structure, as it features 10 nodes along the upper edge of the structure as opposed to the seven specified in Figure 5.32(b). The volumes obtained by all three methods are close to Michell's lower bound for a five-load span with total length $L_{span} = 6$ ($V_{span,5} = 34.2284$ (Lewiński, Sokół & Graczykowski 2019)).

Case Study 5.3: Long Cantilever Structure

Case Study 5.2 revisits the long cantilever problem extensively studied in Chapter 2. The domain is rectangular with width of 20 units and height of 5 units (cf. Figure 2.3). A vertical load of 10 acting downward is applied at the top right of the domain, and two support points are applied at the bottom and top at the left side. The Young's modulus is $E = 10^5$ and the Poisson ratio is $\nu = 0.3$. The initial map is a minimal graph containing the two support points and the load. In the cellular division stage of the ALS, the TPTD (cf. Section 5.2.5) is used to select the pair of markers that, when a new edge is created between them, most reduces the compliance of the structure. At the end of each cellular division stage, colinear edges with the same label are merged at vertices of degree two (i.e., vertices that have only two incident edges) to

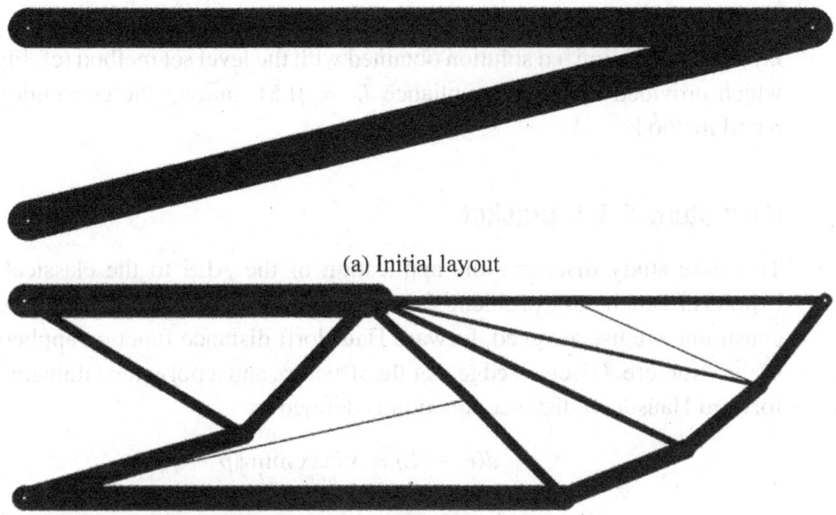

(a) Initial layout

(b) Optimized layout after three cycles; $\hat{J}_c = 0.42$; $v_f = 0.30$

Figure 5.33 Results obtained for the long cantilever problem using the ALS.

reduce the number of available candidate markers at the next stage. The optimization is initiated using one individual with the initial layout shown in Figure 5.33(a) and develops that individual via the ALS defined in Example 5.7, coupled with the TPTD and directed cellular dynamics. For convenience, the ALS definition is repeated here:

$$\Sigma = \{\mathtt{A}\}, \tag{5.28}$$

$$\omega = \mathtt{AAA}, \tag{5.29}$$

$$P = \{[+\overrightarrow{\mathtt{A}}][-\overrightarrow{\mathtt{A}}][+\overleftarrow{\mathtt{A}}][-\overleftarrow{\mathtt{A}}]\overrightarrow{\mathtt{A}}[+\overrightarrow{\mathtt{A}}][-\overrightarrow{\mathtt{A}}][+\overleftarrow{\mathtt{A}}][-\overleftarrow{\mathtt{A}}]\overrightarrow{\mathtt{A}}[+\overrightarrow{\mathtt{A}}][-\overrightarrow{\mathtt{A}}][+\overleftarrow{\mathtt{A}}][-\overleftarrow{\mathtt{A}}]\}. \tag{5.30}$$

The single rule (5.30) divides all edges in half and creates candidate edges for cellular division at all vertices on a face. As mentioned, the TPTD is used to choose the pair of markers for cellular division. After cellular division, the directed cellular dynamics stage performs shape and size optimization on the new layout using the Sequential Quadratic Programming (SQP) method (Schittkowski 1983). The objective of the problem is to minimize the compliance of the structure, with constraints related to volume, collision, and containment (see Kobayashi, Canfield & Kolonay (2021) for details).

The optimized solution at the end of the three cycles is shown in Figure 5.33(b). This solution obtained with the ALS is markedly different from the optimized solution obtained with SIMP (cf. Figure 2.4), and significantly more efficient: the compliance for ALS is $\hat{J}_c = 0.42$, whereas for SIMP, the compliance is $\hat{J}_c = 0.56$. Specifically, comparing the two values of compliance, we note that the optimized solution obtained using ALS is 25% smaller, though the comparison of these values should be done with the caveat that the design domain constrains the pixels in the SIMP model, while the design domain constrains the skeleton in the ALS.

It is noteworthy that the final layout obtained with the ALS closely resembles the layout of the optimized solution obtained with the level set method (cf. Figure 2.17(c)), which provided the best compliance $\hat{J}_c = 0.51$, among the conventional, gradient-based methods.

Case Study 5.4: L-Bracket

This case study discusses the application of the ALS to the classical, non-convex L-bracket benchmark problem shown in Figure 5.34. To implement the non-convex constraint, we use a signed, forward Hausdorff distance function applied to the edges of the structure. Given an edge e of the structure and a polygonal domain L, the signed, forward Hausdorff distance function is defined as

$$d(e \rightarrow L) = s \max_{p \in e} \min_{q \in L} \|p - q\| ,$$

where the sign s is positive if the point p on e that maximizes the distance from L lies outside of the domain L, negative if p lies inside the domain L, and zero if p lies on the boundary of L. The initial layout used for this problem is shown in Figure 5.35(a) and features a minimal graph containing the support and load points and an additional barrier edge dividing the non-convex domain L into two convex polygons. Fixed boundary conditions are applied to the two points at the top of the L-bracket, and a single load is applied at the right tip (Jantos et al. 2019); the point at

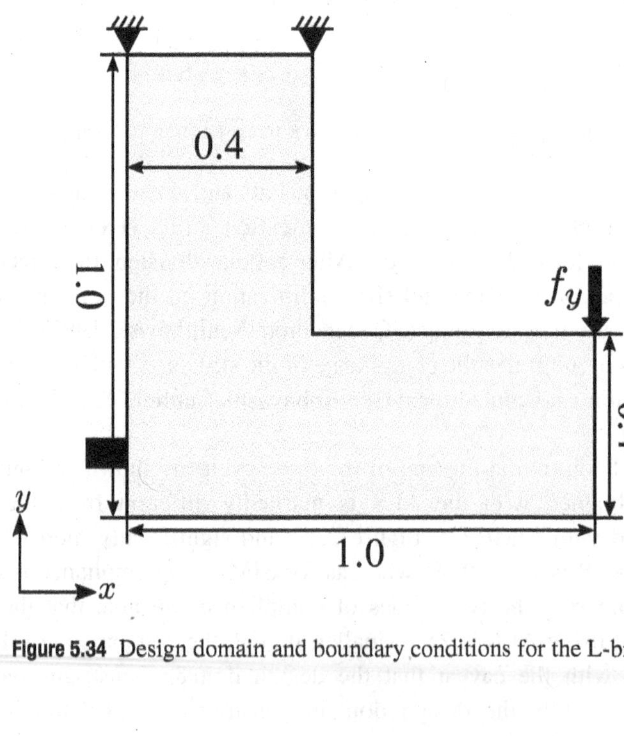

Figure 5.34 Design domain and boundary conditions for the L-bracket problem.

(a) Initial layout for the ALS; $\hat{J}_c = 397.8$, $v_f = 0.187$

(b) ALS solution after one developmental cycle; $\hat{J}_c = 158.7$, $v_f = 0.256$

(c) TTOM solution

Figure 5.35 Comparison between L-bracket solutions generated using the ALS and density-based TTOM approaches (Jantos, Hackl & Junker 2019). Reprinted from (Kobayashi, Canfield & Kolonay 2021) with permission.

Figure 5.36 Optimized layout after the second developmental cycle; $\hat{J}_c = 135.5$, $v_f = 0.256$. Reprinted from (Kobayashi, Canfield & Kolonay 2021) with permission.

the re-entrant corner of the domain is kept fixed during the optimization. The objective is to minimize the compliance of the frame, subject to a constraint on the area of the frame, where the target area for this problem is 0.256. The rules of the ALS are kept fixed and equal to the rules in example Equation 5.7, and cellular division is performed using the TPTD (cf. Section 5.2.5) with constraints on the minimum length of a new edge and minimum angle between the new edge and the existing edges (0.1 and 15°, respectively). Shape and sizing optimization are also employed in a manner identical to the previous test case.

The optimized layout obtained after the first developmental cycle is shown in Figure 5.35(b). This topology agrees well with the reference layout obtained using the thermodynamic topology optimization method (TTOM) reported in Jantos, Hackl & Junker (2019). Considering a second cycle results in the topology shown in Figure 5.36, which features additional structural members and a compliance reduction of 14.6% compared to the optimized layout after the first cycle.

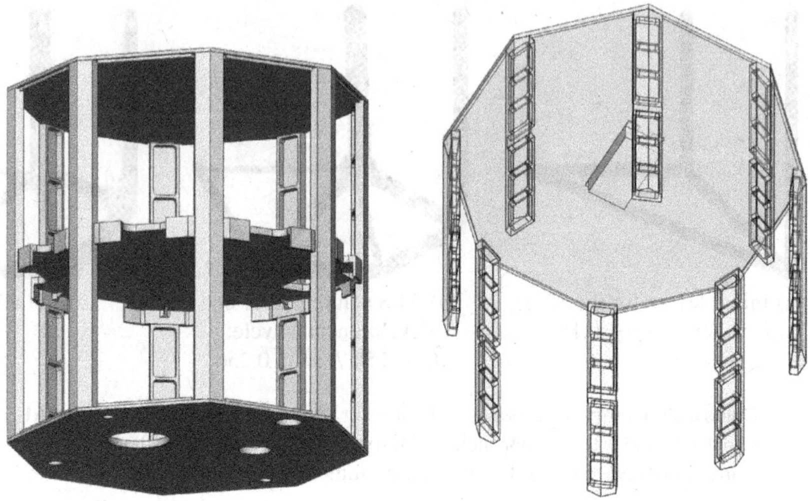

(a) The zenith deck (top)

(b) Top deck, the ribs, and the subsystem (with transparency) as seen from below

Figure 5.37 Structural frame of the HawaiiSat-1.

Case Study 5.5: Satellite Panel Design

Case Study 5.4 examines an application of the ALS to the design of a structural component of a satellite (Nunes 2010). Concretely, we apply the methodology to the simultaneous optimization of subsystem placement and the mass of an external plate of a satellite under launch conditions. This problem addresses part of one of the requirements in the design of the HawaiiSat-1 satellite.[2] The structural component to be optimized, designated as the *nominal zenith deck* or simply the *top deck*, is shown in Figure 5.37 together with the subsystem that requires placement. The top deck is part of the satellite structure and is structurally reinforced by eight ribs, while the subsystem is an Inertial Measurement Unit (IMU) used for navigation and control of the satellite.

Panel Layout

The geometry of the top deck is shown in Figure 5.38 and consists of an octagon with an internal rectangular area where the IMU is placed. The objective is to optimize the layout of the top deck such that its mass is minimized while meeting a set of stress and deformation constraints. Specifically, the maximum displacement of the deck is required to be less than 1 mm; applying a safety factor of 2, this constraint is modified to 0.5 mm or 500 μm. The structure consists of aluminum 6061-T6 (cf. Table 5.4), which was chosen for HawaiiSat-1 due to its high strength and good workability. This alloy

2 www.hsfl.hawaii.edu/missions/hawaiisat-1/.

Table 5.4 Material properties used for the satellite structure design problem.

Aluminum 6061-T6	
Young's modulus (GPa)	68.9
Poisson ratio	0.33
Density (kg/m^3)	2,700
Shear modulus (MPa)	25.84
Yield strength (MPa)	241

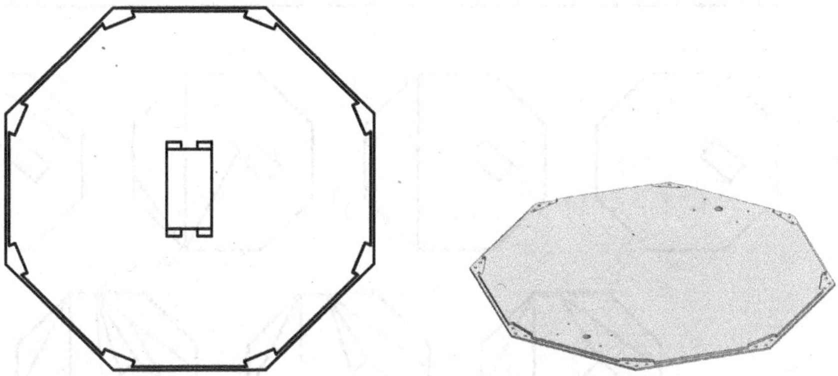

(a) Panel topology (b) Baseline panel design with a weight of 9.37 kg

Figure 5.38 Geometry of the panel.

has a yield stress of $\sigma_Y = 241$ MPa and ultimate tensile stress of $\sigma_{UTS} = 290$ MPa.[3] To maintain structural integrity and again considering a safety factor of 2, a stress constraint of $\sigma_{max} = 120.5$ MPa is applied.

The panel is constructed using a single shell and a frame of beams. The shell is divided into the *global shell* (the main structural support of the panel) and the *subsystem shell* (the portion of the panel that supports the subsystem). The beams are arranged on top of the *global shell* using the resulting layout generated by the ALS as exemplified in Figure 5.39. The beams themselves have square cross-sections and are divided in two categories: internal and external. Internal beams are designated as all beams created during the cellular developmental process, while external beams are defined as those present in the initial layout (i.e., the edges of the octagon). Thus, the design variables for this case study include the top deck topology (determined using

[3] Other typical values for the yield and ultimate tensile stresses are $\sigma_Y = 275$ MPa and $\sigma_{UTS} = 310$ MPa. For safety, the smaller values were chosen.

Table 5.5 Global constants used in the case study that represent physical terms.

Constants	Value	Description
g	9.80665	Gravitational acceleration (m/s^2)
A_{panel}	0.281605	Static area of the planar panel (m^2)
A_{IMU}	0.009158	Static area of the planar IMU (m^2)
L_{edge}	0.2415	Length of the side edges in (m)
m_{IMU}	0.299	Weight of the IMU (kg)
f_z	−10.00	Vertical load (g)
f_x	8.75	Lateral load (g)

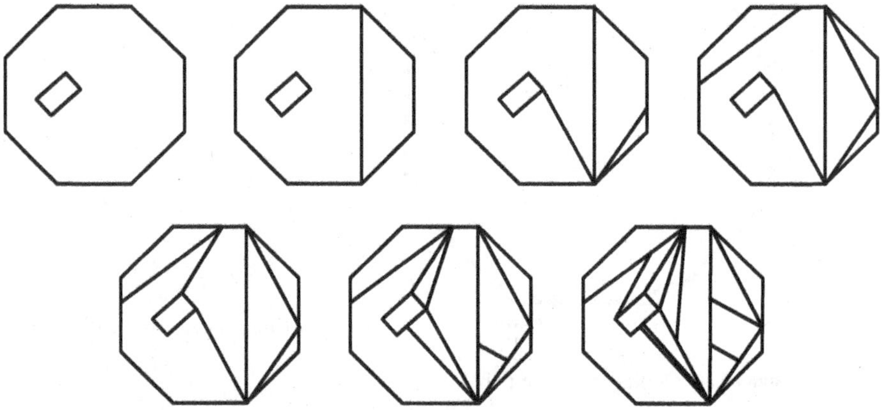

Figure 5.39 Example of the first six steps of the cellular division process using the ALS for the zenith deck of the HawaiiSat-1.

the ALS), the plate thicknesses of both the main region and the IMU regions of the panel, the lengths and square cross-sectional areas of both the external and internal beams, and the subsystem placement. Fixed boundary conditions are applied to the eight vertices of the octagon panel, while the edges of the panel are free to deform. During the cellular development process, constraints were placed on the minimum edge length (2% of the characteristic length of the panel, $\approx$ 0.6 m); minimum cell area (2% of the characteristic area of the panel, $\approx$ 0.28 m^2); and minimum angle (10°, for manufacturing purposes). Finite element modeling and analyses were performed using COMSOL and assumed the physical constants provided in Table 5.5. For reference purposes, a baseline design is established (cf. Figure 5.38(b)), which features a panel thickness of 12.5 mm and a weight of 9.37 kg (excluding the weight of the IMU).

Optimization

A series of optimizations was performed using a variety of different population sizes and maximum generation numbers. Herein, results that consider a population size

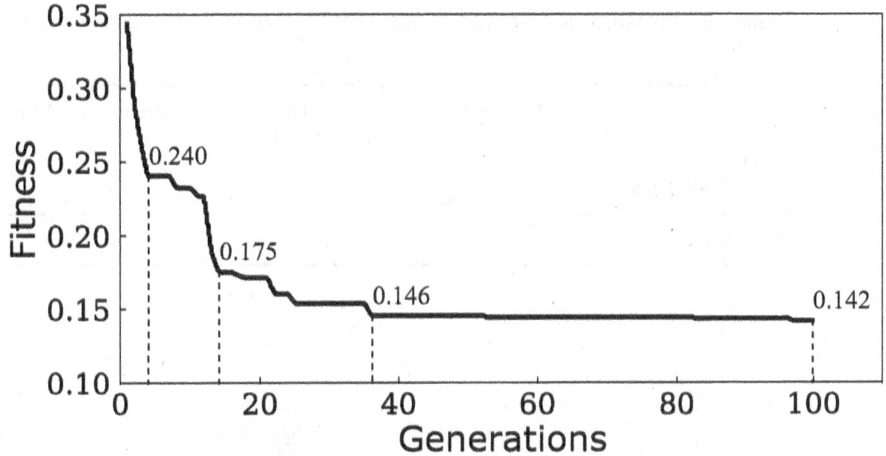

(a) Fitness values as a function of generations

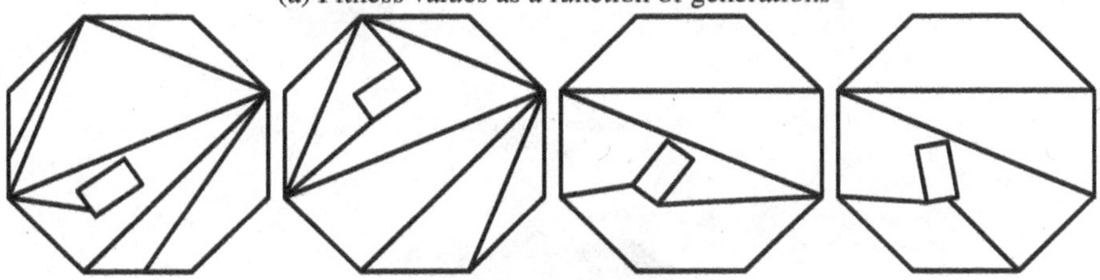

(b) Topologies that correspond to the selected fitness values in the plot above

Figure 5.40 Zenith deck optimization results for a 200-individual population evolved for 100 generations.

of 200 for 100 generations will be reported. Figure 5.40 illustrates the improvement in weight reduction as the population evolves; the final optimized mass of 1.280 kg represents a reduction of 86% compared to the mass of the baseline. This result follows the usual pattern of genetic optimization with jumps in the fitness improvement followed by long "diffusive" periods of stasis. Also shown in Figure 5.40 are key topological changes in the best-performing individuals during optimization. Note that these structures become topologically simpler as the optimization proceeds, with the final result consisting of a relatively simple geometry that differs from conventional truss-like designs.

Figure 5.41 shows the results of the finite element analysis for the optimized layout. Note that the stress levels obtained are much smaller than the allowable stress for this alloy, which indicates that stiffness (via the displacement constraint) drives this design rather than strength (stress constraint). For validation purposes, a 3-D SOLIDWORKS model of the optimized structure was constructed, and the displacement and stress distribution simulations obtained using both COMSOL and SOLIDWORKS were compared. While the COMSOL model uses 1-D beams and 2-D plate finite elements, the SOLIDWORKS model features 3-D continuum elements; this increase in

Table 5.6 Comparison between results from COMSOL and SOLIDWORKS for the most optimized structure.

	Mass (kg)	Max. Displacement (μm)	Max. Stress (MPa)
COMSOL	1.280	473.28	33.6
SOLIDWORKS	1.202	461.91	42.2
Difference	6%	2%	26%

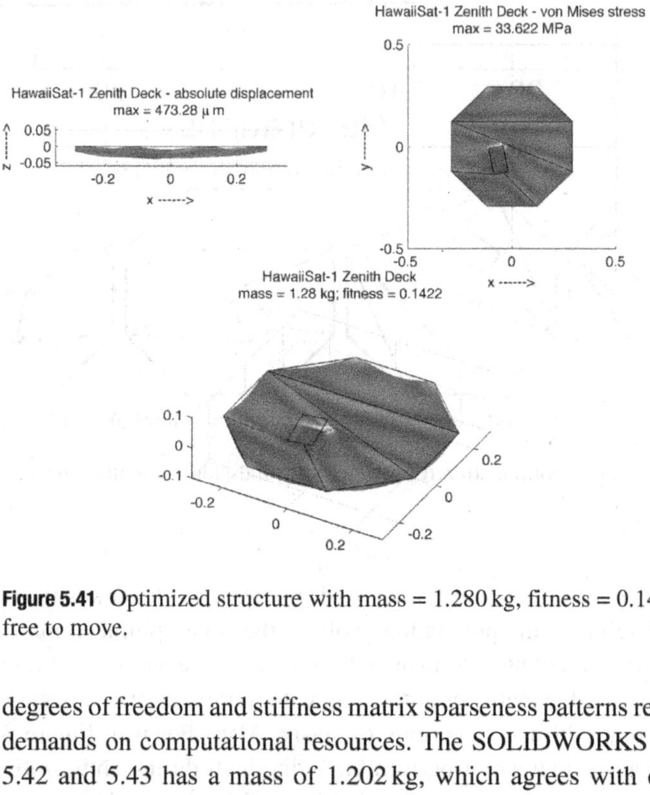

Figure 5.41 Optimized structure with mass = 1.280 kg, fitness = 0.1422, and subsystem that is free to move.

degrees of freedom and stiffness matrix sparseness patterns results in drastically higher demands on computational resources. The SOLIDWORKS model shown in Figures 5.42 and 5.43 has a mass of 1.202 kg, which agrees with optimized mass obtained using COMSOL, but is slightly lower due to a more accurate modeling of the joints. The values of the maximum displacement and maximum von Mises stress predicted by both models are compared in Table 5.6 and show good agreement, validating the use of the more efficient COMSOL model for the optimization (the stress values differ due to modeling and mesh differences, but they are still far from the allowable stresses).

5.4 Conclusions

This chapter presented and demonstrated a method for the design and optimization of engineering structures via the processes observed during cellular development of organisms. Based on these results, the following conclusions can be drawn:

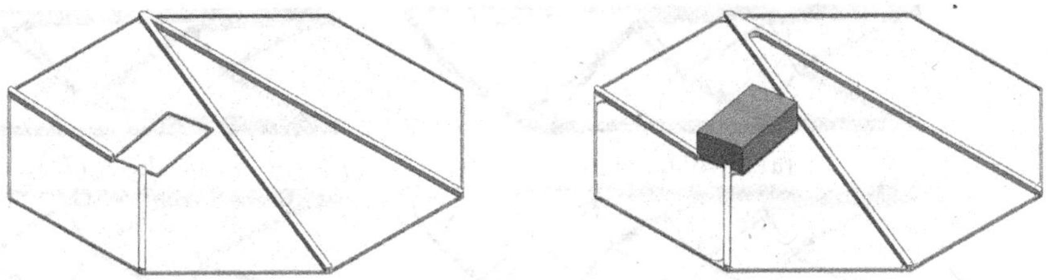

Figure 5.42 SOLIDWORKS model for the best individual. Raw model on the left and finalized model with chamfers on the right.

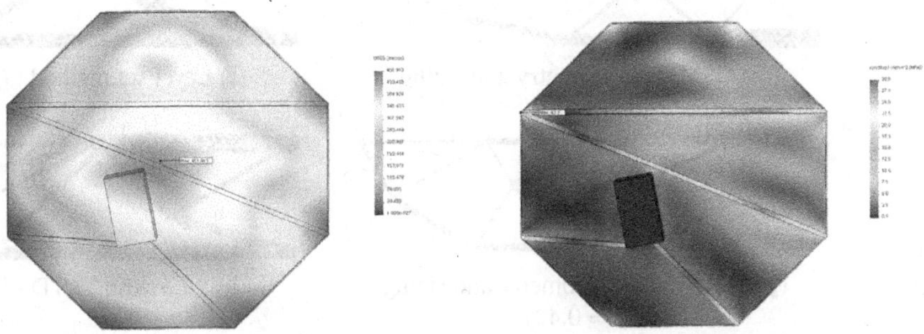

Figure 5.43 Optimized structure. Mass = 1.202 kg, Disp = 461.9 μm, von Mises Stress = 42.2 MPa.

- The developmental approach of the ALS lends itself to the creation of frame structures when searching for optimal designs. This improves the efficacy of the evolutionary programming in discovering and exploring features in the design space, allowing ALS to create full populations of effective, realizable designs. In contrast, evolutionary programming methods based on the conventional pixelation of the design domain must discover connected load paths via trial and error. Furthermore, when realizable designs are discovered using conventional pixel-based methods, the features guaranteeing feasibility are not guaranteed to be passed to the offspring.
- The guiding of the search toward frame structures greatly enhances the exploration of the design space in the evolutionary programming. That enhancement, in turn, allows for a smaller population of designs and a smaller number of generations to be used. While convergence toward a local optima in the design space may remain slow if the optimization relies solely on the GA, the exploitation of local optimized solutions can be accelerated via directed cellular dynamics. This approach takes advantage of the smoothness of the functional with respect to shape parameters to create an efficient search method capable of finding local optimal solutions.
- The use of the directed search enhances the search for the optimal design by developing a basin of attraction to the optimized solution: any individual design that is

(a) SIMP ($\hat{J}_c = 0.56$)

(b) SERA ($\hat{J}_c = 0.54$)

(c) Homogenization ($\hat{J}_c = 0.50$)

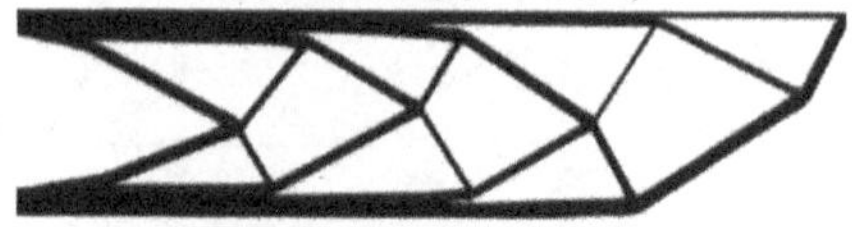

(d) EA-SIMP ($\hat{J}_c = 0.55$)

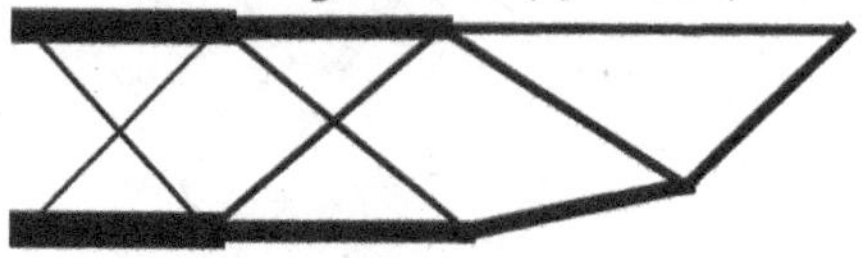

(e) GSM with geometry and sizing ($\hat{J}_c = 0.46$)

(f) Level set method ($\hat{J}_c = 0.51$)

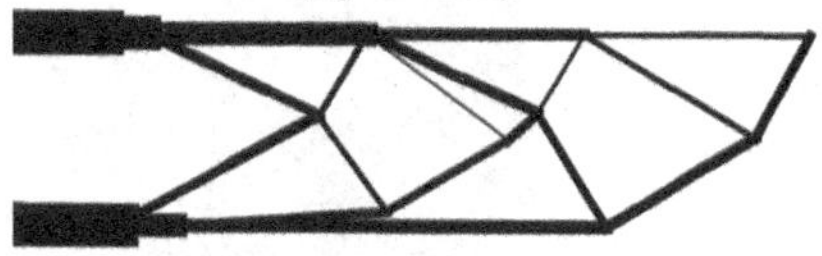

(g) SPIDRS with geometry and sizing ($\hat{J}_c = 0.42$)

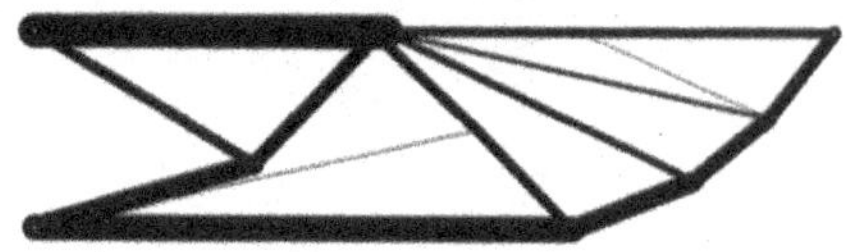

(h) ALS with TPTD ($\hat{J}_c = 0.42$)

Figure 5.44 Summary of solutions to the topology optimization problem of Figure 2.3, including those obtained from the various methods reviewed in Chapter 2 and the most capable of SPIDRS and ALS implementations.

sufficiently close to the optimized solution can be driven toward it by the directed search stage.

- The final component presented in this chapter is the use of the TPTD to decide when/where to divide a cell along a pair of markers. With this method, a population of one individual has been shown to be capable of providing solutions to benchmark problems that perform as well as or better than other optimized solutions available in the literature.
- Compared to the popular methods, such as SIMP and the level set method (cf. Chapter 2), the cellular developmental method greatly reduces the number of design variables and allows for easy control of the size and aspect ratio of frame members and the angles between them. In addition, the optimized frame requires neither additional postprocessing nor *a priori* knowledge of the desired resolution of the frame features or size of the design domain.
- As illustrated in Figure 5.44, the solution obtained for the long cantilever benchmark problem using the TPTD features a smaller compliance than all the gradient-based methods studied in Chapter 2 and one comparable to the most capable SPIDRS-generated solution detailed in Chapter 4.

- Optimization of a satellite zenith panel shows that the methodology developed in this chapter can be successfully employed in the simultaneous optimization of structural topology and subsystem placement.

Having presented methodologies for topology optimization in this and previous chapters, we now turn our attention to demonstrating the potential of these methodologies through real-world, multidisciplinary problems.

6 Multiphysical Applications

A motivating factor in the development of a bioinspired topology optimization approach is the need to consider multiphysical design problems, or problems that involve more than one simultaneously occurring physical field (mechanical, thermal, aerodynamic, electromagnetic, etc.), especially those for which no prior intuition exists. As discussed in Chapter 1, the complex couplings between multiple physical fields can complicate the calculation of the derivatives needed for classical topology optimization approaches and often result in additional design objectives and/or constraints, further hindering rigorous exploration of the design space using gradient-based algorithms. Having demonstrated the efficacy of the proposed approaches using benchmark structural topology optimization problems, this chapter will summarize several applications of these methods to multiphysical design problems. This is by no means an exhaustive survey, as this remains an active area of research; however, the variety of applications considered should convey the ease with which each L-system interpretation approach can be adapted and recast for nonstructural design problems.

6.1 Turtle Graphics Interpretation

Chapter 3 demonstrated that a turtle graphics interpretation of L-system encodings is unsuitable for structural topology optimization. However, design problems considering other physical mechanisms could represent applications for which the branched topologies generated by this interpretation method may be advantageous. In particular, work performed by Kobayashi (2010) explored the use of the L-system/turtle graphics approach in bioinspired problems involving fluid transport and heat transfer. The findings of this work are summarized here, with the goal of simply demonstrating the relevance of the method to this class of problems rather than commenting on evolutionary biology and the selection of any particular designs.

6.1.1 Fluid Transport via Leaf Venation

The first problem examines the evolution of the venation of a cordate leaf, which is modeled here as a cordioid: $r = a(1 + \cos\theta), \theta \in [0, 2\pi]$, where (r, θ) are polar coordinates centered at the leaf root. The leaf consists of a porous material usually

Table 6.1 Description and relation/assumed values for the variables in Equation 6.1.

Variable	Description	Relation	Assumed Value
p	Dim. pressure	$(P - P_0)/\Delta P$	0 (root)
P	Pressure	–	–
P_0	Root pressure	–	–
ΔP	Pressure drop	$\dot{m}\nu/\delta K_M$	–
$\dot{m}$	Fluid transpiration rate	–	0 (leaf border)
ν	Kinematic viscosity	–	–
δ	Leaf thickness	–	–
K_M	Membrane permeability	–	–
κ	Dim. permeability	–	1 (membrane), 1,200 (veins)
τ	Dim. transpiration rate	–	1 (membrane), 0 (veins)

modeled via Darcy's law (Roth-Nebelsick et al. 2001) that can be split into two parts: the high permeability veins that facilitate efficient mass flow and the compact membrane, which enhances the diffusion of nutrients and transpiration. In nondimensional form, Darcy's law for the fluid transport across a leaf is written as

$$\nabla \cdot (\kappa \nabla p) + \tau = 0, \tag{6.1}$$

with variables defined in Table 6.1. The multiobjective optimization problem seeks to design the best layout of the highly permeable vein that simultaneously minimizes two objectives: leaf mass and the total pressure drop over the leaf. The vein layout is modeled using an L-system coupled with a turtle graphics interpretation (cf. Chapter 3); the L-system consists of an alphabet of four letter characters ($\boldsymbol{\Sigma} = \{\mathtt{A}, \mathtt{B}, \mathtt{C}, \mathtt{D}\}$) and therefore four production rules, a two-character axiom, and four levels of recursion. The turtle graphics interpretation requires the variables δ_{α_i}, δ_w, and δ_θ (cf. Table 3.1). Results were obtained using a population size of 200 for 100 generations, with the COMSOL Multiphysics® suite (COM 2008) used to compute the two objective functions. Both functions are penalized to eliminate designs where the percentage of vein mass with respect to the total mass is less than 3% and greater than 15%, ensuring meaningful solutions.

Figure 6.1 shows a portion of the approximate Pareto frontier and the aggregate population from various generations, with a number of Pareto-optimal topologies designated by squares on the frontier, also illustrated on the right to better demonstrate the effect of the conflicting design requirements. It is noticeable that designs with more mass feature vein topologies that are highly bifurcated. This is due to the high permeability ratio between the veins and the membrane, which allows fluid to be carried along the veins with little effort compared to through the membrane and causes the pressure in areas along the veins to remain consistent with that of the root value, provided the veins are not too thin. Therefore, the leaf venation tends to bifurcate to increase its effective coverage area and reduce the pressure drop across the leaf. Designs with smaller vein mass, however, do not have secondary bifurcations, as the

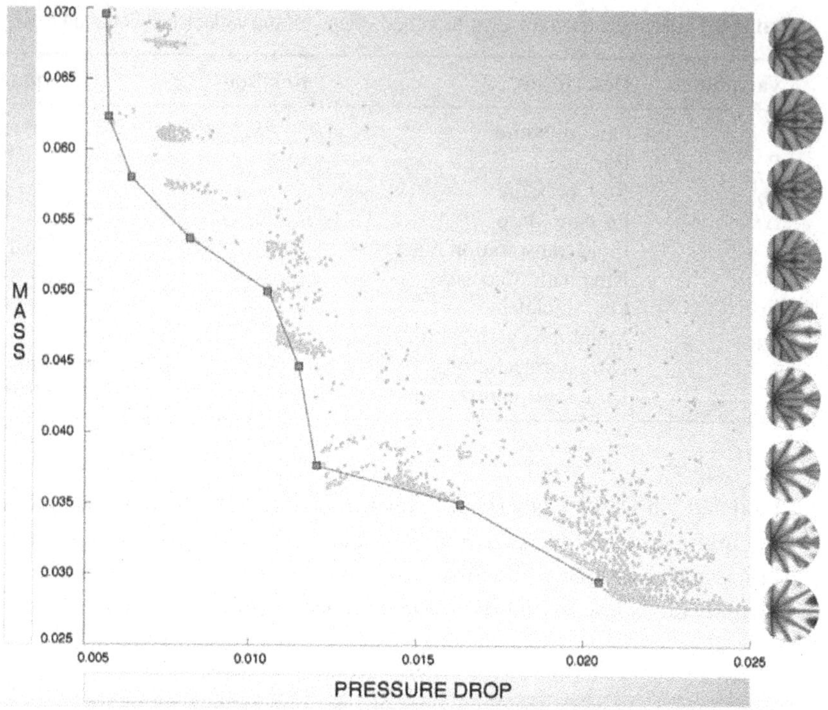

Figure 6.1 Approximate Pareto frontier for the evolution of cordate leaf venation with the objectives of minimizing mass and pressure drop across the leaf along with several topologies of interest. Adapted from Kobayashi (2010) with permission.

increase in the pressure drop near extremely thin veins does not compensate for the relatively small increase in coverage area.

6.1.2 Fluid Transport for Cooling Systems

Inspired by the results of Section 6.1.1, a second problem considers the optimum layout design of a porous medium for cooling. Given that the cooling of microchips is a major hurdle to the continued applicability of Moore's Law (i.e., the number of transistors in an integrated circuit will double every two years (Mack 2011)), this represents a real-world design problem that has been widely studied in previous works (Bruns 2007, Dirker & Meyer 2013, Coffin & Maute 2016). Here, the analogy between the transport phenomenon in plant leaves and cooling systems represents a compelling but complex bioinspiration for the design of fluidic systems to cool central processing units.

The design domain consists of a thin, axisymmetric disk of porous medium occupying a circular region of radius R. The disk is heated from below at a constant rate q'', while cooling fluid enters from the center of the disk and leaves the domain at the outer radius R. As before, the porous medium consists of two different components: a highly permeable material (i.e., the "veins") and a less permeable material (i.e., the

Table 6.2 Description and relation/assumed values for the variables in Equation 6.2.

Variable	Description	Relation	Assumed Value
p	Dim. pressure	$(P - P_0)/\Delta P$	0 (outlet)
P	Pressure	–	–
ΔP	Pressure drop	$U_{in}\mu\sqrt{A}/K_M$	–
P_0	Outlet pressure	–	–
U_{in}	Fluid velocity	–	1 (inlet)
μ	Fluid viscosity	–	–
A	Surface area	–	–
K_M	Membrane permeability	–	–
τ	Dim. transpiration rate	–	1 (membrane), 0 (veins)
κ	Dim. permeability	–	1 (membrane), 10^4 (veins)
$\mathbf{v}$	Dim. velocity	$\mathbf{V}/U_{in}$	–
$\mathbf{V}$	Velocity	–	–
ϵ	Porosity	–	0.1 (membrane), 0.6 (veins)
Pe	Peclet number	$U_{in}\sqrt{A}/\alpha_f$	100
α_f	Fluid thermal diffusivity	–	–
θ_f	Dim. fluid temperature	$(T_f - T_0)/T_R$	0 (inlet)
θ_s	Dim. solid temperature	$(T_s - T_0)/T_R$	0 (inlet)
T_f	Fluid temperature	–	–
T_s	Solid temperature	–	–
T_0	Inlet temperature	–	–
T_R	Reference temperature	$q''A/\delta k_f$	–
q''	Heat rate	–	–
δ	Disk thickness	–	–
k_f	Fluid thermal conductivity	–	–
k_s	Solid thermal conductivity	–	–
Nu	Nusselt number	$h_{sf}\sqrt{A}/k_f$	4
h_{sf}	Interstitial heat transfer coeff.	–	–
$\tilde{k}$	Thermal conductivity ratio	k_s/k_f	500
a^*_{sf}	Reduced specific surface area	$a_{sf}\sqrt{A}$	100
a_{sf}	Specific surface area	–	–
Bi	Biot number	$h\sqrt{A}/k_s$	10

"membrane"). Darcy's law is again used to model both the fluid flow in the porous medium and the transport of energy in the medium for heat transfer. The full model for this problem is defined in nondimensional form as

$$\begin{aligned} &\nabla \cdot (\kappa \nabla p) + \tau = 0, \\ &\epsilon Pe \mathbf{v} \cdot \nabla \theta_f = \nabla \cdot \left(\epsilon \nabla \theta_f\right) + Nu\left(\theta_s - \theta_f\right) + \epsilon, \\ &\nabla \cdot \left[(1-\epsilon)\tilde{k}\nabla \theta_s\right] - Nu\left(\theta_s - \theta_f\right) + 1 - \epsilon = 0, \end{aligned} \tag{6.2}$$

with variables defined in Table 6.2. As boundary conditions, we define a negligible diffusive flux of fluid energy $\partial\theta_f/\partial r = 0$ and a convection heat transfer $\partial\theta_s/\partial r = Bi(\theta_f - \theta_s)$ in the solid. The multiobjective topology optimization problem seeks to design the best layout of the highly permeable vein that minimizes both the

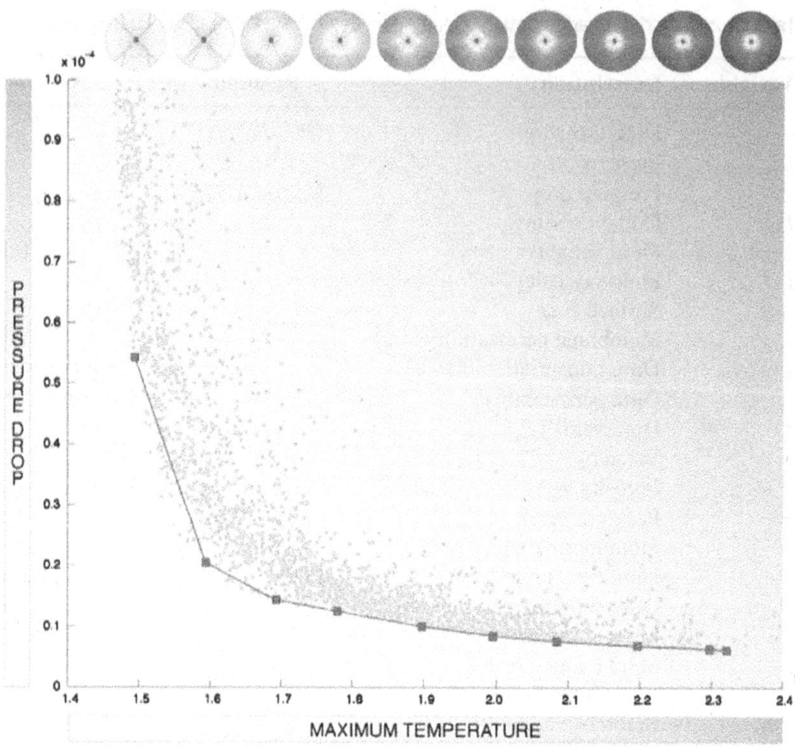

Figure 6.2 Approximate Pareto frontier for the evolution of a cooling system with the objectives of minimizing maximum temperature and pressure drop across a porous medium along with several topologies of interest. Adapted from Kobayashi (2010) with permission.

maximum temperature in the membrane (essential to the correct and efficient operation of electronics) and the pressure drop in the fluid between the inlet and outlet (increases power consumption in the cooling system). The vein layout is modeled using the same L-system/turtle graphics interpretation discussed in Section 6.1.1, with results obtained using a population size of 200 for 100 generations and analysis again completed using the COMSOL Multiphysics® suite.

A portion of the resulting approximate Pareto frontier and aggregate population is illustrated in Figure 6.2, along with several Pareto-optimal topologies along the frontier as designated by squares. The high thermal conductivity of the membrane relative to the veins quickly equalizes the temperature in the annular direction such that a low vein area decreases the maximum temperature. However, because the membrane is more compact, this decrease in temperature is accompanied by an increase in the pressure-drop. Large vein areas result in the opposite effect, with the transition between the two states resulting in a smooth increase in both vein thickness and coverage. Interestingly, the results indicate that major reductions in the pressure drop are predicted along the portion of the approximate Pareto frontier that corresponds with a relatively small temperature increase (i.e., at maximum temperatures of less than 1.6×10^{-3} in Figure 6.2).

6.2 SPIDRS

The graph-based interpretation of parametric L-system encodings associated with the SPIDRS algorithm in Chapter 4 demonstrated improved performance for purely structural topology optimization problems when compared to the turtle graphics approach, especially due to its effectiveness in generating well-connected convex cells. This approach has also shown promise in design problems that consider additional physical mechanisms beyond merely forces, deflections, stresses, and strains due to its fundamental role as a network design tool. Here, we will consider two examples: a straightforward structural insulation problem that combines both structural and heat transfer analysis, and a more complex electrical circuit problem that exemplifies potential applications in the field of electromagnetics, with strong future potential to combine structural and electromagnetic design toward multifunctional and sensory structures.

6.2.1 Structural Insulation

The goal of this multiobjective problem is to design a structure capable of acting as an insulator (i.e., minimizing thermal conductivity) while also maximizing its resistance to both axial and shear loads. Note the clear paradoxes in this problem, as the numerous connectivities critical to structural stiffness directly and adversely increase thermal conductivity as well. Here, we consider the periodic structure shown in Figure 6.3, which consists of small unit cells with heights and widths of 100 mm each. The initial (null) structure and subsequently generated structural members feature an assumed square cross-section (e.g., x–z and y–z planes) of 1 mm × 1 mm and consist of isotropic aluminum with a specified elastic modulus of $E = 68$ GPa, Poisson's ratio of $\nu = 0.36$, density of $\rho = 2,700\,\text{kg/m}^3$, and bulk thermal conductivity of $K_{bulk} = 210\,\text{W/m}\cdot\text{K}$. The initial state of the graph is defined as $F = \{[1,4,3,2],[1,2,3,4]\}$, with geometric symmetry boundary conditions placed on the planes associated with the half-edges e_{12} and e_{34}. Note that structural members along these edges of symmetry are removed from the final graph topology to prevent predetermining aspects of the topology, which will prove critical to the final results.

For this problem, three individual thermal-mechanical conditions are considered, as shown in Figure 6.4. First, starting from an initial temperature of $T_0 = 298$ K, a temperature of $T_1 = 318$ K is applied to the right side of the structure, while the left side is held constant at T_0 (Figure 6.4(a)). The thermal conductivity of the structure is then calculated as

$$K = \frac{q_t L}{T_1 - T_0},$$

where q_t is the steady-state heat flux, or rate per unit area at which heat flows through the structure as computed via finite element analysis, and L is the distance between the two walls of the structure. Next, a fully fixed boundary condition (i.e., $u_x = u_y = 0$) is applied to the left side of the structure, and an applied distributed force of $f_x = -0.001$ N/mm is applied to the right side (Figure 6.4(b)). The resulting output dis-

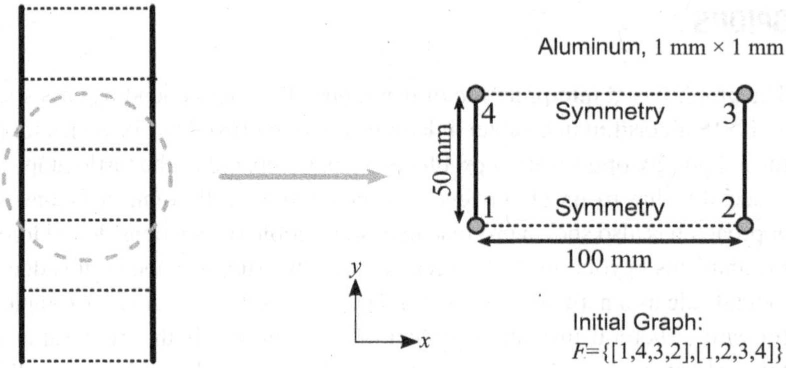

Figure 6.3 Initial graph and boundary conditions associated with the structurally stiff insulator design problem.

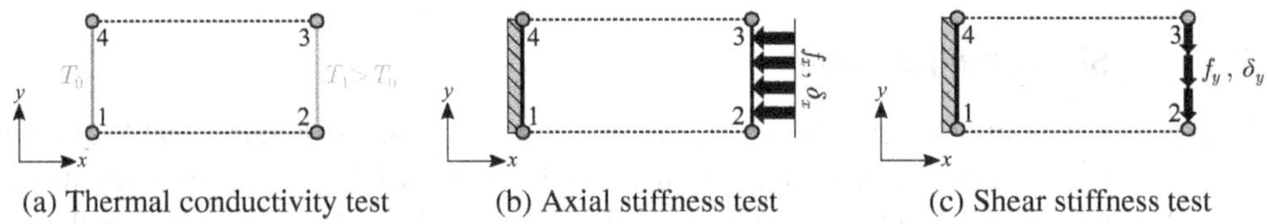

Figure 6.4 Illustration of the three load cases for the structurally stiff insulator design problem.

placement δ_x is then measured to provide an indication of the axial stiffness k_{axial} of the structure. Finally, with the fully fixed boundary condition retained, an applied distributed force of $f_y = -0.001$ N/mm is applied to the right side (Figure 6.4(c)), with the resulting output displacement δ_y providing an indication of the shear stiffness k_{shear}. Each load case is considered using a static and fully linear analysis step.

The optimization problem illustrated in Figure 6.4 is considered using the topology optimization framework discussed in Chapter 4 for 500 generations, a population size of 100 individuals, and three parametric L-system recursions. For each design, the resulting thermal conductivity, axial stiffness, and shear stiffness values are normalized by those associated with a completely solid aluminum structure ($\hat{K} = K/-2.1$ W/m·K, $\hat{k}_{axial} = k_{axial}/705$ kN/m, and $\hat{k}_{shear} = k_{shear}/96.8$ kN/m). The resulting approximate Pareto frontier in three dimensions is shown in Figure 6.5. The goal of minimizing the thermal conductivity of the structure favors designs with minimal load paths that are as long as possible (which will result in low structural stiffness), while the two stiffness objectives to be maximized call for the creation of numerous load paths (which will increase the thermal conductivity of the design). These trends are observable in the topologies shown in Figure 6.6 (denoted by gray squares in Figure 6.5), which also depicts the temperature contours calculated when analyzing the load case illustrated in Figure 6.4(a). Of special interest is the best insulating design shown in Figure 6.6(a), wherein structural members are constructed in a "zigzagged" pattern, making the total integrated path length between the two walls of the structure much greater than that seen in other designs, and thus greatly reducing the thermal conductivity of the part.

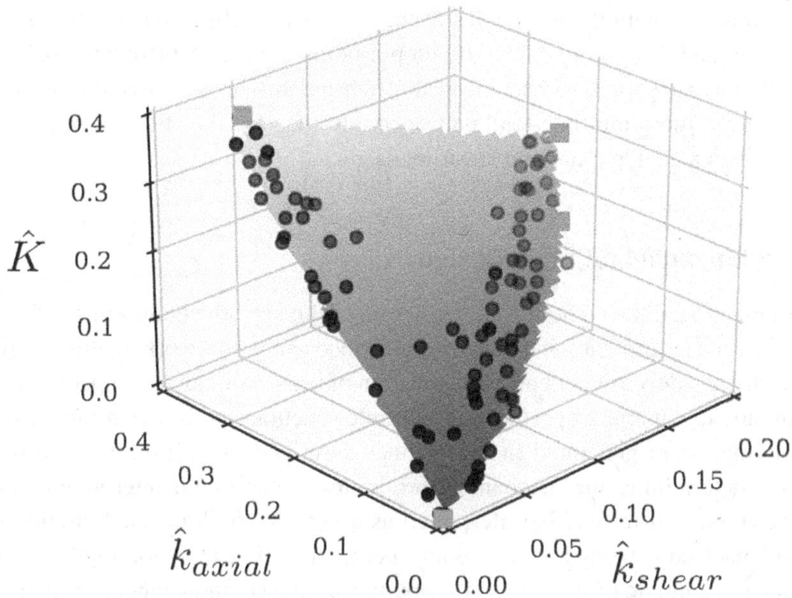

Figure 6.5 Three-objective Pareto frontier associated with the structurally stiff insulator design problem illustrated in Figures 6.3 and 6.4.

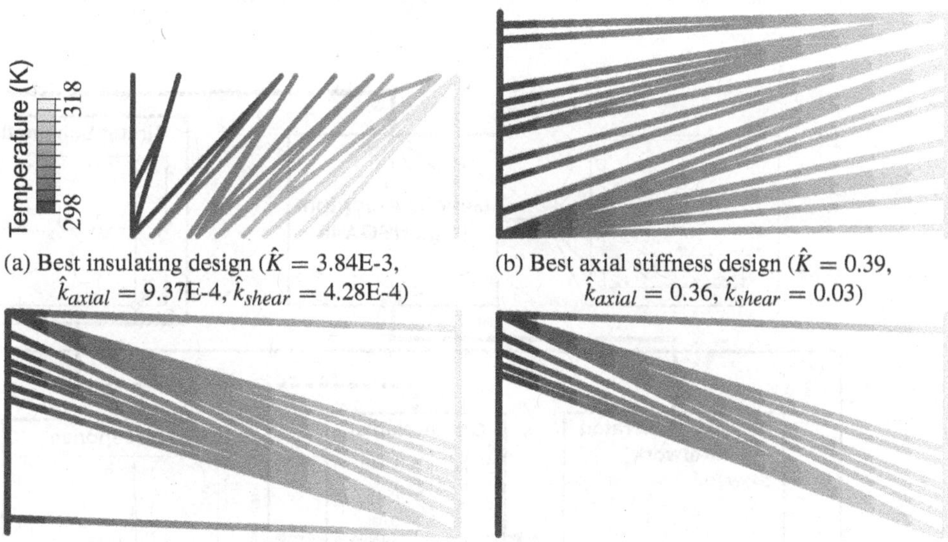

(a) Best insulating design ($\hat{K} = 3.84\text{E-3}$, $\hat{k}_{axial} = 9.37\text{E-4}$, $\hat{k}_{shear} = 4.28\text{E-4}$)

(b) Best axial stiffness design ($\hat{K} = 0.39$, $\hat{k}_{axial} = 0.36$, $\hat{k}_{shear} = 0.03$)

(c) Best shear stiffness design ($\hat{K} = 0.28$, $\hat{k}_{axial} = 0.24$, $\hat{k}_{shear} = 0.20$)

(d) Design closest to the utopia point ($\hat{K} = 0.21$, $\hat{k}_{axial} = 0.17$, $\hat{k}_{shear} = 0.16$)

Figure 6.6 Examples of SPIDRS-generated topologies along the Pareto frontier associated with the structurally stiff insulator design problem, shown here with temperature contours calculated when analyzing the load case illustrated in Figure 6.4(a).

Contrast this with the topologies of the best axial stiffness design, best shear stiffness design, and design closest to the utopia point (Figures 6.6(b)–6.6(d)), which feature multiple load paths for heat to flow through and thus an increased thermal conductivity. However, these multiple load paths serve to increase the stiffness of these topologies when subjected to axial and shear loads.

6.2.2 Electromagnetic/Circuit Design

As mentioned throughout Chapter 4, one of the key advantages of SPIDRS is its suitability to any application that can be described using a network. In this example, we will discuss the development and optimization of resistor–inductor–capacitor (RCL) circuit topologies capable of performing multiple functions as shown in Figure 6.7. Network topologies are generated in the manner described in Section 4.3.4; however, rather than determining the mechanical properties of the graph interpreted as a structure, the graph will instead be interpreted as a network of connected circuit components, with each edge being assigned both a component type (resistor, inductor, or capacitor) and a magnitude of the relevant property (resistance, inductance, or capacitance). The connectivity between components and the definitions of the components themselves are then translated into a circuit netlist, which contains all necessary information to perform conventional RCL circuit analysis (Nagel 1975). As the netlist is constructed,

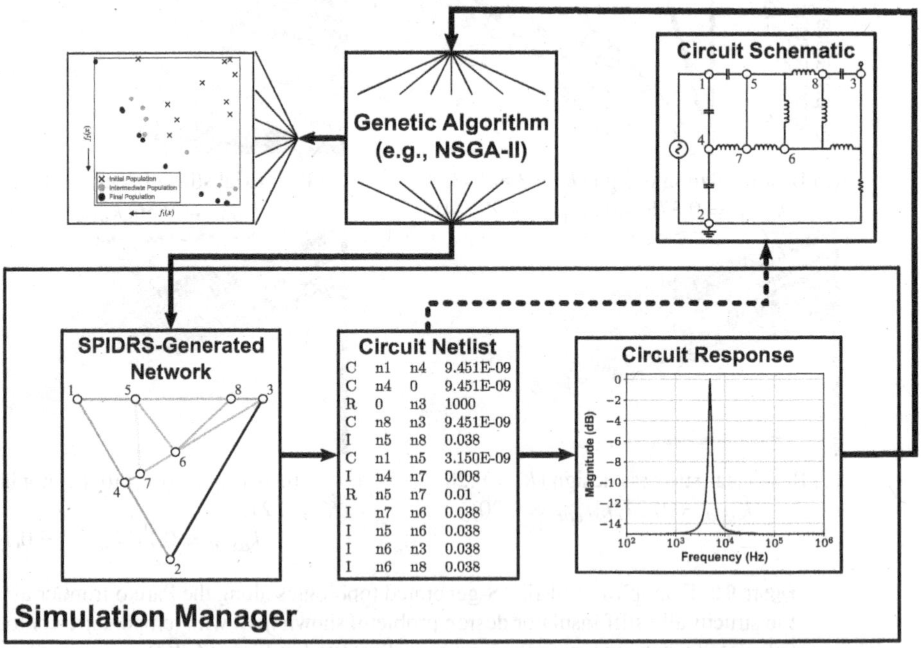

Figure 6.7 Flowchart depicting how the SPIDRS algorithm introduced in Chapter 4 can be used for electrical circuit design problems.

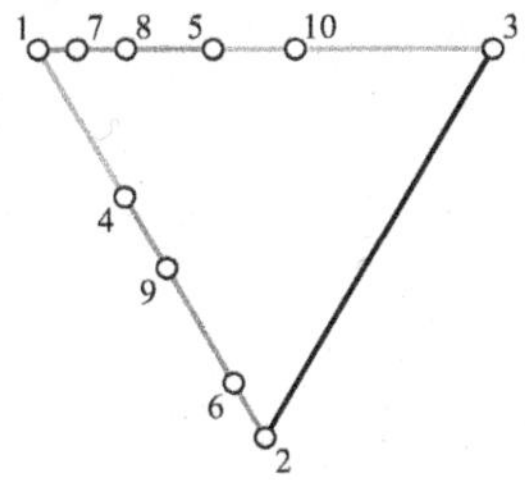

Circuit Netlist

I	n1	n4	0.341
C	n4	n9	3.740E-09
C	n9	n6	4.400E-09
C	n6	0	4.400E-09
R	0	n3	2000
I	n10	n3	0.974
I	n5	n10	0.974
C	n8	n5	1.562E-09
C	n7	n8	1.562E-09
C	n1	n7	1.562E-09

(a) Initial circuit topology

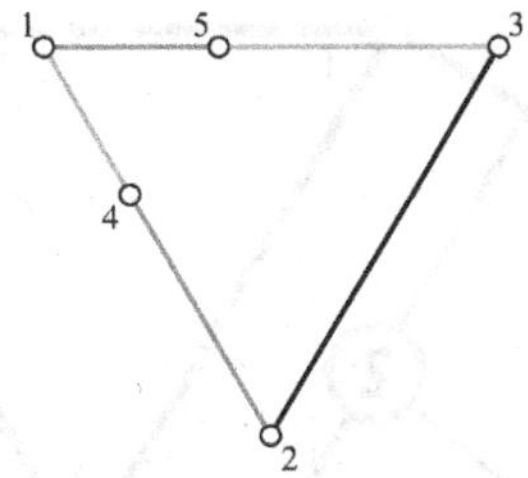

Circuit Netlist

I	n1	n4	0.341
C	n4	0	1.385E-09
R	0	n3	2000
I	n5	n3	1.947
C	n1	n5	5.206E-10

(b) Reduced circuit topology

Figure 6.8 Example of how SPIDRS-generated circuits can be simplified by combining components that are in series.

we simplify the circuit by combining any like components in series into an equivalent replacement component with a new equivalent property value per the following:

$$R_{eq} = R_1 + \cdots + R_N, \quad \frac{1}{C_{eq}} = \frac{1}{C_1} + \cdots + \frac{1}{C_N}, \quad I_{eq} = I_1 + \cdots + I_N,$$

where N is the number of components in series. An example is illustrated in Figure 6.8, where the 10 components in a circuit are reduced to 5. The simplified netlist is then passed to the Ahkab circuit simulator (Venturini 2015), a Python module that performs alternating current (AC) analysis, which returns the critical performance metrics (e.g., simulated frequency responses) for each circuit design that will be used by the genetic algorithm (GA) to evolve the population (cf. Section 1.3.3).

To demonstrate the design of an electrical circuit using SPIDRS, we consider a single band-pass filter (BPF) that rejects frequencies outside a specified range, with rejection being defined here as a power reduction of −3 dB or more. The goal is to match the response of a BPF adapted from Kuphaldt (1988) (cf. Figure 6.10(c)) with a center frequency of 5 kHz by minimizing both the error in the goal circuit response and individual circuit response and the number of circuit components required. We desire an error function that accounts for the decreasing importance of predictive accuracy as the Bode magnitude (i.e., the magnitude of the transfer function V_{out}/V_{in}) approaches the noise floor (set conservatively at −15 dB), and which neglects altogether inaccuracies below this level. Therefore, we choose an error function for the BPF that is weighted such that

$$\varepsilon_{BPF} = \sum_{i=1}^{N} \sqrt{\left(m_i^g - m_i^d\right) W_i \left(m_i^d; m_i^g\right)},$$

where m_i^g is the goal Bode magnitude at the ith frequency, m_i^d is the Bode magnitude of the current circuit design at the ith frequency, and W is a dynamic weighting function constructed such that all errors calculated below −15 dB are neglected while errors near the 0 dB goal peak are strongly penalized. For example,

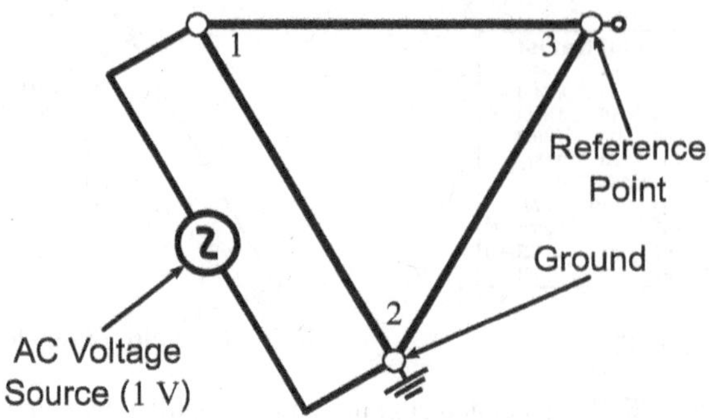

Figure 6.9 Boundary conditions and initial graph used for the two BPF design problems (cf. Figures 6.10–6.13).

$$W_i = \begin{cases} 0; & m_i^d < m_i^g = -15\,\text{dB}, \\ \left(m_i^g + 15\,\text{dB}\right)^2; & m_i^g > m_i^d > -15\,\text{dB}, \\ \left(m_i^d + 15\,\text{dB}\right)^2; & m_i^d > m_i^g. \end{cases}$$

Since $m_i^g \geq -15\,\text{dB}\ \forall\, i = 1 \ldots N$, this can be written compactly as

$$W_i = \left[\max\left(m_i^g, m_i^d\right) + 15\,\text{dB}\right]^2. \tag{6.3}$$

The initial graph that represents the circuit is illustrated in Figure 6.9 and considers a three-node network with an AC voltage source connected between nodes 1 and 2, a grounded condition at node 2, and a reference point located at node 3; the response of each SPIDRS-generated circuit is evaluated across $N = 500$ evenly distributed frequencies ranging from 10^2–10^6 Hz. The maximum possible values that can be assigned to the resistors, capacitors, and inductors in each SPIDRS-generated circuit are 20 kΩ, 10 nF, and 1 H, respectively. The optimization of the single BPF considers a population size of 100 individuals evolved for 1,000 generations. Figure 6.10 shows the reduced network and associated netlist, the corresponding circuit schematic, and the Bode plot for the SPIDRS-designed circuit (topology and component values) that best matches the specified goal response. The simple circuit utilizes seven components and features a center frequency of 5.05 kHz and a bandwidth (i.e., the difference between the upper and lower frequencies the filter will pass) of approximately 550 Hz, which agrees well with the specified circuit response (cf. Figure 6.10(c)), especially at meaningful power levels. Notice that the component between nodes 1 and 7 has an inductance of 1.25 H, which is greater than the maximum possible assigned value; this is a result of the circuit reduction technique previously discussed (cf. Figure 6.8), as this component represents four inductors in series with magnitudes all below 1 H.

For a more difficult design problem, we next consider a dual BPF, or a circuit that passes two distinct ranges of frequencies. An arbitrary and challenging Bode plot response is selected to represent the goal with one center frequency of ≈4 kHz and another of ≈25 kHz as given in Table 6.3 and shown in Figure 6.13(c). The optimization objectives are redefined to minimize the error between the goal response and

Table 6.3 Select data points from the goal response for the dual BPF design problem.

Peak 1		Peak 2	
Frequency (kHz)	Bode Mag. (dB)	Frequency (kHz)	Bode Mag. (dB)
2.500	−15.00	12.00	−15.00
3.041	−10.61	16.92	−10.43
3.631	−2.568	21.80	−2.955
3.715	−1.599	22.91	−1.522
3.802	−0.875	23.99	−0.592
3.890	−0.517	25.12	−0.350
4.074	−0.995	26.30	−0.870
4.169	−1.658	27.54	−1.914
4.266	−2.445	28.18	−2.532
5.695	−10.69	39.55	−10.51
7.500	−15.00	60.00	−15.00

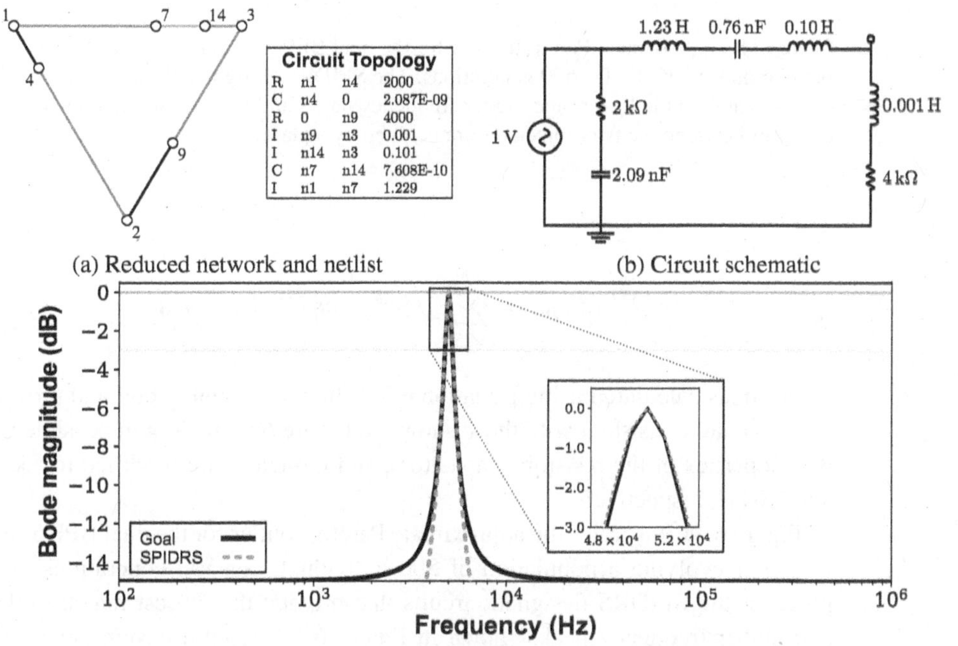

(a) Reduced network and netlist (b) Circuit schematic

(c) Bode plot

Figure 6.10 SPIDRS-designed single BPF with a center frequency closest to 5 kHz.

individual circuit response across both the first half (10^2–10^4 Hz) and the second half (10^4–10^6 Hz) of the considered frequency range, defined as

$$\varepsilon^1_{DBPF} = \sum_{i=1}^{250} \sqrt{(m_i^g - m_i^d)} W_i \left(m_i^d; m_i^g\right)$$

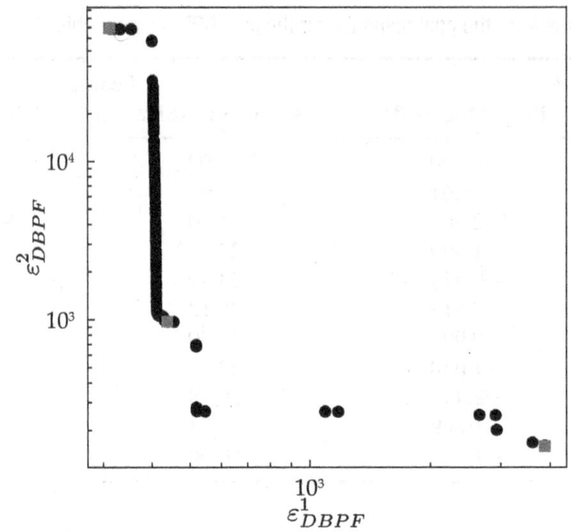

Figure 6.11 Approximate Pareto frontier for the dual BPF design problem after considering a population size of 500 for 500 generations. The SPIDRS-designed circuits that provide the closest match to the lower- and higher-frequency peaks and the circuit that represents the best trade-off between the two objectives are denoted by squares.

and

$$\varepsilon^2_{DBPF} = \sum_{i=251}^{500} \sqrt{(m_i^g - m_i^d)W_i\left(m_i^d; m_i^g\right)},$$

where W_i is calculated using Equation 6.3. While the initial circuit and provided voltage input are consistent with those shown in Figure 6.9, maximum possible values for the properties of the resistors, capacitors, and inductors are modified to 5 kΩ, 10 nF, and 0.04 H, respectively.

Figure 6.11 illustrates the approximate Pareto frontier for the dual BPF design problem after evolving a population of 500 individuals for 500 generations, with Bode plots for the SPIDRS-designed circuits that provide the closest match to the lower- and higher-frequency peaks shown in Figure 6.12. From a design perspective, it is also interesting to observe the SPIDRS-designed circuit that represents the best trade-off between the two objectives, the reduced network and netlist, corresponding circuit schematic, and Bode plot which is illustrated in Figure 6.13. This 12-component circuit has center frequencies of ≈4 kHz and ≈25 kHz, although it is clear from Figure 6.13(c) that the resulting response does not match the specified goal as closely as that of the single BPF example (cf. Figure 6.10(c)), particularly in regards to the bandwidth of the higher frequency range. However, the ability of the SPIDRS algorithm to generate a design capable of matching such a challenging, arbitrary, and potentially useful response to the accuracy demonstrated further proves the potential of this approach and enables the future consideration of more complex, multiphysical problems especially

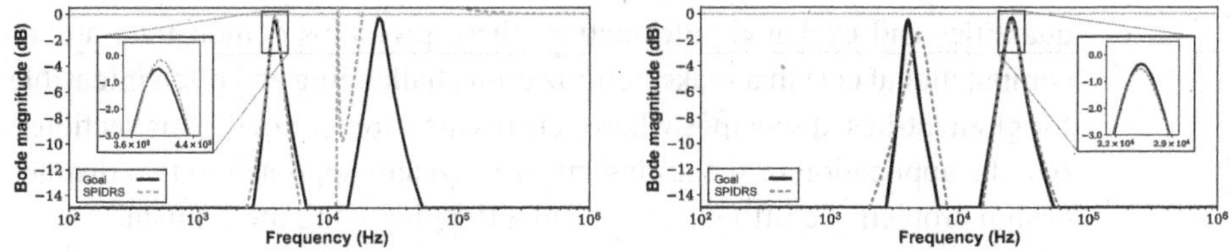

(a) Individual 1 (b) Individual 500

Figure 6.12 Bode plots for the SPIDRS-designed circuits that provide the closest match to the lower- and higher-frequency peaks.

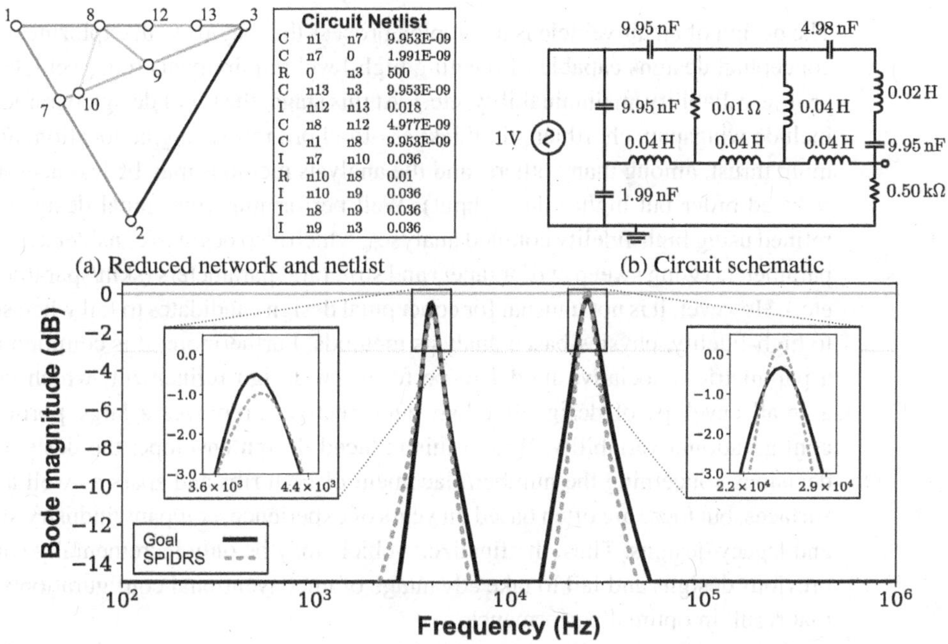

(a) Reduced network and netlist (b) Circuit schematic

(c) Bode plot

Figure 6.13 SPIDRS-generated dual BPF (Individual 471) that represents the best trade-off between matching the two arbitrarily specified center frequencies of ≈4 kHz and ≈25 kHz.

as mechanical and electromagnetic responses become highly coupled (e.g., as in the fields of soft robotics and wearable sensors).

6.3 Arrangement L-system

In addition to the promising structural performance observed for topologies generated using the arrangement L-system mentioned in Chapter 5, this approach has also been applied to aeroelastic design problems that consider the interaction between structural and aerodynamic forces. Due to the need for two different solvers to calculate physical quantities and exchange information, these problems demonstrate a complexity and computational cost that make optimization challenging and often infeasible, especially for gradient-based schemes where sensitivities are required. This section will summarize the application of the arrangement L-system approach to two distinct aeroelastic design problems: a lifting surface and a flapping wing mechanism.

6.3.1 Aeroelastic Lifting Surface

The design of an air vehicle is a complex process that begins with exploring a series of conceptual designs capable of meeting high-level requirements (e.g., vehicle mission profile, reliability, maintainability, etc.). At this stage, the set of design parameters may include wingspan, chord length, thickness-to-chord ratios, engine location, and maximum thrust, among many others, and the analysis methods may be less accurate (e.g., reduced order but higher throughput). Well-performing conceptual designs are then refined using high-fidelity coupled analyses, which also begin to consider aerodynamic parameters (wing sweep and/or taper) and structural parameters (skin/spar thicknesses, etc.). However, it is not unusual for conceptual design candidates to fail when subjected to high-fidelity, physics-based analysis methods. Furthermore, it is common to create a parametric associative model to perform the design refinement, which conserves a small envelope of design freedom but effectively removes a large portion of the configurational variability. Within this reduced design envelope, the designer makes decisions concerning the number/placement of both ribs and spars as well as control surfaces, but these are often based on years of experience, company/industry standards, and legacy designs. Thus, the finalized vehicle may be only incrementally superior to previous designs and fail to take advantage of unconventional configuration strategies that result in optimal performance.

With the above in mind, this section summarizes a novel optimization methodology that combines aerodynamic shape, structural sizing, structural topology, and subsystem topology (e.g., control surface number, size, and location) to design an air vehicle lifting surface with limited designer intuition and intervention (Kolonay & Kobayashi 2015). Specifically, this approach utilizes a bi-level optimization algorithm (cf. Figure 6.14) with the GA-based outer loop determining the structural layout of the lifting surface using the ALS, while also searching over the variables governing aerodynamic shape and control surface number, size, and location. A gradient-based inner loop then performs weight minimization using structural sizing variables (i.e., skin/panel thickness and spar/rib cap and post areas; cf. Section 4.4.1). The outer loop optimization problem is solved by NSGA-II (Deb et al. 2000) with the Automated STRuctural Optimization System (ASTROS) (Johnson & Neill 1988), providing structural analysis via linear finite elements and aerodynamic analysis using midplane linear potential theory and doublet lattice theory for steady and unsteady aerodynamics, respectively. For the inner loop, ASTROS utilizes the gradient-based μ-dot algorithm (Vanderplaats & Moses 1973) with analytic gradient calculations.

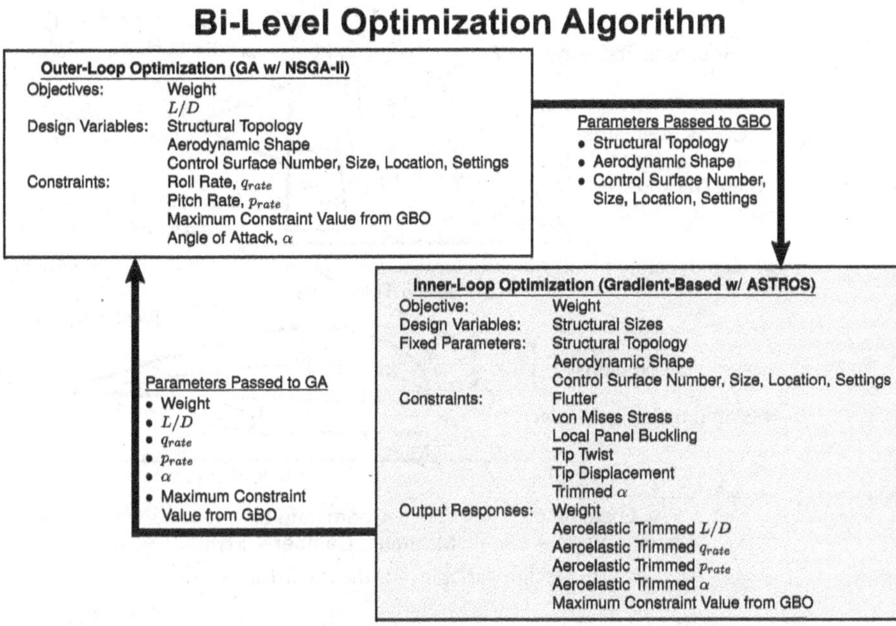

Figure 6.14 Bi-level optimization algorithm data flow. Adapted from Kolonay and Kobayashi (2015) with permission.

To demonstrate the proposed methodology, a multiobjective design problem is considered that seeks to minimize structural weight while maximizing the lift-to-drag ratio L/D (representing efficient aerodynamic performance) for generic fighter aircraft wing performing various maneuvers at a Mach number $M = 0.7$ subject to structural and aerodynamic constraints (cf. Figures 6.14 and 6.15). The approximate Pareto frontier generated for the multiobjective design problem after considering a population size of 200 evolved for 100 generations is illustrated in Figure 6.16, along with the entire feasible population after 100 generations. Figure 6.17 depicts the aerodynamic planform, underlying structural topology, and control surface size and locations for four distinct designs ordered along the Pareto frontier. These designs progress along the Pareto frontier, where design 1 represents a light structure with poor aerodynamic performance and design 4 represents a heavy structure with the most efficient aerodynamic performance. For designs 2 and 3, the active constraints include both top skin panel buckling during a 6g pullup maneuver and flutter constraints, while for design 4, the active constraints include skin panel buckling, flutter, and wingtip displacement. Additionally, in designs 2–4, a counterintuitive trend emerges: As the aspect ratio of these wings increases, the number of structural members in the topology decreases. Finally, considering the wing planform of each design, it is noticeable that design 1 corresponds to a typical high-performance fighter lifting surface, while designs 2 and 3 resemble "birdlike" wings. Design 4, on the other hand, is an example of a relatively straight high-aspect ratio wing planform. For a more detailed discussion of these results, see Kolonay and Kobayashi (2015).

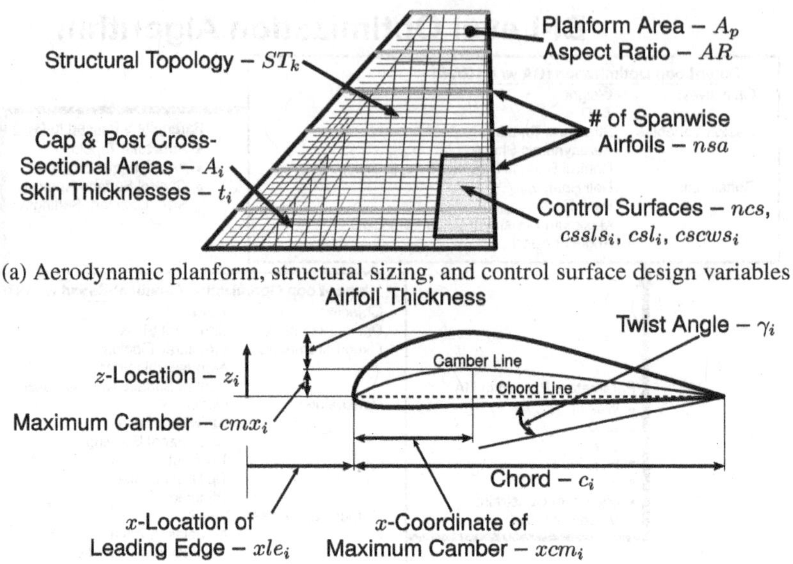

(a) Aerodynamic planform, structural sizing, and control surface design variables

(b) Design variables for the ith airfoil section

Figure 6.15 Design variables for the aeroelastic lifting surface design problem. Adapted from Kolonay and Kobayashi (2015) with permission.

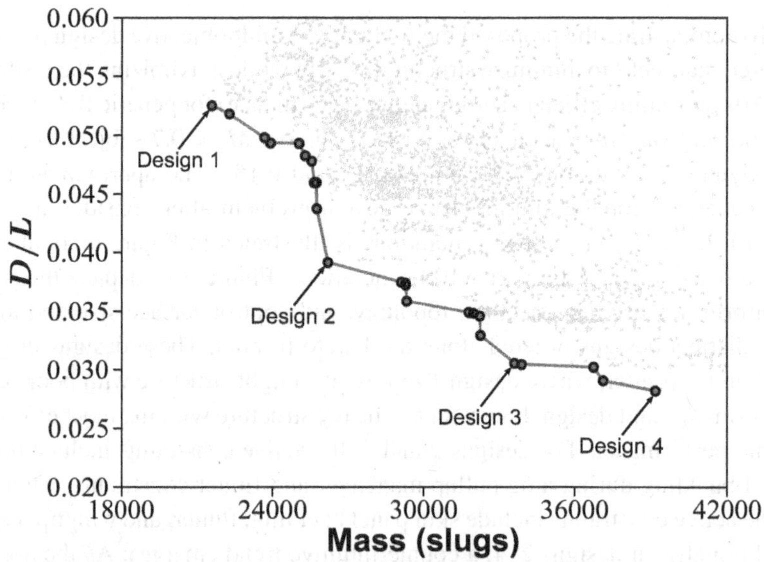

Figure 6.16 Pareto frontier and all feasible individuals in the population after 100 generations generated using the bi-level optimization algorithm described in Figure 6.14. Adapted from Kolonay and Kobayashi (2015) with permission.

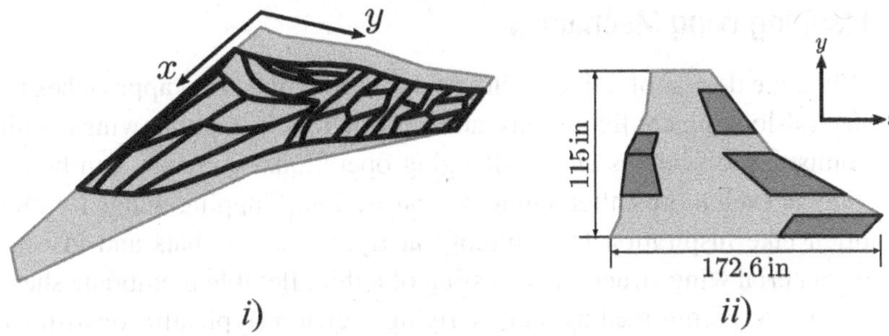

(a) Design 1

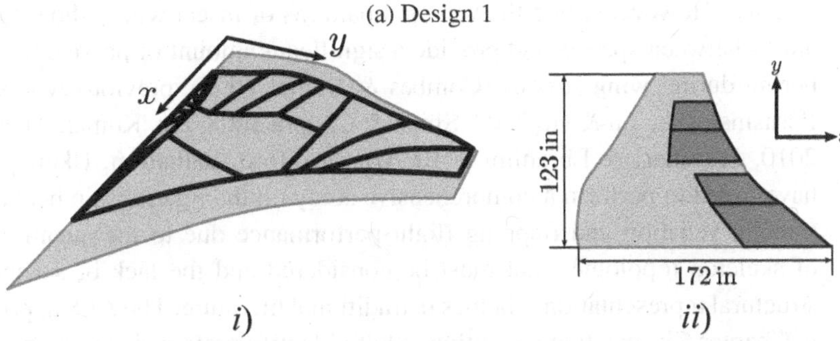

(b) Design 2

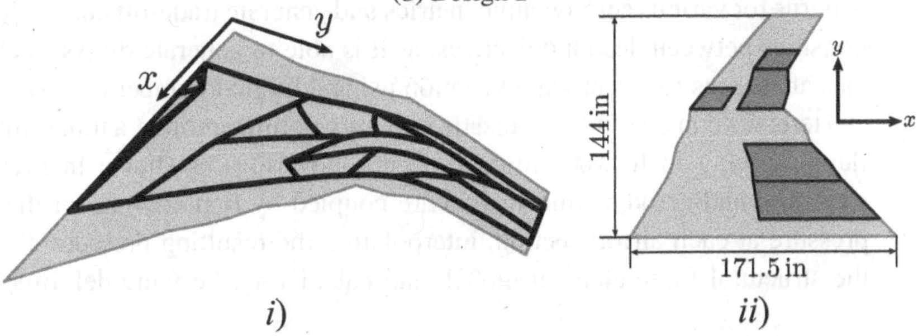

(c) Design 3

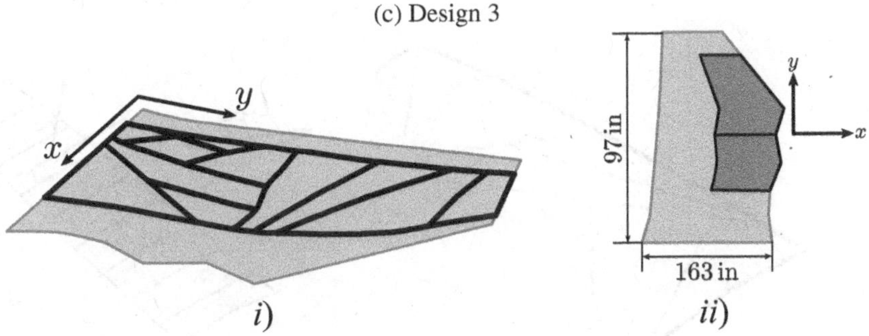

(d) Design 4

Figure 6.17 Illustrations of the (i) structural topology and (ii) control surface size/location and aerodynamic planform for the Pareto optimal designs denoted in Figure 6.16. Adapted from Kolonay and Kobayashi (2015) with permission.

6.3.2 Flapping Wing Mechanism

While the design of a fixed wing using these bioinspired approaches resulted in some natural-looking configurations, an investigation of flapping wings, with their multiple competing objectives and challenging operational contexts, can be expected to provide an even more enlightening demonstration. Flapping wings for micro air vehicles often take inspiration from biological flyers such as bats and insects and feature a bi-material wing structure consisting of a thin, flexible membrane sheet (latex, Mylar, nylon, etc.) reinforced by an underlying skeleton (typically consisting of carbon fiber battens). However, while the venation patterns of insect wings show substantial variations between species and provide a significant amount of passive twist and/or cambering during wing strokes (Combes & Daniel 2003), previous works (Ho, Nassef, Pornsinsirirak, Tai & Ho 2003, Singh & Chopra 2008, Hu, Kumar, Abate & Albertani 2010, Mazaheri & Ebrahimi 2010, Wu, Stanford, Sällström, Ukeiley & Ifju 2011) have failed to perform a comprehensive study on the relationship between membrane skeletal venation and flapping flight performance due to the quantity and diversity of skeleton topologies that must be considered and the lack of sufficiently flexible structural representation schemes in traditional literature. The ALS approach discussed in Chapter 5 is an attractive option to both identify preferred membrane wing venation patterns for various aerodynamic metrics and generate trade-off curves that define relationships between design objectives, as it is able to generate diverse, cell-like topologies analogous to insect wing venation using a limited number of design variables.

Here, we consider the aeroelastic topology optimization of a thin cambered flexible flapping wing in forward flight (i.e., non-hovering) as shown in Figure 6.18. The structural and aerodynamic solvers are coupled by first solving for the aerodynamic pressure at each airfoil section, interpolating the resulting pressure distribution onto the structural finite element model, and calculating the wing deformation to update

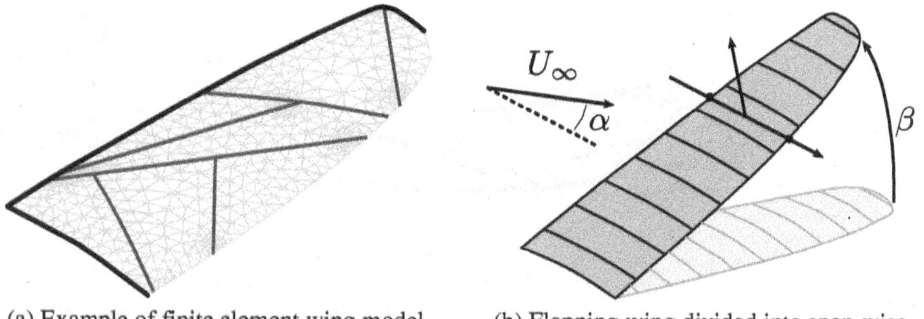

(a) Example of finite element wing model, with the membrane discretized into triangular elements and wing venation (dark gray), root, leading edge, and tip (black) represented as beam elements

(b) Flapping wing divided into span-wise sections

Figure 6.18 Structural and aerodynamic configurations for the flapping wing design problem. Adapted from Stanford, Beran & Kobayashi (2012) with permission.

the shape of each airfoil. Structurally, the membrane of the wing is discretized into a series of triangular elements, while the root, leading edge, tip, and generated battens that represent the venation of the wing are represented as linear Euler–Bernoulli beam elements. Aerodynamically, the wing is divided into a series of span-wise sections consisting of a parabolic arc airfoil with a peak camber of 2% of the local chord, no twist, and no dihedral. Aerodynamic pressures are calculated for each section by solving for the no-penetration condition using the airloads theory of Peters, Hsieh & Torrero (2007); this method allows for coupling with an unsteady wake model and accounts for large rigid body motion and small elastic deformations. Additionally, induced flow from the trailing wake is calculated using an inflow theory (Peters, Karunamoorthy & Cao 1995) that has been utilized for flexible aircraft with rigid body motions (Patil & Hodges 2006). For additional information, the reader is referred to Stanford et al. (2012).

For this study, the wing is defined to have a root chord of 0.16 m, a tip chord of 0.04 m, and a wing length of 0.4 m (root to tip). The membrane of the wing (E = 2 MPa, $\nu = 0.5$) is 0.1 mm thick, while the root, leading edge, and tip (E = 300 GPa, $\nu = 0.34$) are assigned a cross-section of 2 mm × 5 mm; the battens generated by the ALS (E = 300 GPa, $\nu = 0.34$) have an assumed cross-section of 0.8 mm × 3 mm. The wing is subjected to an external flow U_∞ = 10 m/s (corresponding to a Reynolds number $Re = 10{,}000$) at a static angle-of-attack $\alpha = 4^\circ$ and is flapped up and down through a single Euler angle $\beta = 30^\circ$ at a frequency ω = 40 rad/s. To develop the topology of the wing, the ALS is first applied to a square geometric domain before being mapped to the space defined by the planform of the vehicle (cf. Figure 6.19). The edges of the four outer walls are mapped to the wing planform to become the leading edge, wing root, trailing edge, and wingtip, while the internal edges are redrawn. Note that each internal edge could also be mapped to the wing planform, but this would result in curved wing battens/venation that are not amenable to membrane wing fabrication (Wu et al. 2011).

The goal of this multiobjective optimization problem is to determine the Pareto frontier that maximizes average lift and thrust while minimizing power. For a more detailed discussion on the optimization process and resulting topologies, the reader is referred to Stanford, Beran & Kobayashi (2012); here, we will simply comment on the three single-objective optimization problems explored and the resulting design trends. Each problem considers a population size of 100 for 100 generations to determine the corresponding optimal designs for maximizing thrust, minimizing power, and maximizing lift, which are determined via coefficients given as

$$C_T = -2F_x/\left(\rho_\infty U_\infty^2 S\right),$$
$$C_P = 2P/\left(\rho_\infty U_\infty^3 S\right),$$
$$C_L = -2F_y/\left(\rho_\infty U_\infty^2 S\right),$$

where F_x and F_y are cycle-averaged aerodynamic forces generated over a single wing normal and parallel to the freestream vector (cf. Figure 6.18(b)), respectively; P is

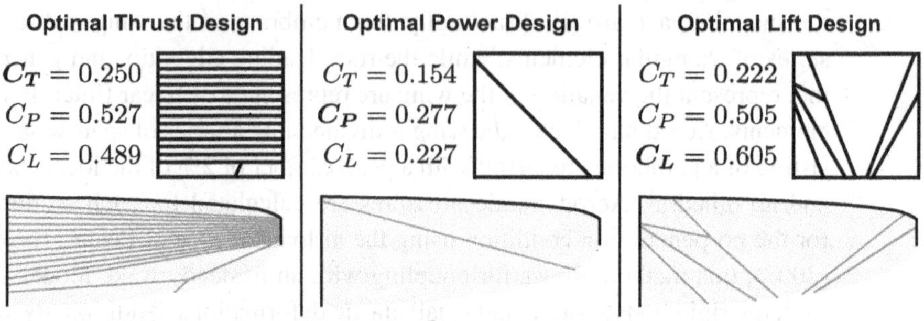

Figure 6.19 Resulting topologies for the single-objective optimization problems associated with each design objective. Adapted from Stanford, Beran & Kobayashi (2012) with permission.

the total power input (the sum of aerodynamic power, strain energy rate, and kinetic energy rate); ρ_∞ is the flow density; and S is the wing planform area. The resulting designs and corresponding objective coefficients are given in Figure 6.19. It is immediately noticeable that the topology of the peak thrust design is oriented for the wing to behave in a beam-like fashion and leverage spanwise bending (driven by inertial and aerodynamic loading) to increase the local lift vector, and thus the resulting thrust. Conversely, the minimum power topology is very sparse and more flexible in terms of chord-wise bending. This lack of structural support in the membrane results in adaptive washout at the trailing edge, causing chordwise wing deformation to attempt to align each wing section with the flow to reduce aerodynamic resistance, and thus the power draw. The maximum lift design is a combination between the two and features a large membrane region reinforced by several battens oriented diagonally; the non-reinforced triangular region of the membrane inflates during the downstroke, thereby increasing the camber and resulting lift force. In a broader sense, there appears to be an unclear trade-off between thrust and lift, as the high-thrust design also has higher lift and vice versa due to the large aerodynamic force production required for both. However, power requirements for flapping can be lowered by reducing aerodynamic resistance and thus clearly conflict with lift/thrust generation.

6.4 Conclusions

This chapter summarizes several efforts toward applying various L-system topology optimization approaches to design problems that require the consideration of multiple physical fields. The ability to address such problems was a key motivating factor behind the development of these methods and represents a clear advantage over the more traditional topology optimization approaches existing in practice and reviewed in Chapter 2.

Figure 6.20 illustrates the various single- and multiphysical problems considered using L-system topology optimization throughout this book; while Chapters 3–5 focused on proof-of-concept structural problems, in this final chapter, we have

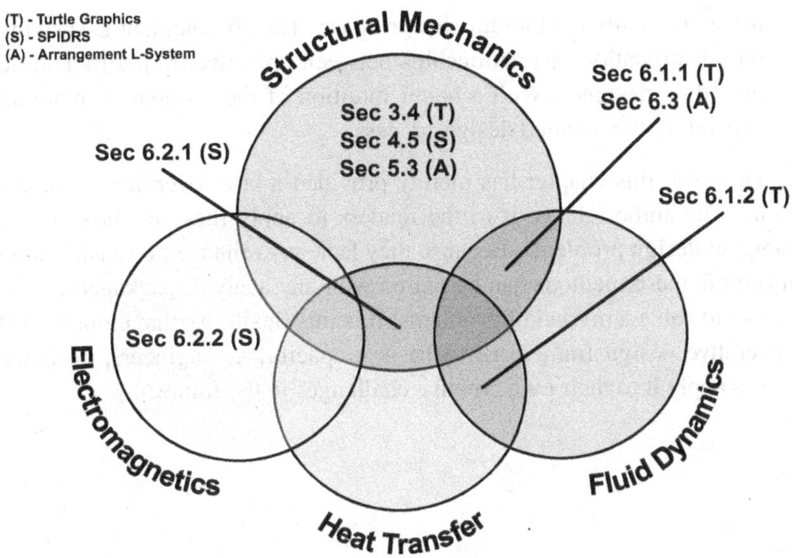

Figure 6.20 Overview of the various single and multiphysical problems considered using L-system topology optimization in this book, as organized by physical mechanism(s) considered.

demonstrated the ease with which the proposed methods can be recast to consider additional or overlapping physical fields. Considering these example problems and the results obtained, the following conclusions can be drawn:

- While it was previously demonstrated that a turtle graphics L-system interpretation approach is ill-suited for structural design (cf. Chapter 3), the branched topologies generated by the approach have utility when considering heat transfer and/or fluid flow. Here, the bifurcations that prohibited the creation of efficient structural load paths serve to expand the effective coverage area of a given topology, which is crucial when design objectives seek to minimize pressure drops or temperatures.
- The fundamental role of the SPIDRS algorithm as a network design tool makes it extremely adaptable to a variety of problems. While able to consider a simple combined structural/heat transfer problem, perhaps the more interesting contribution here is the ability of the same underlying algorithm to consider the design of electrical circuits, a class of problems that is rarely addressed using topology optimization. In particular, the ability to add additional attribute information, and therefore tune the complexity of the algorithm to the complexity of the problem, as well as match specified responses rather than single scalar objective values, will serve to further broaden its applicability in future efforts.
- The arrangement L-system is particularly well-suited for developing unconventional topological configurations, especially when considering complex aeroelastic design problems. Additionally, the ease with which this approach can be integrated with shape, sizing, and subsystem optimization allows designers to better and

more efficiently explore the design space. The arrangement L-system also enables the identification of relationships between objective trade-offs and design trends, providing designers with a better intuition of the problem at hand and hopefully expediting the detailed design process.

However, this chapter has merely provided a brief overview of potential applications. The authors leave it to the readers to apply these methods to a much wider range of design problems. Because they lack any reliance on sensitivities during optimization, these methods can be paired with any analysis package(s) of interest to the reader to solve a myriad of problems. It is this flexibility that empowers the proposed generative design framework to be so impactful as engineers, designers, and even artists apply it to their own creative challenges in the future.

Appendix A Vector Spaces

This appendix defines a vector space and summarizes its properties that are relevant to the book. There is an extensive literature on the subject, and the interested reader may consult, for instance, Axler (2015) for details on the theory of vector spaces.

A.1 Definition of Vector Space and Linear Maps

A vector space is an abstraction of $\mathbb{R}^n$. Specifically, in this book, we concern ourselves with real vector spaces only, that is, vector spaces defined over $\mathbb{R}$.

A vector space generalizes $\mathbb{R}^n$ to spaces where two operations, namely, addition and scalar multiplication, are defined. An **addition** on a set V is a mapping that assigns an element $\mathbf{u} + \mathbf{v} \in V$ for all pairs $\mathbf{u}, \mathbf{v}$ in V. A **scalar multiplication** is a mapping that assigns the element $a\mathbf{v}$ in V for all $a \in \mathbb{R}$ and $\mathbf{v} \in V$. A set V, along with an addition on V and a scalar multiplication on V, defines a **vector space** over the reals if the following properties hold:

- $\mathbf{u} + \mathbf{v} = \mathbf{v} + \mathbf{u}$, for all $\mathbf{u}, \mathbf{v} \in V$;
- $(\mathbf{u} + \mathbf{v}) + \mathbf{w} = \mathbf{u} + (\mathbf{v} + \mathbf{w})$, for all $\mathbf{u}, \mathbf{v}, \mathbf{w} \in V$;
- There is an element $0 \in V$, such that $\mathbf{v} + 0 = \mathbf{v}$, for all $\mathbf{v} \in V$;
- For each $\mathbf{v} \in V$, there exists $\mathbf{u} \in V$, such that $\mathbf{v} + \mathbf{u} = 0$;
- $(ab)\mathbf{v} = a(b\mathbf{v})$, for all $a, b \in \mathbb{R}$ and all $\mathbf{v} \in V$;
- $1\mathbf{v} = \mathbf{v}$, for all $\mathbf{v} \in V$;
- $a(\mathbf{u} + \mathbf{v}) = a\mathbf{u} + a\mathbf{v}$, for all $a \in \mathbb{R}$ and all $\mathbf{u}, \mathbf{v} \in V$; and
- $(a + b)\mathbf{v} = a\mathbf{v} + b\mathbf{v}$, for all $a, b \in \mathbb{R}$ and all $\mathbf{v} \in V$.

The prototypical example of a vector space is $\mathbb{R}^n$, for some $n \in \mathbb{N}$. Another example of a vector space is the set of polynomials with real coefficients, $\mathcal{P}(\mathbb{R})$.

A **linear combination** of a list $\mathbf{v}_1, \ldots, \mathbf{v}_m$ of vectors in V is a vector of the form

$$a_1\mathbf{v}_1 + \cdots + a_m\mathbf{v}_m,$$

where $a_1, \ldots, a_m \in \mathbb{R}$. Every vector space V has a **basis**, that is, a subset $\{\mathbf{v}_i\}_{i\in I}$, with I an index set and $\mathbf{v}_i \in V$ for all $i \in I$, such that each vector $\mathbf{v} \in V$ can be written uniquely as a linear combination as follows:

$$\mathbf{v} = \sum_{j\in J} a_j \mathbf{v}_j,$$

where J is a finite subset of I. The numbers $a_j, j \in J$, are called the components of the vector $\mathbf{v}$ with respect to the basis $\mathbf{v}_i$, $i \in I$. If I is a finite set, we say V is a finite-dimensional vector space and call the number of elements of I the **dimension** of V; for instance, the dimension of $\mathbb{R}^n$ is n, or, in symbols, $\dim \mathbb{R}^n = \mathrm{n}$. The set of polynomials $\mathcal{P}(\mathbb{R})$ is an example of an infinite-dimensional vector space.

The list $\mathbf{e}_1, \ldots, \mathbf{e}_n$ of vectors in $\mathbb{R}^n$ is defined such that the ordered tuple $\mathbf{e}_i$, $i \in \{1, \ldots, n\}$ has all entries equal to zero, except for its ith entry, which is equal to one, forming a special basis for $\mathbb{R}^n$ called **standard basis**.

A mapping $\mathbf{T}\colon V \to W$ between two vector spaces V and W is a **linear map** if

$$\mathbf{T}(\mathbf{u} + \mathbf{v}) = \mathbf{T}\mathbf{u} + \mathbf{T}\mathbf{v},$$

for all $\mathbf{u}, \mathbf{v} \in V$, and

$$\mathbf{T}(a\mathbf{v}) = a\mathbf{T}\mathbf{v},$$

for all $a \in \mathbb{R}$ and all $\mathbf{v} \in V$. The space of linear maps between two vector spaces V and W is itself a vector space and it is denoted $L(V, W)$.

Integration and differentiation are examples of linear maps between $\mathcal{P}(\mathbb{R})$ and $\mathcal{P}(\mathbb{R})$. Linear maps between $\mathbb{R}^n$ and $\mathbb{R}^m$, $\mathbf{T}\colon \mathbb{R}^n \to \mathbb{R}^m$, are mappings that can be written as follows:

$$\mathbf{T}\mathbf{v} = \mathbf{A}\mathbf{v},$$

for all $\mathbf{v} \in \mathbb{R}^n$, where $\mathbf{A}$ is an $m \times n$ matrix, that is,

$$\mathbf{A} = \begin{bmatrix} a_{11} & \cdots & a_{1n} \\ \vdots & \vdots & \vdots \\ a_{m1} & \cdots & a_{mn} \end{bmatrix},$$

and where, by convention, vectors are represented by column matrices.

As a matter of fact, fixing a basis $\mathbf{e}_1, \ldots, \mathbf{e}_n$ of V and a basis $\mathbf{f}_1, \ldots, \mathbf{f}_m$ of W, where $\dim V = n$ and $\dim W = m$, there is a unique m by n matrix, $\mathbf{A}$, such that

$$\mathbf{T}\mathbf{e}_j = \sum_{i=1}^{m} A_{ij}\mathbf{f}_i,$$

for $j = 1, \ldots, n$. We say that the matrix $\mathbf{A}$ represents the linear map $\mathbf{T}$ with respect to the bases $\mathbf{e}_1, \ldots, \mathbf{e}_n$ of V and $\mathbf{f}_1, \ldots, \mathbf{f}_m$ of W in the following sense: for any vector $\mathbf{v} \in V$ and $\mathbf{w} = \mathbf{T}\mathbf{v}$, we have

$$w_i = \sum_{j=1}^{n} A_{ij} v_j,$$

$i = 1, \ldots, m$, where $\mathbf{w} = w_1\mathbf{f}_1 + \cdots + w_m\mathbf{f}_m$ and $\mathbf{v} = v_1\mathbf{e}_1 + \cdots + v_n\mathbf{e}_n$.

The transpose of a matrix $\mathbf{A}$, denoted $\mathbf{A}^T$, is the matrix obtained interchanging the rows and columns of $\mathbf{A}$, that is,

$$A_{ij}^T = A_{ji},$$

$i = 1, \ldots, m, j = 1, \ldots, n$.

The space of linear maps between V and $\mathbb{R}$, $L(V, \mathbb{R})$ is an important vector space called the **dual space** of V, and it is denoted $V^* = L(V, \mathbb{R})$; the elements of V^* are called **linear functionals**. Given a basis $\mathbf{v}_1, \ldots, \mathbf{v}_n$ of V, where $\dim V = n$, the functionals $\phi^1, \ldots, \phi^n$, defined such that

$$\phi^i(v_j) = \delta^i_j,$$

where δ^i_j is the **Kronecker delta**

$$\delta^i_j = \begin{cases} 1 & \text{if } i = j, \\ 0 & \text{otherwise,} \end{cases}$$

form a basis of V^* called the **dual basis** of $\mathbf{v}_1, \ldots, \mathbf{v}_n$.

A.2 Norm in a Vector Space

Each vector $\mathbf{v} = (v_1, \ldots, v_n)$ in $\mathbb{R}^n$ has a length defined as follows:

$$|\mathbf{v}| = \sqrt{v_1^2 + \cdots + v_n^2}.$$

A norm of a vector generalizes the concept of length of a vector in $\mathbb{R}^n$ to vector spaces. Concretely, a **norm** on a vector space V is a function $\|\cdot\| : V \to \mathbb{R}_+$, such that the following properties hold:

- $\|\mathbf{v}\| \geq 0$, for all $\mathbf{v} \in V$ and $\|\mathbf{v}\| = 0 \Leftrightarrow \mathbf{v} = 0$;
- $\|a\mathbf{v}\| = |a| \, \|\mathbf{v}\|$, for all $a \in \mathbb{R}$ and all $\mathbf{v} \in V$;
- $\|\mathbf{u} + \mathbf{v}\| \leq \|\mathbf{u}\| + \|\mathbf{v}\|$, for all $\mathbf{u}, \mathbf{v} \in V$.

The last property is called the triangular inequality since it generalizes relation between the sides of a triangle in $\mathbb{R}^2$.

The most common norms on $\mathbb{R}^n$ are the following:

- $\|\mathbf{v}\|_1 = |v_1| + \cdots + |v_n|$;
- $\|\mathbf{v}\|_2 = \sqrt{v_1^2 + \cdots + v_n^2}$; and
- $\|\mathbf{v}\|_\infty = \max(|v_1|, \ldots, |v_n|)$.

Each of these norms is an example of the general p-norm:

$$\|\mathbf{v}\|_p = \sqrt[p]{|v_1|^p + \cdots + |v_n|^p},$$

where p is a real number greater or equal to one, and

$$\|\mathbf{v}\|_\infty = \lim_{p\to\infty} \|\mathbf{v}\|_p,$$

for all $\mathbf{v} \in V$. For a linear map, $\mathbf{T}$, between two normed vector spaces V and W, the **subordinate norm** of $\mathbf{T}$ is defined as follows:

$$\|\mathbf{T}\| = \sup_{0\neq\mathbf{v}\in\mathbf{V}} \frac{\|\mathbf{T}\mathbf{v}\|_W}{\|\mathbf{v}\|_V}.$$

In finite-dimensional normed vector spaces, we have the remarkable result that all norms are equivalent; that is, given two norms $\|\cdot\|_\alpha$ and $\|\cdot\|_\beta$ on a normed vector space V, there exist positive constants, $0 < c \leq C$, such that

$$c\,\|\mathbf{v}\|_\alpha \leq \|\mathbf{v}\|_\beta \leq C\,\|\mathbf{v}\|_\alpha,$$

for all $\mathbf{v} \in V$. Equivalent norms means, in particular, that converging sequences and their limits on a finite dimensional normed space are independent of the norm; that is, a sequence is converging on a finite-dimensional normed space if, and only if, the sequence is converging with respect to all norms, or, what is the same due to the equivalence, if, and only if, the sequence is converging with respect to some norm on V.

A.3 Inner Product in a Vector Space

The geometry of $\mathbb{R}^n$ is intimately related to the dot product on $\mathbb{R}^n$:

$$\mathbf{u}.\mathbf{v} = u_1 v_1 + \cdots + u_n v_n,$$

for all $\mathbf{u}, \mathbf{v} \in \mathbb{R}^n$, $\mathbf{u} = (u_1, \ldots, u_n)$ and $\mathbf{v} = (v_1, \ldots, v_n)$. For instance, the length of a vector, $\mathbf{v} \in \mathbb{R}$, is defined as follows:

$$|\mathbf{v}| = \sqrt{\mathbf{v}.\mathbf{v}}.$$

The concept of orthogonality between two vectors $\mathbf{u}, \mathbf{v} \in \mathbb{R}$ requires that:

$$\mathbf{u} \perp \mathbf{v} \Leftrightarrow \mathbf{u}.\mathbf{v} = 0.$$

Inner products abstract the properties of the dot product to vector spaces. Indeed, an **inner product**,$\langle\cdot|\cdot\rangle : V \times V \to \mathbb{R}$ on a vector space V is a function with the following properties:

- $\langle\mathbf{u}|\mathbf{v}\rangle = \langle\mathbf{v}|\mathbf{u}\rangle$, for all $\mathbf{u}, \mathbf{v} \in V$;
- $\langle a\mathbf{u} + b\mathbf{v}|\mathbf{w}\rangle = a\,\langle\mathbf{u}|\mathbf{w}\rangle + b\,\langle\mathbf{v}|\mathbf{w}\rangle$, for all $\mathbf{u}, \mathbf{v}, \mathbf{w} \in V$;
- $\langle\mathbf{v}|\mathbf{v}\rangle \geq 0$, for all $\mathbf{v} \in V$, and $\langle\mathbf{v}|\mathbf{v}\rangle = 0 \Leftrightarrow \mathbf{v} = 0$.

Examples of inner product vector spaces are provided by $\mathbb{R}^n$ with the dot product, and $\mathcal{P}(\mathbb{R})$ with the following inner product:

$$\langle p|q\rangle = \int_a^b p(x)q(x)dx,$$

for all $p, q \in \mathcal{P}(\mathbb{R})$, where $a, b \in \mathbb{R}$, $a < b$ are fixed constants.

Similarly to the dot product, the inner product defines a norm and orthogonality on V. Indeed, for the norm, we have

$$\|\mathbf{v}\| = \sqrt{\langle \mathbf{v}|\mathbf{v}\rangle},$$

for all $\mathbf{v} \in V$; while for orthogonality, given two vectors, $\mathbf{u}, \mathbf{v} \in V$, in an inner product vector space V, we say $\mathbf{u}$ and $\mathbf{v}$ are orthogonal to each other, denoted $\mathbf{u} \perp \mathbf{v}$, if $\langle \mathbf{u}|\mathbf{v}\rangle = 0$.

A result of great importance in inner product vector spaces is the Riesz representation theorem, which we state next for finite-dimensional vector spaces – the result for infinite-dimensional vector spaces requires topological considerations that lie beyond the scope of this book; the interested reader may consult Kreyszig (1978) for the statement and proof of the Riesz representation theorem in the general context of functional analysis.

Riesz Representation Theorem for Finite-Dimensional Spaces. Suppose V is a finite-dimensional inner product vector space, and $\phi \in V^*$ is a linear functional. Then, there exists a unique vector $\mathbf{u} \in V$, such that

$$\phi(\mathbf{v}) = \langle \mathbf{v}|\mathbf{u}\rangle\,,$$

for all $\mathbf{v} \in V$ and $\|\mathbf{u}\|_V = \|\phi\|_{V^*}$, with respect to the norm defined by the respective inner product. In fact, the vector $\mathbf{u} \in V$ is given by

$$\mathbf{u} = \phi(\mathbf{e}_1)\mathbf{e}_1 + \cdots + \phi(\mathbf{e}_n)\mathbf{e}_n,$$

where $\mathbf{e}_1, \ldots, \mathbf{e}_n$ is any orthonormal basis of V (which always exists).

One example of application of the Riesz theorem is in the definition of the gradient of a differentiable function. Indeed, let f be a differentiable function on $\mathbb{R}^n$. Then the differential of $f\colon \mathbb{R}^n \to \mathbb{R}$ at a point $\mathbf{u} \in \mathbb{R}^n$ is the linear functional, $df_{\mathbf{u}}$ defined as follows:[1]

$$df_{\mathbf{u}}(\mathbf{v}) = \left.\frac{\partial f}{\partial x_1}\right|_{\mathbf{u}} v_1 + \cdots + \left.\frac{\partial f}{\partial x_n}\right|_{\mathbf{u}} v_n,$$

for all $\mathbf{v} \in \mathbb{R}^n$, where $x_1, \ldots, x_n$ is a Cartesian coordinate system for $\mathbb{R}^n$. Then, the gradient vector field, ∇f, is the vector field that at each point $\mathbf{u}$ is such that

[1] In calculus books, the typical notation for the differential is $df = \frac{\partial f}{\partial x_1}dx_1 + \cdots + \frac{\partial f}{\partial x_n}dx_n$. The two definitions are compatible if one notices that $dx_1|_{\mathbf{u}}, \ldots, dx_n|_{\mathbf{u}}$ can be interpreted as an orthonormal basis of the dual $(\mathbb{R}^n)^*$ at the point $\mathbf{u} \in \mathbb{R}^3$. Note that to fully account for the dependency of the differential df on the point in general contexts requires knowledge of tensor analysis that falls outside the scope of this book.

$$df_{\mathbf{u}}(\mathbf{v}) = \langle \nabla f(\mathbf{u}) | \mathbf{v} \rangle ,$$

for all $\mathbf{v} \in \mathbb{R}^n$, where the inner product is the dot product. The existence and uniqueness of the gradient thus is ensured by the Riesz representation theorem, and the formula thus obtained for the gradient agrees with its usual formula,

$$\nabla f = \left(\frac{\partial f}{\partial x_1}, \ldots, \frac{\partial f}{\partial x_1} \right).$$

This definition of the gradient of a function may seem unconventional, but it is helpful in understanding the gradient of functionals in more general contexts, such as the computation of the gradient in the level set method studied in Chapter 2.

A.4 Tensors

Besides vectors, many theories in mechanics, such as elasticity theory, require knowledge of tensors and tensor fields. Thus, this section defines tensors and briefly describes tensor operations relevant to this book. The reader interested in tensor analysis may consult, for instance, Sokolnikoff (1951), for a classical exposition of the subject, or Wasserman (2009), for a modern account of the matter. Finally, for simplicity, in this section, we assume all vector spaces to be finite-dimensional.

We start with the definition of multilinear maps, which extends the concept of a linear map defined above. Thus, given vector spaces $V_1, \ldots, V_n$, and W, a map $f\colon V_1 \times \cdots \times V_n \to W$ is **multilinear** if it is linear with respect to each of its arguments, when all remaining arguments are kept fixed; that is, if for each index $i = 1, \ldots, n$ and for any fixed list of $n - 1$ vectors $\mathbf{v}_1, \ldots, \mathbf{v}_{n-1}$, the map

$$V_i \to W, \quad \mathbf{v} \mapsto f(\mathbf{v}_1, \ldots, \mathbf{v}_{i-1}, \mathbf{v}, \mathbf{v}_{i+1}, \ldots, \mathbf{v}_{n-1})$$

is linear. An example of a multilinear map is the inner product on a vector space, which a bilinear map $V \times V \to \mathbb{R}$.

A **tensor** of type $\binom{p}{q}$, $\mathbf{t}$, is a multilinear mapping

$$\mathbf{t}\colon \underbrace{V \times \cdots \times V}_{q \text{ times}} \times \underbrace{V^* \times \cdots \times V^*}_{p \text{ times}} \to \mathbb{R}.$$

Within this context, a vector is a $\binom{1}{0}$ tensor, a linear functional is a $\binom{0}{1}$ tensor, and the inner product is a $\binom{0}{2}$ tensor. Important tensors in linear elasticity are the stress tensor, σ, of type $\binom{2}{0}$, the small strain tensor, ϵ, of type $\binom{0}{2}$, and the elasticity tensor, $\mathbf{E}$, of type $\binom{4}{0}$. This definition highlights the vital property of tensors: They are defined independently of the basis of the vector spaces. It is this independence from a basis that allows physical meaning to be attached to tensors.

The space of $\binom{p}{q}$ tensors is a vector space denoted $T^p_q(V)$. To find a basis for this space, we define the **tensor product** $\mathbf{t} \otimes \mathbf{s} \in T^{p+a}_{q+b}$ between two tensors $\mathbf{t} \in T^p_q$ and $\mathbf{s} \in T^a_b$ as follows:

$$(\mathbf{t} \otimes \mathbf{s})(\mathbf{v}_1, \ldots, \mathbf{v}_q, \mathbf{w}_1, \ldots, \mathbf{w}_b, \alpha^1, \ldots, \alpha^p, \beta^1, \ldots, \beta^a) = \mathbf{t}(\mathbf{v}_1, \ldots, \mathbf{v}_q, \alpha^1, \ldots, \alpha^q) \\ \mathbf{s}(\mathbf{w}_1, \ldots, \mathbf{w}_b, \beta^1, \ldots, \beta^a),$$

for all $\mathbf{v}_1, \ldots, \mathbf{v}_q, \mathbf{w}_1, \ldots, \mathbf{w}_b \in V$ and for all $\alpha^1, \ldots, \alpha^p, \beta^1, \ldots, \beta^a \in V^*$. With the tensor product, let $\mathbf{e}_1, \ldots, \mathbf{e}_n$ be a basis of V and $\phi^1, \ldots, \phi^n$ its dual basis, then the set

$$\left\{ \mathbf{e}_{i_1} \otimes \cdots \otimes \mathbf{e}_{i_p} \otimes \phi^{j_1} \otimes \cdots \right\},$$

where $\dim V = n$, forms a basis for T^p_q. With this basis, any tensor $\mathbf{t} \in T^p_q$ can be written as the following linear combination:

$$\mathbf{t} = \sum t^{i_1 \ldots i_p}_{j_1 \ldots j_q} \mathbf{e}_{i_1} \otimes \cdots \otimes \mathbf{e}_{i_p} \otimes \phi^{j_1} \otimes \cdots \otimes \phi^{j_q},$$

where the summation is over the indices $i_1, \ldots, i_p, j_1, \ldots, j_q \in \{1, \ldots, n\}$.

As a concrete example, we consider the expansion for the stress tensor in three dimensions, namely,

$$\begin{aligned} \boldsymbol{\sigma} = \sigma^{11}\mathbf{e}_1 \otimes \mathbf{e}_1 + \sigma^{12}\mathbf{e}_1 \otimes \mathbf{e}_2 + \sigma^{13}\mathbf{e}_1 \otimes \mathbf{e}_3 + \\ \sigma^{21}\mathbf{e}_2 \otimes \mathbf{e}_1 + \sigma^{22}\mathbf{e}_2 \otimes \mathbf{e}_2 + \sigma^{23}\mathbf{e}_2 \otimes \mathbf{e}_3 + \\ \sigma^{31}\mathbf{e}_3 \otimes \mathbf{e}_1 + \sigma^{32}\mathbf{e}_3 \otimes \mathbf{e}_2 + \sigma^{33}\mathbf{e}_3 \otimes \mathbf{e}_3, \end{aligned}$$

where $\mathbf{e}_1, \mathbf{e}_2, \mathbf{e}_3$ form a basis for $\mathbb{R}^3$.

There are numerous natural identifications among tensor spaces, one of which is of particular interest to elasticity, namely, $T^4_0 \simeq L(T^0_2, T^2_0)$. With this identification, Hooke's law can be written in a short and elegant manner as follows:

$$\boldsymbol{\sigma} = \mathbf{E}\boldsymbol{\epsilon},$$

where the elasticity tensor $\mathbf{E} \in T^4_0$ is understood in this relation as a linear operator $\mathbf{E} \in L(T^0_2, T^2_0)$, transforming the small strain $\boldsymbol{\epsilon} \in T^0_2$ into the stress tensor $\boldsymbol{\sigma} \in T^2_0$. In turn, this physical relation can be conveniently written in compact form, using indicial notation and the tensor components, as follows:

$$\sigma^{ij} = \sum_{k=1}^{n} \sum_{l=1}^{n} E^{ijkl} \epsilon_{kl},$$

where $n = 1, 2, 3$ is the dimension of the space.

Finally, we define another useful operation between tensors, namely, the **double contraction,** $\mathbf{t} : \mathbf{s} \in \mathbb{R}$, between a tensor $\mathbf{t} \in T^2_0$ and a tensor $\mathbf{s} \in T^0_2$:

$$\mathbf{t} : \mathbf{s} = \sum_{i=1}^{n} \sum_{j=1}^{n} t^{ij} s_{ji}.$$

Using the double contraction, the strain energy density, S, can be written as

$$S = \frac{1}{2}\boldsymbol{\sigma} : \boldsymbol{\epsilon} = \frac{1}{2}\mathbf{E}\boldsymbol{\epsilon} : \boldsymbol{\epsilon},$$

or

$$S = \frac{1}{2}\sum_{i=1}^{n}\sum_{j=1}^{n}\sum_{k=1}^{n}\sum_{l=1}^{n} E^{ijkl}\epsilon_{ij}\epsilon_{kl}.$$

Appendix B Classical Methods in Optimization

This appendix briefly discusses classical methods in optimization that are referred to in this book. For a deeper and broader analysis of these methods, the interested reader is directed to Papalambros and Wilde (2017) and Fletcher (1987).

B.1 Simplex Method

The feasible space of linear programming problems is a polytope, with a solution occurring at one of its extremum points (Fletcher 1987). The simplex method is an iterative approach to linear programming problems that starts at one extremum point of the polytope and progresses through such points until convergence, if a solution exists, by reducing (not increasing) the cost function at each step.

Without loss of generality, the simplex method starts with the following standard form of linear programming (SFLP) problems:

Find $\boldsymbol{x} \in \mathbb{R}^n$ that minimizes the cost function,

$$f(\boldsymbol{x}) = \boldsymbol{c}^T\boldsymbol{x}, \tag{B.1}$$

subject to linear constraints

$$\boldsymbol{A}\boldsymbol{x} = \boldsymbol{b}, \quad \boldsymbol{x} \geq 0, \tag{B.2}$$

where $\boldsymbol{A}$ is a full rank $m \times n$ matrix, and $m \leq n$ (typically, $m < n$).

The simplex method generates a sequence of feasible points $\boldsymbol{x}^0, \boldsymbol{x}^1, \ldots,$ which terminates in a finite number of iterations at a solution. In each iteration, the solution space is partitioned into basic and nonbasic variables: $\boldsymbol{x}^T = [\boldsymbol{x}_B^T \, \boldsymbol{x}_N^T]$, where $\boldsymbol{x}_B$ and $\boldsymbol{x}_N$ refer to the basic and nonbasic variables, respectively. Similarly, the matrix $\mathbf{A}$ is partitioned as $\mathbf{A} = [\mathbf{A}_B : \mathbf{A}_N]$, where the base matrix $\mathbf{A}_B$ is an $m \times m$, and $\mathbf{A}_N$ is $m \times (n - m)$. With this partition, we can write

$$[\mathbf{A}_B : \mathbf{A}_N] \begin{bmatrix} \boldsymbol{x}_B \\ \boldsymbol{x}_N \end{bmatrix} = \mathbf{A}_B\boldsymbol{x}_B + \mathbf{A}_N\boldsymbol{x}_N = \mathbf{b}.$$

Matrix $\mathbf{A}$ being of full rank, we can choose $\mathbf{A}_B$ to be non-singular, and compute

$$\boldsymbol{x}_B = \mathbf{A}_B^{-1}(\mathbf{b} - \mathbf{A}_N\boldsymbol{x}_N) = \hat{\mathbf{b}} - \mathbf{A}_B^{-1}\mathbf{A}_N\boldsymbol{x}_N,$$

where $\hat{\mathbf{b}} = \mathbf{A}_B^{-1}\mathbf{b}$. This partition allows for the parameterization of the solution $\boldsymbol{x}$ as a function of $\boldsymbol{x}_N$. Each feasible point in the simplex sequence is such that

$$\boldsymbol{x}^k = \begin{bmatrix} \hat{\mathbf{b}}^k \\ 0 \end{bmatrix} \geq 0,$$

so that $\hat{\mathbf{b}}^k \geq 0$. Such a feasible point, with $\boldsymbol{x}_N = 0$, is called a basic feasible solution (BFS).

Using the parameterization of the solution $\boldsymbol{x}$ as a function of $\boldsymbol{x}_N$, we obtain the reduced cost function as

$$\begin{aligned} f(\boldsymbol{x}_N) &= \mathbf{c}_B^T\boldsymbol{x}_B + \mathbf{c}_N^T\boldsymbol{x}_N, \\ &= \mathbf{c}_B^T(\hat{\mathbf{b}} - \mathbf{A}_B^{-1}\mathbf{A}_N\boldsymbol{x}_N) + \mathbf{c}_N^T\boldsymbol{x}_N, \\ &= \hat{f} + \hat{\mathbf{c}}_N^T\boldsymbol{x}_N, \end{aligned}$$

where

$$\hat{\mathbf{c}}_N = \mathbf{c}_N - \mathbf{A}_N^T\pi,$$

with

$$\pi = \mathbf{A}_B^{-1}\mathbf{c}_B.$$

The coefficients $\hat{\mathbf{c}}_N$ are called **reduced costs** at the basic feasible solution. Because the feasible points must be positive ($\boldsymbol{x} \geq 0$), if the reduced costs are positive ($\hat{\mathbf{c}}_N \geq 0$) then $f(\boldsymbol{x}_N) = \hat{f} + \hat{\mathbf{c}}^T\boldsymbol{x}_N \geq \hat{f}$, and we conclude that the feasible iterate $\boldsymbol{x}^k$ is a global minimum to the LP problem and the simplex algorithm stops. Otherwise, there is an index $q \in N$, such that $\hat{c}_q < 0$, and the cost function can be reduced along x_q. In this case, simplex will swap a basic index with a nonbasic index; for the non-basic index, it is common to choose the index q with minimum reduced cost, that is,

$$q = \arg\min_{i \in N} \hat{c}_i.$$

To choose the basic index, we note that a change in x_q will affect all basic variables according to

$$\boldsymbol{x}_B = \hat{\mathbf{b}} - \mathbf{A}_B^{-1}\mathbf{a}_q x_q, \tag{B.3}$$
$$= \hat{\mathbf{b}} + \mathbf{d}x_q, \tag{B.4}$$

where $\mathbf{a}_q$ is the qth column of $\mathbf{A}$ and $\mathbf{d} = -\mathbf{A}_B^{-1}\mathbf{a}_q$ is the derivative of $\boldsymbol{x}_B$ with respect to x_q. It follows at once from Equation B.4 that if $\mathbf{d} \geq 0$, then $f(\boldsymbol{x})$ can be decreased without limit and the LP problem has no solution. Otherwise, a change in x_q along a direction $i \in B$ for which $d_i < 0$ will reduce the associated x_i until it becomes zero and $x_q = \hat{b}_i/d_i$. The first basic variable x_p to become null when x_q varies is chosen for swapping with the nonbasic variable x_q; that is,

$$p = \arg\min_{\substack{i \in B \\ d_i < 0}} \frac{\hat{b}_i}{-d_i}.$$

The procedure described thus far is not self-starting, since it requires a BFS $\boldsymbol{x}^0$. For this reason, this procedure is called the Phase II of the simplex method. In the Phase I of the simplex method, we determine an initial BFS satisfying the constraints. One simple procedure to obtain such initial BFS is the *method of artificial variables*, where artificial variables $r_i, i \in \{1,\ldots,m\}$ are defined such that $\mathbf{r} = \mathbf{b} - \mathbf{A}\boldsymbol{x} \in \mathbb{R}^m$ (i.e., $\mathbf{r}$ represents the residual of the equation $\mathbf{A}\boldsymbol{x} = \mathbf{b}$). In this method, reversing signs if needed, we assume that $\mathbf{b} \geq 0$; the ancillary problem (AP) can then be solved as

Find $\boldsymbol{r} \in \mathbb{R}^m$ *that minimizes* $\sum_{i=1}^m r_i$ *subject to*

$$\boldsymbol{A}\boldsymbol{x} + \boldsymbol{r} = \boldsymbol{b}, \quad \boldsymbol{x} \geq 0, \quad \boldsymbol{r} \geq 0. \tag{B.5}$$

Let $\boldsymbol{x}'$ and $\mathbf{r}'$ solve the AP. If $\mathbf{r}' = 0$, then $\boldsymbol{x}'$ is a BFS for the LP problem; otherwise, there is no feasible solution to the LP problem.

Since the ancillary problem is an LP, Phase II of the simplex method can be used to compute a solution to the AP. Indeed, by noting that $\mathbf{r} = \mathbf{b}$ and $\boldsymbol{x} = 0 \Rightarrow \mathbf{A}\boldsymbol{x} + \mathbf{r} = \mathbf{A}0 + \mathbf{b} = \mathbf{b}$, a BFS for the AP problem is immediately available by taking the artificial variables to be basic (that is, $\mathbf{r}^0 = \mathbf{b}$ and $\boldsymbol{x}^0 = 0$). The analysis above shows that Phase II of the simplex method determines the solution to the LP problem. The steps of the Phase II of the simplex method are summarized in Algorithm 8.

Algorithm 8 Simplex Method: Phase II

Require:

1. LP problem in SFLP: Minimize $f(\boldsymbol{x}) = \mathbf{c}^T\boldsymbol{x}$
$\boldsymbol{x} \in \mathbb{R}^n : \mathbf{A}\boldsymbol{x} = \mathbf{b}, \boldsymbol{x} \geq 0$, where $\mathbf{A}$ is full rank

2. An initial BFS (Phase I)

repeat
  Choose $B \subset I = \{1,\ldots,n\}$ and $N = I\backslash B$, such that $\boldsymbol{x} = \boldsymbol{x}_B$ is the initial BFS
  Define $\mathbf{A} = [\mathbf{A}_B : \mathbf{A}_N]$, $\boldsymbol{x}^T = (\boldsymbol{x}_B^T, \boldsymbol{x}_N^T)$, $\mathbf{c}^T = (\mathbf{c}_B^T, \mathbf{c}_N^T)$ and $\pi = \mathbf{A}_B^{-T}\mathbf{c}_B$
  Compute the reduced costs $\hat{\mathbf{c}}_N = \mathbf{c}_N - \mathbf{A}_N^T\pi$
  if $\hat{c}_i \geq 0, \forall i \in N$ **then**
    $\boldsymbol{x} = \hat{\mathbf{b}} = \mathbf{A}_B^{-1}\mathbf{b}$ is an optimal BFS; stop
  else
    Let $q = \arg\min_{i \in N} \hat{c}_i$
    Compute $\mathbf{d} = -\mathbf{A}_B^{-1}\mathbf{a}_q$
    if $\mathbf{d} \geq 0$ **then**
      The problem has no solution; stop
    else
      Let $p = \arg\min_{i \in B \,\&\, d_i < 0} \frac{\hat{b}_i}{-d_i}$
    end if
    Swap the indices p and q
  end if
until a stopping criterion is met

B.2 Steepest Descent Method

The steepest descent method is one of the oldest and simplest methods in nonlinear programming. The method was first proposed by Cauchy in the nineteenth century (Cauchy 1847) for the solution of a nonlinear equation defined by a positive function and is based on the idea that in a sufficiently small neighborhood of a point, a differentiable function will decrease along the direction opposite to the gradient at that point. In fact, many modern optimization methods use a line search as a step in their algorithms (Fletcher 1987), which follows Cauchy's idea.

Specifically, to accelerate and stabilize their search for an optimal solution, these methods employ a line search along a direction $\mathbf{s}^k$. As part of the line search, a step size $\alpha_k \in \mathbb{R}$ is determined such that the updated design variables $\boldsymbol{x}^{k+1}$ along $\mathbf{s}^k$,

$$\boldsymbol{x}^{k+1} = \boldsymbol{x}^k + \alpha_k \mathbf{s}^k,$$

minimize the objective function along that line such that

$$\alpha_k = \arg\min_{\alpha>0} f(\boldsymbol{x}^k + \alpha \mathbf{s}^k),$$

where f is the nonlinear objective function. As mentioned, in the steepest descent method the search direction is the opposite to the gradient at $\boldsymbol{x}^k$: $\mathbf{s}^k = -\nabla f(\boldsymbol{x}^k)$. A popular choice for α_k for unconstrained minimization problems is that proposed in Barzilai and Borwein (1988). Instead of solving the line minimization problem, these authors compute α_k from a least-square approximation to quasi-Newton methods yielding

$$\alpha_k = \frac{\mathbf{d}^{k-1^T} \mathbf{d}^{k-1}}{\mathbf{d}^{k-1^T} \boldsymbol{y}^{k-1}},$$

where $\mathbf{d}^{k-1} = \boldsymbol{x}^k - \boldsymbol{x}^{k-1}$ and $\boldsymbol{y}^{k-1} = \nabla f(\boldsymbol{x}^k) - \nabla f(\boldsymbol{x}^{k-1})$.

As defined, the steepest descent method is simple to implement and easy to understand. However, it can exhibit a slow, linear convergence behavior, which discourages its widespread use in applications. Despite its practical limitations, steepest descent still serves as a simple model to introduce and understand more advanced methods. For more information on this method the interested reader is directed to Papalambros and Wilde (2017) and Fletcher, Qi, Teo and Yang (2005).

Appendix C Classical Simplex-Based Approach to Ground Structure Optimization

This appendix demonstrates how the simplex optimization method (cf. Appendix B) can be applied to simple ground structure-based optimization problems such as those introduced in Section 2.3.2. For further details, the interested reader is directed to Hemp (1973).

Equilibrium of internal and external forces on the ground structure can be written as follows:

$$\sum_{i=1}^{m} \mathbf{T}_i \gamma_i + \mathbf{F} = 0, \qquad \text{(C.1)}$$

where $\mathbf{F} \in \mathbb{R}^n$ is the external force vector, $\mathbf{T}_i$ is the member force, and $\gamma_i \in \mathbb{R}^n$ is the direction cosine vector of member i given by

$$[\gamma_i]_p = \begin{cases} \cos(\phi_{i,k}) & \text{if } p = j_{i,1,k} \text{ for some } k \in \{1,\ldots,d\}, \\ -\cos(\phi_{i,k}) & \text{if } p = j_{i,2,k} \text{ for some } k \in \{1,\ldots,d\}, \\ 0 & \text{otherwise,} \end{cases}$$

where $p \in \{1,\ldots,n\}$, $\phi_{i,k} \in [0, 2\pi]$ denotes the angle between the kth coordinate axis and the ith rod, and $j_{i,1,k}, j_{i,2,k} \in \{1,\ldots,Nd\}$ denote the nodal displacement coordinate of the starting and ending nodes of the ith rod. To build the stiffness matrix, $\mathbf{K}$, of the ground structure, we assume small displacements of the nodes and linear elasticity, resulting in the following equality:

$$\mathbf{K}(\mathbf{A}) = \sum_{i=1}^{m} A_i \frac{E_i}{\ell_i} \gamma_i \gamma_i^T \in \mathbb{R}^{n \times n},$$

where $\mathbf{A} \in \mathbb{R}_+^m$ is the vector of cross-sectional areas of the rods such that $A_1,\ldots,A_m \geq 0$, $E_i > 0$ is the Young's modulus, and $\ell_i > 0$ is the length of the ith rod. With the stiffness matrix, the equilibrium of internal and external forces in Equation C.1 can be written as follows:

$$\mathbf{K}(\mathbf{A})\mathbf{u} = \mathbf{F},$$

where $\mathbf{u}$ denotes the nodal displacement vector $\mathbf{u} \in \mathbb{R}^n$.

The corresponding mathematical programming problem (MPP) for the ground structure can then be stated as follows:

Find the cross-sectional areas $A_1, \ldots, A_m$ and displacements $u \in \mathbb{R}^n$, which minimize the compliance functional,

$$J_c = \mathbf{F}^T \mathbf{u},$$

subject to equilibrium

$$\mathbf{K}(\mathbf{A})\mathbf{u} = \mathbf{F},$$

volume constraint

$$\sum_{i=1}^{m} A_i \ell_i \leq V,$$

and positivity $A_i \geq 0$, $i \in \{1, \ldots, m\}$.

To obviate the nonlinearity of the equilibrium equation, the equivalent dual formulation of the MPP is considered (Achtziger 2007):

Find a virtual displacement $\bar{u} \in \mathbb{R}^n$ that maximizes the virtual work,

$$W = \mathbf{F}^T \bar{\mathbf{u}},$$

subject to virtual strain constraints

$$-1 \leq \frac{\sqrt{E_i}}{\ell_i} \gamma_i^T \bar{\mathbf{u}} \leq 1.$$

If $\bar{\mathbf{u}}^*$ is a solution to the dual MPP problem with corresponding Lagrange multipliers $\lambda^*, \mu^* \in \mathbb{R}_+^m$, then $\mathbf{F}^T \bar{\mathbf{u}}^* > 0$ and $(\mathbf{A}^*, \mathbf{u}^*)$ is a global minimum of the original primal MPP (*ibidem*), with

$$\mathbf{u}^* = \frac{1}{V}(\mathbf{F}^T \bar{\mathbf{u}}^*)\bar{\mathbf{u}}^*, \tag{C.2}$$

$$A_i^* = \frac{V(\lambda_i^* + \mu_i^*)}{\mathbf{F}^T \bar{\mathbf{u}}}, \quad i \in \{1, \ldots, m\}. \tag{C.3}$$

The dual formulation has the advantage of being a linear programming (LP) problem, which can be efficiently solved using LP methods (Fletcher 1987). For concreteness, we select the simplex method for the solution of the dual problem and present a summary of this method in Appendix B.

As a guiding example of classical truss layout optimization, we consider the ground structure with four joints and five bars as depicted in Figure C.1. The load F and the non-null displacements u_1, u_2, u_3, u_4 of the joints 3 and 4 are shown, where joints 1 and 2 are pinned. The same material is used for all rods. Under these conditions, the dual LP problem for this example can be written as follows:

Find virtual displacements

$$\bar{u}_1, \bar{u}_2, \bar{u}_3, \bar{u}_4 \in \mathbb{R}_+$$

that minimize the virtual work

$$W = F\bar{u}_2, \quad F > 0 \tag{C.4}$$

subject to the following virtual strain constraints:

$$-\frac{\sqrt{E}\bar{u}_1}{l} \leq 1, \tag{C.5}$$

$$-\frac{\sqrt{E}\bar{u}_1}{l} \geq -1, \tag{C.6}$$

$$\frac{\sqrt{E}(\bar{u}_2 - \bar{u}_4)}{l} \leq 1, \tag{C.7}$$

$$\frac{\sqrt{E}(\bar{u}_2 - \bar{u}_4)}{l} \geq -1, \tag{C.8}$$

$$\frac{\sqrt{E}\bar{u}_3}{l} \leq 1, \tag{C.9}$$

$$\frac{\sqrt{E}\bar{u}_3}{l} \geq -1, \tag{C.10}$$

$$\frac{\sqrt{E}(\bar{u}_3 - \bar{u}_4)}{2l} \leq 1, \tag{C.11}$$

$$\frac{\sqrt{E}(\bar{u}_3 - \bar{u}_4)}{2l} \geq -1, \tag{C.12}$$

$$\frac{\sqrt{E}(\bar{u}_2 - \bar{u}_1)}{2l} \leq 1, \tag{C.13}$$

$$\frac{\sqrt{E}(\bar{u}_2 - \bar{u}_1)}{2l} \geq -1. \tag{C.14}$$

To use the simplex method, the LP problem (Equations C.4–C.14) has to be rewritten into a standard form of linear programming (SFLP) problem as introduced in Appendix B. Here, this entails rewriting the maximization problem in Equation C.4 as an equivalent minimization problem, and the inequality constraints (Equations C.5–C.14) into equivalent equality constraints. The former is accomplished by stating

$$\max_{\bar{u}} W(\bar{u}) = -\min_{\bar{u}} -W(\bar{u}).$$

As for the inequality constraints, they are first rewritten as "$\leq$" inequalities with a positive right hand side to simplify finding an initial basic feasible solution (BFS) and then converted into equality constraints by introducing **slack variables**. For instance,

$-\frac{\sqrt{E}\bar{u}_1}{l} \leq 1$ is equivalent to $-\frac{\sqrt{E}\bar{u}_1}{l} + s_1 = 1$, where $s_1 \geq 0$ is a slack variable, while $-\frac{\sqrt{E}\bar{u}_1}{l} \geq -1$ is rewritten as $\frac{\sqrt{E}\bar{u}_1}{l} \leq 1$, which is equivalent to $\frac{\sqrt{E}\bar{u}_1}{l} + t_1 = 1$ provided $t_1 \geq 0$. Applying this procedure to the remaining inequalities results in the following LP problem in standard form:

Find virtual displacements

$$\bar{u}_1, \bar{u}_2, \bar{u}_3 \in \mathbb{R}_+$$

and slack variables

$$s_i, t_i \in \mathbb{R}_+, \quad i = 1, \ldots, 4$$

that minimize

$$z = -F\bar{u}_2, \quad F > 0$$

subject to the following virtual strain constraints:

$$\begin{aligned}
-\frac{\sqrt{E}\bar{u}_1}{l} + s_1 &= 1, \\
\frac{\sqrt{E}\bar{u}_1}{l} + t_1 &= 1, \\
\frac{\sqrt{E}(\bar{u}_2 - \bar{u}_4)}{l} + s_2 &= 1, \\
-\frac{\sqrt{E}(\bar{u}_2 - \bar{u}_4)}{l} + t_2 &= 1, \\
\frac{\sqrt{E}\bar{u}_3}{l} + s_3 &= 1, \\
-\frac{\sqrt{E}\bar{u}_3}{l} + t_3 &= 1, \\
\frac{\sqrt{E}(\bar{u}_3 - \bar{u}_4)}{2l} + s_4 &= 1, \\
-\frac{\sqrt{E}(\bar{u}_3 - \bar{u}_4)}{2l} + t_4 &= 1, \\
\frac{\sqrt{E}(\bar{u}_2 - \bar{u}_1)}{2l} + s_5 &= 1, \\
-\frac{\sqrt{E}(\bar{u}_2 - \bar{u}_1)}{2l} + t_5 &= 1.
\end{aligned}$$

Defining

$$\boldsymbol{x} = (\bar{u}_1, \bar{u}_2, \bar{u}_3, \bar{u}_4, s_1, t_1, s_2, t_2, s_3, t_3, s_4, t_4, s_5, t_5),$$

we have

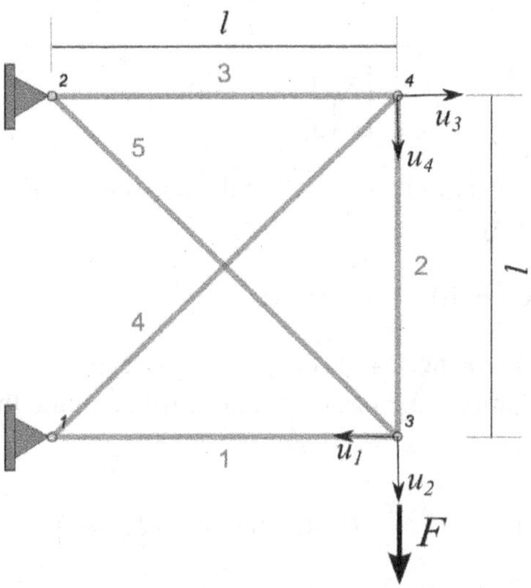

Figure C.1 Ground structure, displacements, and load for truss example.

$$\mathbf{c} = \begin{bmatrix} 0 & -F & 0 & 0 & 0 & 0 & 0 & 0 & 0 & 0 & 0 & 0 & 0 & 0 \end{bmatrix}^T,$$

$$\mathbf{A} = \begin{bmatrix}
-\frac{\sqrt{E}}{l} & 0 & 0 & 0 & 1 & 0 & 0 & 0 & 0 & 0 & 0 & 0 & 0 & 0 \\
\frac{\sqrt{E}}{l} & 0 & 0 & 0 & 0 & 1 & 0 & 0 & 0 & 0 & 0 & 0 & 0 & 0 \\
0 & \frac{\sqrt{E}}{l} & 0 & -\frac{\sqrt{E}}{l} & 0 & 0 & 1 & 0 & 0 & 0 & 0 & 0 & 0 & 0 \\
0 & -\frac{\sqrt{E}}{l} & 0 & \frac{\sqrt{E}}{l} & 0 & 0 & 0 & 1 & 0 & 0 & 0 & 0 & 0 & 0 \\
0 & 0 & \frac{\sqrt{E}}{2l} & 0 & 0 & 0 & 0 & 0 & 1 & 0 & 0 & 0 & 0 & 0 \\
0 & 0 & -\frac{\sqrt{E}}{l} & 0 & 0 & 0 & 0 & 0 & 0 & 1 & 0 & 0 & 0 & 0 \\
0 & 0 & \frac{\sqrt{E}}{l} & -\frac{\sqrt{E}}{l} & 0 & 0 & 0 & 0 & 0 & 0 & 1 & 0 & 0 & 0 \\
0 & 0 & -\frac{\sqrt{E}}{l} & \frac{\sqrt{E}}{l} & 0 & 0 & 0 & 0 & 0 & 0 & 0 & 1 & 0 & 0 \\
-\frac{\sqrt{E}}{2l} & \frac{\sqrt{E}}{2l} & 0 & 0 & 0 & 0 & 0 & 0 & 0 & 0 & 0 & 0 & 1 & 0 \\
\frac{\sqrt{E}}{2l} & -\frac{\sqrt{E}}{2l} & 0 & 0 & 0 & 0 & 0 & 0 & 0 & 0 & 0 & 0 & 0 & 1
\end{bmatrix}, \tag{C.15}$$

and

$$\mathbf{b} = \begin{bmatrix} 1 & 1 & 1 & 1 & 1 & 1 & 1 & 1 & 1 & 1 \end{bmatrix}^T. \tag{C.16}$$

A glance at Equations C.15–C.16 reveals the benefit of converting "≥" into "≤" inequalities with a positive right-hand side: It allows for an easy computation of the initial BFS. Indeed, choosing the slack variables as basic variables yields an identity matrix for $\mathbf{A}_B$, and the positive $\mathbf{b}$ yields the needed positive $\hat{\mathbf{b}}$. By selecting the base set $B = \{5, \ldots, 14\}$ and $N = \{1, 2, 3, 4\}$, we obtain $\mathbf{A}_B = \mathsf{Id}_8$, $\boldsymbol{x}_N^0 = 0$, and

$$\boldsymbol{x}_B^0 = \hat{\mathbf{b}} = \mathbf{A}_B^{-1}\mathbf{b} = \mathsf{Id}_8 \cdot \mathbf{b} = \begin{bmatrix} 1 & 1 & 1 & 1 & 1 & 1 & 1 & 1 & 1 & 1 \end{bmatrix}^T \geq 0,$$

from which it follows that

$$\boldsymbol{x}^0 = \begin{bmatrix} \boldsymbol{x}_B^0 \\ \boldsymbol{x}_N^0 \end{bmatrix}$$

is an initial BFS as desired. The cost function at $\boldsymbol{x}^0$ evaluates to $z = -\mathbf{F}\bar{\mathbf{u}}_2 = 0$, while the reduced cost vector is given by

$$\hat{\mathbf{c}}_N = \begin{bmatrix} 0 & -F & 0 & 0 \end{bmatrix}^T.$$

Since $\hat{c}_2 = -F < 0$, the index in the nonbasic set, $q = 2$, is selected for swapping. To determine the swapping index among the basic variables, we compute the derivative vector $\mathbf{d}$, given by

$$\mathbf{d} = \frac{\sqrt{E}}{l} \begin{bmatrix} 0 & 0 & -1 & 1 & 0 & 0 & 0 & 0 & -\frac{1}{2} & \frac{1}{2} \end{bmatrix},$$

so that $p = \arg\min_{i \in B \,\&\, d_i < 0} \frac{\hat{b}_i}{-d_i} = 3$. Swapping the indices $q = 2$ and $p = 3$, we obtain the new basic and nonbasic sets $B = \{5, 6, 2, 8, 9, 10, 11, 12, 13, 14\}$ and $N = \{1, 7, 3, 4\}$. After three additional iterations of Algorithm 8 in Appendix B, the simplex converges to the following solution of the dual LP problem:

$$\begin{bmatrix} \bar{u}_1^* & \bar{u}_2^* & \bar{u}_3^* & \bar{u}_4^* \end{bmatrix} = \frac{l}{\sqrt{E}} \begin{bmatrix} 1 & 3 & 0 & 2 \end{bmatrix},$$

and $W^* = \frac{3Fl}{\sqrt{E}}$. The reduced cost at the solution is $\hat{c}_N = \frac{Fl}{\sqrt{E}} \begin{bmatrix} 1 & 0 & 0 & 2 \end{bmatrix}$ with a nonbasic set $N = \{6, 7, 3, 13\}$, indicating that $s_2, u_3 \in \mathbb{R}_+$ can be varied at will without affecting the optimized value of W. In other words, there are infinite solutions to the dual LP problem, and other criteria such as stability, buckling, or aesthetics must be invoked to select a design.

To visualize the optimized truss, we determine the geometry and attributes of the optimized layouts. We start with the displacements of optimized solution, $\mathbf{u}^*$, which can be computed at once from Equation C.2, yielding

$$\begin{aligned} \begin{bmatrix} u_1^* & u_2^* & u_3^* & u_4^* \end{bmatrix} &= \frac{lW^*}{V\sqrt{E}} \begin{bmatrix} 1 & 3 & 0 & 2 \end{bmatrix} \\ &= \frac{3Fl^2}{VE} \begin{bmatrix} 1 & 3 & 0 & 2 \end{bmatrix}, \end{aligned}$$

from which the minimum compliance of all optimized designs is calculated as $J_c^* = \mathbf{F}^T\mathbf{u}^* = (W^*)^2/V = \frac{9F^2l^2}{VE}$. To compute the cross-sectional area of the rods, we first compute the force T_i and stress σ_i at each rod, and then obtain their cross-sectional areas from

$$T_i^* = \sigma_i^* A_i^*.$$

By the principle of virtual work,

$$\sum_i T_i \bar{\epsilon}_i l_i = F\bar{u}_2,$$

we obtain

$$-\bar{u}_1\left(T_1+\frac{T_5}{\sqrt{2}}\right)+\bar{u}_2\left(T_2+\frac{T_5}{\sqrt{2}}-F\right)+\bar{u}_3\left(T_3+\frac{T_4}{\sqrt{2}}\right)-\bar{u}_4\left(T_2+\frac{T_4}{\sqrt{2}}\right)=0,$$

which yields the following equilibrium equations:

$$T_1+\frac{T_5}{\sqrt{2}}=0,$$
$$T_2+\frac{T_5}{\sqrt{2}}=F,$$
$$T_3+\frac{T_4}{\sqrt{2}}=0,$$
$$T_2+\frac{T_4}{\sqrt{2}}=0.$$

Solving for T_1, T_3, T_4, T_5 as a function of T_2, we obtain

$$T_1=-F+T_2, \quad (C.17)$$
$$T_3=T_2, \quad (C.18)$$
$$T_4=-\sqrt{2}T_2, \quad (C.19)$$
$$T_5=\sqrt{2}(F-T_2). \quad (C.20)$$

The stresses in rods 1, 2, 3, 4, and 5 can be computed using Hooke's law for small displacements as follows:

$$\sigma_1=E\epsilon_1^*=-\frac{E}{l}u_1^*=-\frac{3Fl}{V}<0, \quad (C.21)$$
$$\sigma_2=E\epsilon_2^*=\frac{E}{l}(u_2^*-u_4^*)=\frac{3Fl}{V}>0, \quad (C.22)$$
$$\sigma_3=E\epsilon_3^*=\frac{E}{l}u_3^*\geq 0, \quad (C.23)$$
$$\sigma_4=E\epsilon_4^*=\frac{E}{2l}(u_3^*-2), \quad (C.24)$$
$$\sigma_5=E\epsilon_5^*=\frac{E}{2l}(u_2^*-u_1^*)=\frac{3Fl}{V}>0. \quad (C.25)$$

In addition to equilibrium, the volume of the structure must equal V, given by

$$V=\sum_i A_i l_i=\sum_i \frac{T_i}{\sigma_i}l_i. \quad (C.26)$$

Substituting Equations C.17–C.20 and C.21–C.25 into Equation C.26, we obtain the following, which constrains the values of T_2 and u_3^*:

$$V=\frac{2E^2(u_3^*)^2V^2-3EFl^2u_3^*V+18F^2l^4}{18EF^2l^2u_3^*-3E^2F(u_3^*)^2V}T_2+V.$$

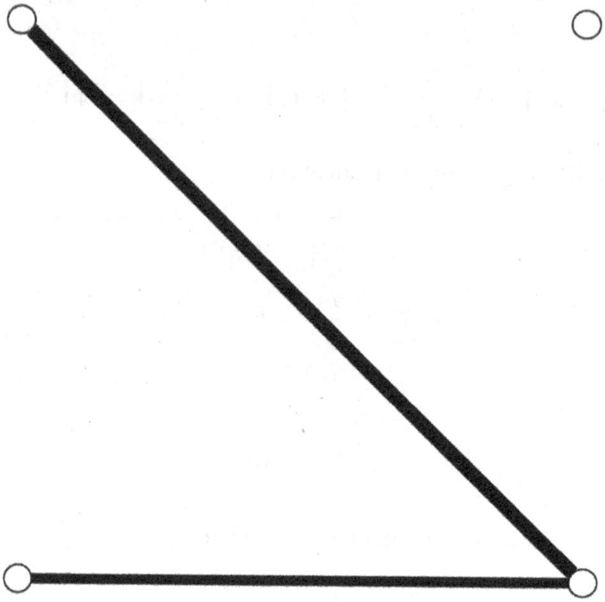

Figure C.2 Five bar problem: Ground structure solution.

Solving this equation, we obtain $T_2 = 0$ or

$$u_3^* = \frac{3Fl^2}{4EV}(1 \pm \mathrm{i}\sqrt{15}),$$

from which we conclude that the only physical solution to the MP is $T_2 = 0$. To compute the cross-sectional areas, we start with Equations C.17–C.20 and determine

$$T_1 = -F, T_2 = 0, T_3 = 0, T_4 = 0, T_5 = \sqrt{2}F.$$

The cross-sectional areas are then computed using $A_i = T_i/\sigma_i$, resulting in the following optimized cross-sectional areas:

$$\begin{aligned} A_1 &= \frac{V}{3l}, \\ A_2 &= 0, \\ A_3 &= 0, \\ A_4 &= 0, \\ A_5 &= \frac{\sqrt{2}V}{3l}. \end{aligned}$$

This indicates that rods 2, 3, and 4 are not part of the optimized layout, since $A_2 = A_3 = A_4 = 0$. With these areas, the optimal structure is fully stressed ($-\sigma_1 = \sigma_5 = 3Fl/V$) and has the layout shown in Figure C.2.

Appendix D Increasing Structural/Networking Feasibility in Turtle-Constructed Graphs

While the turtle graphics interpretation of L-systems is relatively easy to implement, the branched nature of the resulting graphs can lead to a number of challenges. In particular, a prevalence of free ends is rarely of great advantage in many engineering applications. Here, two important approaches for mitigating this disadvantage of the turtle interpretation are provided.

D.1 Branching Structure Boundary Constraints

As discussed in Chapter 3, the creation of structures based on L-system encoded topologies interpreted via a turtle graphics algorithm relies on the spreading of branched segments, potentially within a predetermined bounded domain (cf. Figures 3.6 and 3.8). Segments that would result in the violation of such boundaries must then be actively constrained to only reach the boundaries and potentially connect with them, but not violate them. In practice, this can lead to feasible, if often inefficient, structures and networks as multiple paths connecting exterior segments to the interior may be created. The following provides a brief overview of this constraint.

Consider Figure D.1, which shows the agent (cf. Section 3.2) located at the coordinate (x_1, y_1) inside the design domain, which is rendered in gray and is bounded by a series of segments that include the line between (x_{p1}, y_{p1}) and (x_{p2}, y_{p2}). Now assume that the L-system-generated instructions cause the agent to create a line segment of length L, corresponding to a new location of (x_2, y_2). The first step in this boundary constraint function is to check whether the new location remains inside the boundary, which is performed using the ray casting algorithm (Antonio 1992); if the agent remains within the boundary, the constraint is not violated and the line segment is created as instructed by the L-system. However, suppose that (x_2, y_2) is found to lie outside the domain boundary. Given that the segment is allowed to reach the boundary but not cross it, the intersection point of the boundary and the proposed line segment must be determined, where this point (x_{int}, y_{int}) will be defined as the new segment end point. It can be shown that for a line defined by points (x_1, y_1) and (x_2, y_2) and another line defined by points (x_3, y_3) and (x_4, y_4), the intersection coordinates (x_{int}, y_{int}) between those two lines can be determined by

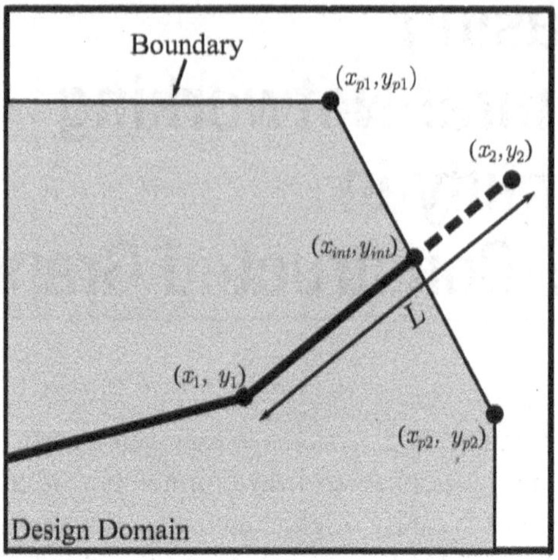

Figure D.1 Illustration of boundary constraints associated with the creation of L-system-generated topologies. Reprinted with permission from (Bielefeldt, Reich, et al. 2019).

$$(x_{int}, y_{int}) = \left(\frac{(x_1y_2 - y_1x_2)(x_3 - x_4) - (x_1 - x_2)(x_3y_4 - y_3x_4)}{(x_1 - x_2)(y_3 - y_4) - (y_1 - y_2)(x_3 - x_4)}, \right. \\ \left. \frac{(x_1y_2 - y_1x_2)(y_3 - y_4) - (y_1 - y_2)(x_3y_4 - y_3x_4)}{(x_1 - x_2)(y_3 - y_4) - (y_1 - y_2)(x_3 - x_4)} \right). \tag{D.1}$$

Substituting in the start and end points for the intersecting line segments (turtle-generated and pre-existing boundary), one obtains the coordinate (x_{int}, y_{int}). The agent state is updated such that it constructs a segment from (x_1, y_2) to (x_{int}, y_{int}), but its heading is left unchanged. Note that, by virtue of Equation D.1 and this algorithm, the agent will not leave (x_{int}, y_{int}) and no new segments will be created until the heading angle is eventually updated so as to drive the agent back into the design domain (cf. Figures 3.7 and 3.9).

D.2 Free-Ended Segment Trimming Algorithm

One clear disadvantage of the branched L-system/turtle approach to topological generation is the necessary existence of line segments with free ends, which do not have the capability of communicating either information or loads with the remainder of the network. Thus, except in the case of fluid-thermal systems (cf. Section 6.1.1), antennas, and other limited cases, free-ended segments often have no engineering significance, artificially inflate structural mass and/or configurational complexity, and may even require increased computational effort as even their null contribution must be calculated (e.g., via elements assigned to such useless segments in solid FEA). Therefore, the implementation of a trimming algorithm to remove free-ended segments is fundamental to the approach taken in this work.

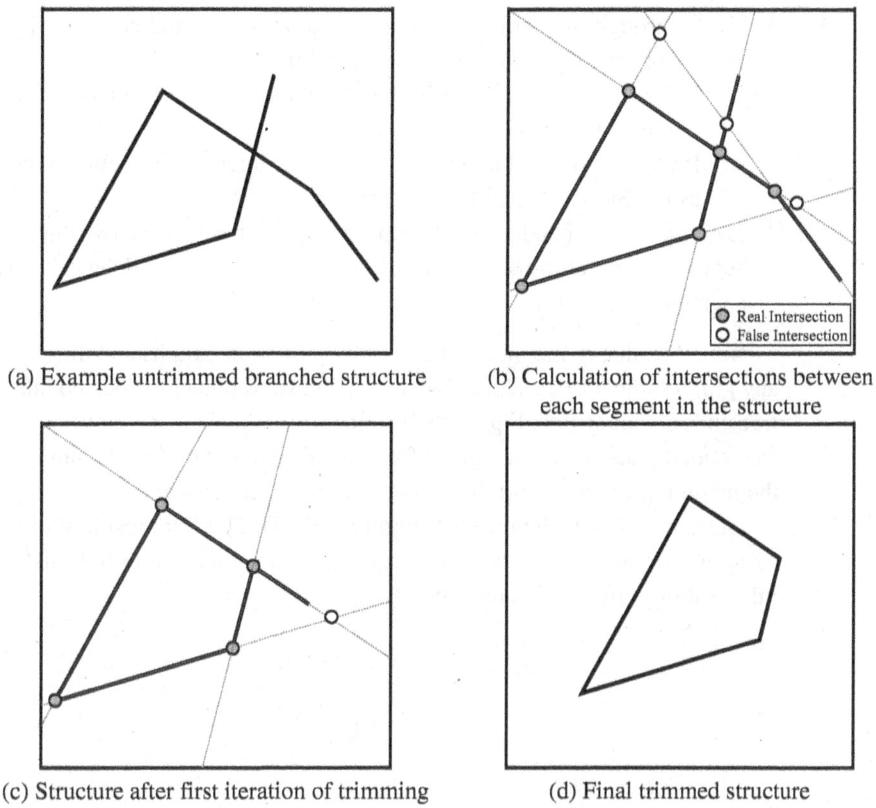

(a) Example untrimmed branched structure

(b) Calculation of intersections between each segment in the structure

(c) Structure after first iteration of trimming

(d) Final trimmed structure

Figure D.2 Schematic illustration of the trimming algorithm as applied to a simple graph. Reprinted with permission from (Bielefeldt, Reich, et al. 2019).

Figure D.2 schematically illustrates the trimming process. As the turtle algorithm was only used for structural examples in Chapter 3, the algorithm used herein actually makes direct use of Abaqus' "auto-trim" feature, although the underlying principles can be translated to any CAD tool. To summarize the approach, the algorithm determines that a segment is free-ended and should be eliminated if one of its two end points is neither (i) shared by another segment, nor (ii) lies somewhere in the middle of (i.e., represents the intersection with) another line segment. Making this determination and removing free-ended segments is a four-step process, summarized here.

1. Cycle through the list of line segments, obtaining and storing the start- and end-points of each;
2. Cycle again through the list of line segments, calculating and storing for each point of intersection between that segment and every other segment using Equation D.1, where many calculated intersection points will represent *extrapolations* from one or both of the considered segments (white dots in Figure D.2(b));

3. Cycle through the list of line segments a final time and check each of the two end points for one of the following two conditions:
   (a) If the endpoint is also an endpoint of another segment as per Step 1, it is not at a free end, and
   (b) If the endpoint is known to be a point of intersection with another line segment as per Step 2, it is not a free end.
4. If *neither* of the conditions of Step 3 are met for any of the two endpoints of a given segment, the "auto-trim" feature of Abaqus is used to delete the free end of the segment in question.

Note that, due to the inherent complexity of some graphs and the evolving nature of any graph subject to the algorithm above, the steps must be repeated until no additional free ends are identified. Figure D.2(c) illustrates this in that a previously analyzed non-free-ended branch may become a free-ended segment during the iterative execution of the trimming process, and the algorithm must iterate over each line segment until the structure remains unchanged (cf. Figure D.2(d)). The final result of this algorithm is a structure (or network) in which all line segments are capable of transferring loads (or information) effectively and robustly.

Appendix E Conversion of Graphs to Volumetric Bodies for Three-Dimensional Analysis and Additive Manufacturing

The bioinspired generative methods introduced throughout this text all rely on a graph-based representation of structures or networks that must then be interpreted in terms of the properties of each nodal connection. In the case of slender structures (i.e., trusses and frames), the nodes in a graph must be given geometric locations, and the line segments between nodes must be assigned sectional properties, both geometric and material-related. Because it is the intent of all structural designs described in this text to represent three-dimensional entities, even if just mathematically and graphically, it may at times be valuable to extend their representations to the truly volumetric and even physical.

To that end, this brief appendix addresses algorithms for post-processing L-system-generated structures into monolithic volumes, which can be suitable for full three-dimensional finite element analysis (e.g., as a means of *numerically validating* reduced-order predictions) and/or for additive manufacturing toward further experimental characterization or end use as functioning structures. Associated codes for these algorithms can be found at ***https://github.com/brentbie/L-system-Topology-Optimization.git***.

E.1 Two-Dimensional Graphs

The conversion of two-dimensional graph-based designs into three-dimensional volumes is relatively straightforward as the extrusion of any closed two-dimensional domain (or collection of such domains) into a three-dimensional filled volume is a common process in all computer-aided drafting (CAD) tools. Each one-dimensional line segment in any two-dimensional graph considered in this work need only to be converted into a two-dimensional (usually high aspect ratio) representation prior to the extrusion step. Specifically, an appropriate algorithm for the full graph-to-volume process could proceed through the following steps:

1. For each line segment in the 2-D (e.g., $x - y$) graph:
   (a) Create two new line segments with length equal to the original, offset from either side of the original by a distance of 1/2 the intended segment thickness;
   (b) At each end of the original line segment, connect the two ends of the new offset line segments to create a closed rectangle;

(a) 3-D (STL) model

(b) Additively manufactured prototype

Figure E.1 Examples of two distinct transversely loaded beam designs developed using the *two-dimensional* turtle algorithm of Chapter 3 showing both digital and physical three-dimensional realizations created using the algorithm provided.

   (c) Delete the original line segment;
   (d) Convert the new closed rectangle to a filled surface or shell.
2. Merge (Boolean union) all resulting rectangular surfaces residing in the original 2-D plane into a single 2-D surface/shell
3. *Optional*: To improve representative fidelity and structural performance, special care should be taken at the intersection of the original edges. Circles or properly rotated squares can be centered at the location of all graph nodes and then merged into the final 2-D surface/shell to create properly reinforced corners/intersections in the resulting part.
4. Extrude the final merged shell in the third Cartesian (e.g., z) direction over the desired distance, usually resulting in a non-simply connected solid domain.

Two examples of such volumes realized using a Python script driving the FreeCAD software package are shown in Figure E.1. Note that all major CAD packages are capable of exporting modeled volumes to `.stl` files, which are suitable for many additive manufacturing processes. An additional three-dimensional realization of an initially two-dimensional graph is shown in Section 3.5, where a physical prototype is shown to operate as designed. Additional examples can be seen in the open literature Bielefeldt, Reich, et al. (2019).

E.2 Three-Dimensional Graphs

The conversion of three-dimensional graphs of connected line segments to monolithic volumes requires additional steps since simple extrusion from a two-dimensional plane is not sufficient. Rather, each individual segment must be converted into a three-dimensional domain that is properly translated and rotated such that its ends are colocated with the appropriate nodes. The collection of all such sub-domains can then be merged into a single part. Note that, because two-dimensional graphs represent a subset of all three-dimensional graphs, the approach described here can be applied to the former (e.g., when non-rectangular and thus non-extrudable cross-sections are desired). An effective algorithm might proceed through the following steps:

1. For each line segment in the 3-D graph:
   (a) Create a three-dimensional representation of the segment based on the desired cross-section using one of at least two approaches:
      - Extrusion of prismatic sections
        (i) Generate a closed/filled two-dimensional representation of the intended cross-section (e.g., rectangular, circular, or otherwise);
        (ii) Extrude this shell/surface in the direction of its normal and with a distance equivalent to the line segment in question, usually resulting in a long slender prismatic shape.

      –OR–
      - Definition of 3-D primitives
        (i) If possible, use CAD package primitive options to directly define a three-dimensional representation of the given edge (e.g., a rectangular prism or cylinder having the desired cross-sectional configuration with the length of the segment in question).

   (b) Rotate the three-dimensional representation such that its long (e.g., neutral) axis is aligned with the original line segment. Note that the angular orientation of the cross-section *about* this longitudinal axis is left to the intentions of the designer;
   (c) Translate the new three-dimensional representation such that one of its ends is colocated with the appropriate node representing the end of the original line segment;
   (d) Delete the original line segment.
2. Merge (Boolean union) all resulting volumetric representations into a single volume, usually resulting in a non-simply connected solid domain;
3. *Optional*: To improve representative fidelity and structural performance, special care should be taken at the intersection of the original edges (i.e., at the nodes). Spheres or properly rotated cubes can be centered at the location of all graph nodes and then merged into the final volume to create properly reinforced corners/intersections in the resulting part.

A CAD model example of such a volume generated from a SPIDRS-created three-dimensional graph is shown in Figure E.2, where edges are assumed to represent slender prismatic domains having circular cross-sections of various diameters. Representations were generated using a Python script implementing the algorithm above (3-D primitive option), which directed the FreeCAD software package. Because circular cross-sections are assumed, the following two considerations apply: (i) only the

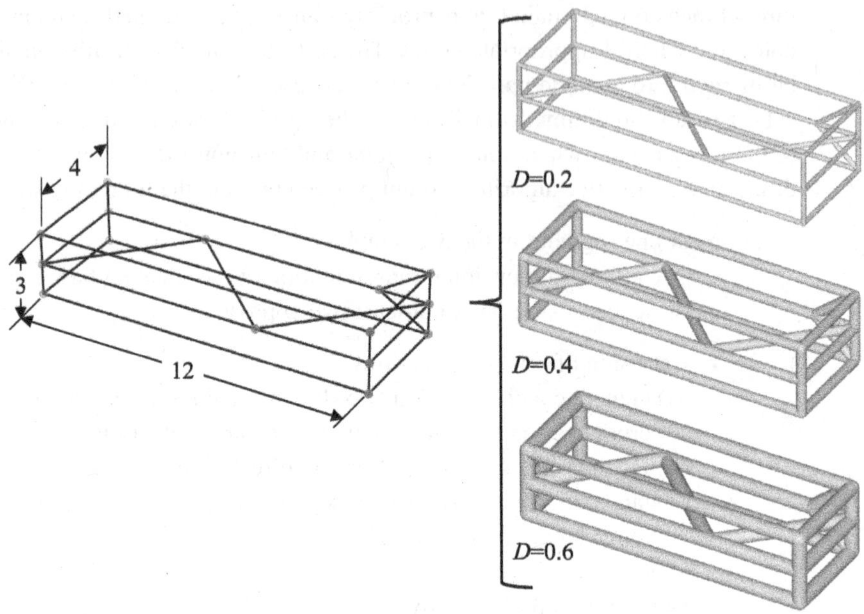

Figure E.2 Example of 3-D SPIDRS-generated frame structure (see Section 4.4.2) in both graph and final volumetric solid forms, where the diameter D of the cylindrical sections can be easily changed. Note that appropriately sized spheres have been merged into the volume at all nodal locations, resulting in smooth and realistic contours in these regions.

three-dimensional orientation of the longitudinal axis of each cylinder need be defined as the angular orientation of the circular cross-section about each long axis is inconsequential and (ii) spheres having the same diameter as the cylinders and centered at each node result in realistic and well-performing segment intersections.

References

Aage, N., Andreassen, E., Lazarov, B. S. & Sigmund, O. (2017), 'Giga-voxel computational morphogenesis for structural design', *Nature* **550**(7674), 84–86.

Abelson, H. & diSessa, A. (1986), *Turtle Geometry: The Computer as a Medium for Exploring Mathematics*, The MIT Press, Cambridge, MA.

Achtziger, W. (2007), 'On simultaneous optimization of truss geometry and topology', *Structural and Multidisciplinary Optimization* **33**, 285–304.

Allaire, G. (2002), *Shape Optimization By the Homogenization Method*, Springer, New York.

Allaire, G., Dapogny, C. & Frey, P. (2011), 'Topology and geometry optimization of elastic structures by exact deformation of simplicial mesh', *Comptes Rendus Mathematique* **349**(17–18), 999–1003.

Allaire, G., Jouve, F. & Toader, A.-M. (2004), 'Structural optimization using sensitivity analysis and a level-set method', *Journal of Computational Physics* **194**, 363–393.

Allaire, G. & Pantz, O. (2006), 'Structural optimization with FreeFem++', *Structural and Multidisciplinary Optimization* **32**, 173–181.

Amstutz, S. & Andrä, H. (2006), 'A new algorithm for topology optimization using a level-set method', *Journal of Computational Physics* **216**(2), 573–588.

Antonio, F. (1992), 'Faster line segment intersection', in D. Kirk, ed., *Graphics Gems III (IBM Version)*, Elsevier, Cambridge, MA, pp. 199–202.

Axler, S. (2015), *Linear Algebra Done Right*, 3rd ed., Springer, New York.

Azid, A., Kwan, A. S. K. & Seetharamu, K. N. (2002), 'A GA-based technique for layout optimization of truss with stress and displacement constraints', *International Journal for Numerical Methods in Engineering* **53**(7), 1641–1674.

Balamurugan, R., Ramakrishnan, C. V. & Singh, N. (2008), 'Performance evaluation of a two stage adaptive genetic algorithm (TSAGA) in structural topology optimization', *Applied Soft Computing* **8**(4), 1607–1624.

Barzilai, J. & Borwein, J. M. (1988), 'Two point step size gradient methods', *IMA Journal of Numerical Analysis* **8**, 141–148.

Bathe, K.-J. (1996), *Finite Element Procedures*, Prentice-Hall, Watertown, MA.

Ben-Tal, A. & Bendsøe, M. P. (1993), 'A new method for optimal truss topology design', *SIAM Journal on Optimization* **3**(2), 322–358.

Bendsøe, M. P. (1989), 'Optimal shape design as a material distribution problem', *Structural and Multidisciplinary Optimization* **1**, 193–202.

Bendsøe, M. P., Ben-Tal, A. & Zowe, J. (1994), 'Optimization methods for truss geometry and topology design', *Structural Optimization* **7**(3), 141–159.

Bendsøe, M. P., Lund, E., Olhoff, N. & Sigmund, O. (2005), 'Topology optimization – Broadening the areas of application', *Control and Cybernetics* **34**(1), 7–35.

Bendsøe, M. P. & Sigmund, O. (2003), *Topology Optimization: Theory, Methods and Applications*, 2nd ed., Springer, New York.

Berdichevsky, V. L. (2009), *Variational Principles of Continuum Mechanics. Part II: Applications*, Springer, New York.

Berg, M., Cheong, O., Kreveld, M., & Overmars, M. (1997), 'Computational geometry', in *Computational Geometry*, Springer, Berlin, pp. 1–17.

Besada, J. L. (2019), 'Math and music, models and metaphors: Alberto Posadas 'tree-like structures', *Contemporary Music Review* **38**(1–2), 107–131.

Bielefeldt, B. R. (2020), 'Multiobjective topology optimization for preliminary design using graph theory and L-system languages', PhD thesis, Texas A&M University.

Bielefeldt, B. R., Akleman, E., Reich, G. W., Beran, P. S. & Hartl, D. J. (2019), 'L-system-generated mechanism topology optimization using graph-based interpretation', *Journal of Mechanisms and Robotics* **11**(2).

Bielefeldt, B. R., Reich, G. W., Beran, P. S. & Hartl, D. J. (2019), 'Development and validation of a genetic L-system programming framework for topology optimization of multifunctional structures', *Computers & Structures* **218**, 152–169.

Bondy, J. A. & Murty, U. S. R. (1976), *Graph Theory with Applications*, vol. 290, Macmillan, London.

Bourdin, B. (2001), 'Filters in topology optimization', *International Journal for Numerical Methods in Engineering* **50**, 2143–2158.

Bourdin, B. & Chambolle, A. (2003), 'Design-dependent loads in topology optimization', *ESAIM: Control, Optimisation and Calculus of Variations* **9**, 19–48.

Bruns, T. E. (2007), 'Topology optimization of convection-dominated, steady-state heat transfer problems', *International Journal of Heat and Mass Transfer* **50**(15–16), 2859–2873.

Cappello, F. & Mancuso, A. (2003), 'A genetic algorithm for combined topology and shape optimizations', *Computer Aided Design* **35**(8), 761–769.

Cauchy, A. (1847), 'Méthode générale pour la résolution des systèmes d'équations simultanées', *Comptes Rendus de l'Académie des Sciences Paris* **25**, 536–538.

Chandrupatla, T. & Belegundu, A. (2021), *Introduction to Finite Elements in Engineering*, 5th ed., Cambridge University Press, Cambridge.

Chapman, C. D., Saitou, K. & Jakiela, M. J. (1994), 'Genetic algorithms as an approach to configuration and topology design', *Journal of Mechanical Design* **116**(4), 1005–1012.

Chen, T.-Y. & Wu, S.-C. (1998), 'Multiobjective optimal topology design of structures', *Computational Mechanics* **21**(6), 483–492.

Choi, K. K. & Kim, N.-H. (2006), *Structural Sensitivity Analysis and Optimization 1: Linear Systems*, Springer Science & Business Media, New York.

Chomsky, N. (1959), 'On certain formal properties of grammars', *Information and Control* **2**(2), 137–167.

Chu, D., Xie, Y., Hira, A. & Steven, G. (1996), 'Evolutionary structural optimization for problems with stiffness constraint', *Finite Elements in Analysis and Design* **21**(4), 239–251.

Chu, D., Xie, Y. & Steven, G. (1998), 'An evolutionary structural optimization method for sizing problems with discrete design variables', *Computers and Structures* **68**, 419–431.

Clark, V. (2019), 'On an evolutionary developmental methodology for pin-joint framework optimization', A thesis submitted to the Graduate Division of the University of Hawai'i at Mānoa in Partial Fulfillment of the Requirements for the Degree of Master of Science in Mechanical Engineering, May 2019.

Clausen, A., Schevenels, M., Lazarov, B. S. & Sigmund, O. (2011), 'Efficient topology optimization in MATLAB using 88 lines of code', *Structural and Multidisciplinary Optimization* **43**, 1–16.

Coello, C. A. C. (2002), *Evolutionary Optimization*, Kluwer, Secaucus.

Coffin, P. & Maute, K. (2016), 'Level set topology optimization of cooling and heating devices using a simplified convection model', *Structural and Multidisciplinary Optimization* **53**(5), 985–1003.

Cohon, J. L. (2003), *Multiobjective Programming and Planning*, Dover Publications, Mineola, NY.

COM (2008), *COMSOL Multiphysics 3.5 User's Manual*.

Combes, S. & Daniel, T. (2003), 'Flexural stiffness in insect wings I. Scaling and the influence of wing venation', *Journal of Experimental Biology* **206**(17), 2979–2987.

Cook, R. D., Malkus, D. S. & Plesha, M. E. (1989), *Concepts and Applications of Finite Element Analysis*, 3rd ed., Wiley, New York.

Das, I. & Dennis, J. E. (1997), 'A closer look at drawbacks of minimizing weighted sums of objectives for Pareto set generation in multicriteria optimization problems', *Structural Optimization* **14**(1), 63–69.

Dassault Systèmes of America Corp. (2014), 'Tosca structure user's guide'.

de Ruiter, M. & van Keulen, F. (2004), 'Topology optimization using a topology description function', *Structural and Multidisciplinary Optimization* **26**(6), 406–416.

Deaton, J. D. & Grandhi, R. V. (2014), 'A survey of structural and multidisciplinary continuum topology optimization: Post 2000', *Structural and Multidisciplinary Optimization* **49**(1), 1–38.

Deb, K. (1999), 'An introduction to genetic algorithms', *Sadhana* **24**(4–5), 293–315.

Deb, K. (2001), *Multi-Objective Optimization Using Evolutionary Algorithms*, John Wiley & Sons, Chichester.

Deb, K., Agrawal, S., Pratap, A. & Meyarivan, T. (2000), 'A fast elitist non-dominated sorting genetic algorithm for multi-objective optimization: NSGA-II', in *International Conference on Parallel Problem Solving from Nature*, Springer, Berlin, pp. 849–858.

Deng, S. & Suresh, K. (2016), 'Multi-constrained 3D topology optimization via augmented topological level-set', *Computers & Structures* **170**, 1–12.

Diaz, A. & Lipton, R. (1997), 'Optimal material layout for 3D elastic structures', *Structural Optimization* **13**(1), 60–64.

Diestel, R. (2010), 'Graph theory – Heidelberg', *Graduate Texts in Mathematics* **173**(451), 5.

Dirker, J. & Meyer, J. P. (2013), 'Topology optimization for an internal heat-conduction cooling scheme in a square domain for high heat flux applications', *Journal of Heat Transfer* **135**(11).

Dunning, P. D., Stanford, B. K. & Kim, H. A. (2015), 'Coupled aerostructural topology optimization using a level set method for 3D aircraft wings', *Structural and Multidisciplinary Optimization* **51**(5), 1113–1132.

Ern, A. & Guermond, J.-L. (2004), *Theory and Practice of Finite Elements*, Springer Verlag, Berlin.

Eschenauer, H. A. & Olhoff, N. (2001), 'Topology optimization of continuum structures: A review', *Applied Mechanics Reviews* **54**(4), 331–390.

Fanjoy, D. W. & Crossley, W. A. (2002), 'Topology design of planar cross-sections with a genetic algorithm: Part 1 – Overcoming the obstacles', *Engineering Optimization* **34**(1), 33–48.

Fantini, P. (2007), 'Effective multiobjective MDO for conceptual design – An aircraft design perspective', PhD thesis, Cranfield University.

Fletcher, R. (1987), *Practical Methods of Optimization*, John Wiley & Sons, West Sussex.

Fletcher, R., Qi, L., Teo, K. & Yang, X. (2005), *Optimization and Control with Applications*, Springer Science and Business Media, New York.

Gain, A. L. & Paulino, G. H. (2013), 'A critical comparative assessment of differential equation-driven methods for structural topology optimization', *Structural and Multidisciplinary Optimization* **48**, 685–710.

Gibbons, A. (1985), *Algorithmic Graph Theory*, Cambridge University Press, Cambridge.

Goldberg, D. E. (1989), *Genetic Algorithms in Search, Optimization, and Machine Learning*, Addison-Wesley, Reading, MA.

Gomes, A. A. & Suleman, A. (2006), 'Application of spectral level set methodology in topology optimization', *Structural and Multidisciplinary Optimization* **31**(6), 430–443.

Gould, S. J. (2002), *The Structure of Evolutionary Theory*, Harvard University Press, Cambridge, MA.

Guan, C. L. & Chun, Y. L. (2009), 'Structural topology optimization using ant colony optimization algorithm', *Applied Soft Computing* **9**(4), 1343–1353.

Guest, J. K. (2009), 'Imposing maximum length scale in topology optimization', *Structural and Multidisciplinary Optimization* **37**(5), 463–473.

Guest, J. K., Asadpoure, A. & Ha, S.-H. (2011), 'Eliminating beta-continuation from Heaviside projection and density filter algorithms', *Structural and Multidisciplinary Optimization* **44**(4), 443–453.

Guest, J. K., Prévost, J. H. & Belytschko, T. (2004), 'Achieving minimum length scale in topology optimization using nodal design variables and projection functions', *International Journal for Numerical Methods in Engineering* **61**(2), 238–254.

Guirguis, D., Hamza, K., Aly, M., Hegazi, H. & Saitou, K. (2015), 'Multi-objective topology optimization of multi-component continuum structures via a Kriging-interpolated level set approach', *Structural and Multidisciplinary Optimization* **51**(3), 733–748.

Ha, S.-H. & Cho, S. (2008), 'Level set based topological shape optimization of geometrically nonlinear structures using unstructured mesh', *Computers & Structures* **86**(13–14), 1447–1455.

Hagishita, T. & Ohsaki, M. (2009), 'Topology optimization of trusses by growing ground structure method', *Structural and Multidisciplinary Optimization* **37**(4), 377–393.

Hamda, H., Jouve, F., Lutton, E., Schoenauer, M. & Sebag, M. (2002), 'Compact unstructured representations for evolutionary design', *Applied Intelligence* **16**(2), 139–155.

Hanan, J. S. (1992), 'Parametric L systems and their application to the modeling and visualization of plants', PhD thesis, University of Regina.

Hare, W., Nutini, J. & Tesfamariam, S. (2013), 'A survey of non-gradient optimization methods in structural engineering', *Advances in Engineering Software* **59**, 19–28.

Hartl, D. J., Bielefeldt, B., Reich, G. W. & Beran, P. S. (2017), 'Multi-fidelity analysis and experimental characterization of muscular-skeletal structures optimized via genetic programming', in *25th AIAA/AHS Adaptive Structures Conference*, p. 1442.

Hemp, W. S. (1973), *Optimum Structures*, Clarendon Press, Oxford.

Hingston, P. F., Barone, L. C. & Michalewicz, Z. (2008), *Design by Evolution: Advances in Evolutionary Design*, Springer Science & Business Media, New York.

Ho, S., Nassef, H., Pornsinsirirak, N., Tai, Y.-C. & Ho, C.-M. (2003), 'Unsteady aerodynamics and flow control for flapping wing flyers', *Progress in Aerospace Sciences* **39**(8), 635–681.

Howell, L. L. (2001), *Compliant Mechanisms*, John Wiley & Sons, New York.

Hu, H., Kumar, A. G., Abate, G. & Albertani, R. (2010), 'An experimental investigation on the aerodynamic performances of flexible membrane wings in flapping flight', *Aerospace Science and Technology* **14**(8), 575–586.

Hughes, J. F., van Dam, A., Mcguire, M., Sklar, D. F., Foley, J. D., Feiner, S. K. & Akeley, K. (2014), *Computer Graphics*, 3rd ed., Addison-Wesley, Upper Saddle River, NJ.

Huttenlocher, D. P., Klanderman, G. A. & Rucklidge, W. J. (1993), 'Comparing images using the hausdorff distance', *IEEE Transactions on Pattern Analysis and Machine Intelligence* **15**(9), 850–863.

Jakiela, M. J., Chapman, C., Duda, J., Adewuya, A. & Saitou, K. (2000), 'Continuum structural topology design with genetic algorithms', *Computer Methods in Applied Mechanics and Engineering* **186**(2–4), 339–356.

Jantos, D. R., Hackl, K. & Junker, P. (2019), 'An accurate and fast regularization approach to thermodynamic topology optimization', *International Journal for Numerical Methods in Engineering* **117**, 991–1017.

Johnson, E. & Neill, D. (1988), Automated structural optimization system (ASTROS). Volume 3. Applications manual', Technical report, Northrop Corp., Hawthorne, CA. Aircraft Div.

Junker, P. & Hackl, K. (2015), 'A variational growth approach to topology optimization', *Structural and Multidisciplinary Optimization* **52**, 293–304.

Kaldate, A., Thurston, D., Emamipour, H. & Rood, M. (2006), 'Engineering parameter selection for design optimization during preliminary design', *Journal of Engineering Design* **17**(4), 291–310.

Kane, C. & Schoenauer, M. (1996), 'Topological optimum design using genetic algorithms', *Control and Cybernetics* **25**(5), 1059–1087.

Katoch, S., Chauhan, S. S. & Kumar, V. (2021), 'A review on genetic algorithm: Past, present, and future', *Multimedia Tools and Applications* **80**, 8091–8126.

Kawamoto, A., Matsumori, T., Yamasaki, S., Nomura, T., Kondoh, T. & Nishiwaki, S. (2011), 'Heaviside projection based topology optimization by a PDE-filtered scalar function', *Structural and Multidisciplinary Optimization* **44**(1), 19–24.

Khetan, A., Lohan, D. J. & Allison, J. T. (2015), 'Managing variable-dimension structural optimization problems using generative algorithms', *Structural and Multidisciplinary Optimization* **52**, 695–715.

Kobayashi, M. H. (2010), 'On a biologically inspired topology optimization method', *Communications in Nonlinear Science and Numerical Simulation* **15**(3), 787–802.

Kobayashi, M. H., Canfield, R. A. & Kolonay, R. M. (2021), 'On a cellular developmental method for layout optimization via multi-fidelity analyses and the two-point topological derivative', *Structural and Multidisciplinary Optimization* **64**, 2343–2360.

Kobayashi, M. H., Pedro, H. T. C., Coimbra, C. & da Silva, A. (2009), 'The formal evolutionary development of low entropy dendritic thermal systems', *Journal of Thermophysics and Heat Transfer* **23**(4), 822–827.

Kobayashi, M. H., Pedro, H. T. C., Kolonay, R. M. & Reich, G. W. (2009), 'On a cellular division method for aircraft structural design', *The Aeronautical Journal* **113**, 821–831.

Kolonay, R. M. & Kobayashi, M. H. (2010), 'Topology, shape, and sizing optimization of aircraft lifting surfaces using a cellular division method', in *13th AIAA/ISSMO Multidisciplinary Analysis Optimization Conference*, p. 9079.

Kolonay, R. M. & Kobayashi, M. H. (2015), 'Optimization of aircraft lifting surfaces using a cellular division method', *Journal of Aircraft* **52**(6), 2051–2063.

Kraft, D. (1994), 'Algorithm 733: Tomp – Fortran modules for optimal control calculations', *ACM Transactions on Mathematical Software* **20**, 262–281.

Kreissl, S., Pingen, G. & Maute, K. (2011), 'An explicit level set approach for generalized shape optimization of fluids with the lattice Boltzmann method', *International Journal for Numerical Methods in Fluids* **65**(5), 496–519.

Kreyszig, E. (1978), *Introductory Functional Analysis with Applications*, John Wiley and Sons, New York.

Kuphaldt, T. R. (1988), *Lessons in Electric Circuits, Volume II – Alternating Current*, Open Book Project.

Langdon, W. B. & Poli, R. (2013), *Foundations of Genetic Programming*, Springer Science & Business Media, New York.

Laurain, A. (2018), 'A level set-based structural optimization code using FEniCS', *Structural and Multidisciplinary Optimization* **58**, 1311–1334.

Lazarov, B. S. & Sigmund, O. (2011), 'Filters in topology optimization based on Helmholtz-type differential equations', *International Journal for Numerical Methods in Engineering* **86**(6), 765–781.

Lazarov, B. S. & Wang, F. (2017), 'Maximum length scale in density based topology optimization', *Computer Methods in Applied Mechanics and Engineering* **318**, 826–844.

Lee, J., Mueller, C. & Fivet, C. (2016), 'Automatic generation of diverse equilibrium structures through shape grammars and graphic statics', *International Journal of Space Structures* **31**(2–4), 147–164.

Lewiński, T., Sokół, T. & Graczykowski, C. (2019), *Michell Structures*, Springer Nature, Gewerbestrasse.

Li, T., Sun, X., Lu, Z. & Wu, Y. (2016), 'A novel multiobjective optimization method based on sensitivity analysis', *Mathematical Problems in Engineering* **2016**, 1–12.

Li, Z., Zuo, L., Kuang, J. & Luhrs, G. (2012), 'Energy-harvesting shock absorber with a mechanical motion rectifier', *Smart Materials and Structures* **22**(2), 025008.

Liang, Q., Xie, Y. & Steven, G. (2000), 'Optimal topology selection of continuum structures with displacement constraints', *Computers & Structures* **77**(6), 635–644.

Liefvendahl, M. & Stocki, R. (2006), 'A study on algorithms for optimization of Latin hypercubes', *Journal of Statistical Planning and Inference* **136**(9), 3231–3247.

Lin, J., Luo, Z. & Tong, L. (2010), 'A new multi-objective programming scheme for topology optimization of compliant mechanisms', *Structural and Multidisciplinary Optimization* **40**(1–6), 241–255.

Lindenmayer, A. (1968*a*), 'Mathematical models for cellular interactions in development I. Filaments with one-sided inputs', *Journal of Theoretical Biology* **18**, 280–299.

Lindenmayer, A. (1968*b*), 'Mathematical models for cellular interactions in development II. Simple and branching filaments with two-sided inputs', *Journal of Theoretical Biology* **18**, 300–315.

Lindenmayer, A. (1974), 'Adding continuous components to L-systems', in G. Rozenberg & A. Salomaa, eds., *L Systems*, Springer, New York, pp. 53–68.

Liu, J. & Ma, Y.-S. (2015), '3D level-set topology optimization: A machining feature-based approach', *Structural and Multidisciplinary Optimization* **52**(3), 563–582.

Liu, K. & Tovar, A. (2014), 'An efficient 3D topology optimization code written in Matlab', *Structural and Multidisciplinary Optimization* **50**(6), 1175–1196.

Liu, X., Yi, W.-J., Li, Q. & Shen, P.-S. (2008), 'Genetic evolutionary structural optimization', *Journal of Constructional Steel Research* **64**(3), 305–311.

Loyola, R. A., Querin, O. M., Jiménez, A. G. & Gordoa, C. A. (2018), 'A sequential element rejection and admission (SERA) topology optimization code written in Matlab', *Structural and Multidisciplinary Optimization* **58**, 1297–1310.

Luo, Z., Tong, L., Wang, M. Y. & Wang, S. (2007), 'Shape and topology optimization of compliant mechanisms using a parameterization level set method', *Journal of Computational Physics* **227**(1), 680–705.

Luo, Z., Wang, M. Y., Wang, S. & Wei, P. (2008), 'A level set-based parameterization method for structural shape and topology optimization', *International Journal for Numerical Methods in Engineering* **76**(1), 1–26.

Mack, C. A. (2011), 'Fifty years of Moore's law', *IEEE Transactions on Semiconductor Manufacturing* **24**(2), 202–207.

Madeira, J. F. A., Pina, H. & Rodrigues, H. C. (2010), 'GA topology optimization using random keys for tree encoding of structures', *Structural and Multidisciplinary Optimization* **40**(1–6), 227–240.

Madeira, J. F. A., Rodrigues, H. C. & Pina, H. (2005), 'Multi-objective optimization of structures topology by genetic algorithms', *Advances in Engineering Software* **36**(1), 21–28.

Mazaheri, K. & Ebrahimi, A. (2010), 'Experimental study on interaction of aerodynamics with flexible wings of flapping vehicles in hovering and cruise flight', *Archive of Applied Mechanics* **80**(11), 1255–1269.

McDonald, N. (1983), *Trees and Networks in Biological Models*, John Wiley and Sons, Chichester.

Messac, A. & Ismail-Yahaya, A. (2001), 'Required relationship between objective function and Pareto frontier orders: Practical implications', *AIAA Journal* **39**(11), 2168–2174.

Messac, A., Sundararaj, G. J., Tappeta, R. V. & Renaud, J. E. (2000), 'Ability of objective functions to generate points on nonconvex Pareto frontiers', *AIAA Journal* **38**(6), 1084–1091.

Michell, A. G. M. (1904*a*), 'The limits of economy of material in frame-structures', *Philosophical Magazine* **47**(8), 589–597.

Michell, A. G. M. (1904*b*), 'LVIII. The limits of economy of material in frame-structures', *The London, Edinburgh, and Dublin Philosophical Magazine and Journal of Science* **8**(47), 589–597.

Miettinen, K. (2012), *Nonlinear Multiobjective Optimization*, Springer Science & Business Media, New York.

Muller, D. E. & Preparata, F. P. (1978), 'Finding the intersection of two convex polyhedra', *Theoretical Computer Science* **7**(2), 217–236.

Nagel, L. W. (1975), 'Spice2: A computer program to simulate semiconductor circuits', PhD dissertation, University of California at Berkeley.

Nakamura, A., Lindenmayer, A. & Aizawa, K. (1986), 'Some systems for map generation', in *The Book of L*, Springer, Berlin, pp. 323–332.

Nocedal, J. & Wright, S. (2006), *Numerical Optimization*, Springer Science & Business Media, New York.

Norato, J. A., Bell, B. K. & Tortorelli, D. A. (2015), 'A geometry projection method for continuum-based topology optimization with discrete elements', *Computer Methods in Applied Mechanics and Engineering* **293**, 306–327.

Nunes, M. A. (2010), 'A biologically inspired method for spacecraft design', MS thesis, University of Hawaii at Manoa.

Oliver, J., Yago, D., Cante, J. & Lloberas-Valls, O. (2019), 'Variational approach to relaxed topological optimization: Closed form solutions for structural problems in a sequential pseudo-time framework', *Computer Methods in Applied Mechanics and Engineering* **355**, 779–819.

Oren-Suissa, M. & Podbilewicz, B. (2007), 'Cell fusion during development', *Trends in Cell Biology* **17**(11), 537–546.

Osher, S. & Fedkiw, R. (2003), *Level Set Methods and Dynamic Implicit Surfaces*, Springer, New York.

Osher, S. & Sethian, J. (1998), 'Fronts propagating with curvature-dependent speed: Algorithms based on Hamilton-Jacobi formulations', *Journal of Computational Physics* **79**(1), 12–49.

Otomori, M., Yamada, T., Izui, K. & Nishiwaki, S. (2015), 'Matlab code for a level set-based topology optimization method using a reaction diffusion equation', *Structural and Multidisciplinary Optimization* **51**(5), 1159–1172.

Papalambros, P. Y. & Wilde, D. J. (2017), *Principles of Optimal Design*, 3rd ed., Cambridge University Press, Cambridge.

Patil, M. J. & Hodges, D. H. (2006), 'Flight dynamics of highly flexible flying wings', *Journal of Aircraft* **43**(6), 1790–1799.

Pedro, H. T. C., Kobayashi, M. H., Coimbra, C. & da Silva, A. (2008), 'On the effectiveness of complex design through an evolutionary approach', *Journal of Thermophysics and Heat Transfer* **22**(1), 115–118.

Pedro, H. T. C. & Kobayashi, M. H. (2011), 'On a cellular division method for topology optimization', *International Journal for Numerical Methods in Engineering* **88**(11), 1175–1197.

Peters, D. A., Hsieh, M.-C. & Torrero, A. (2007), 'A state-space airloads theory for flexible airfoils', *Journal of the American Helicopter Society* **52**(4), 329–342.

Peters, D. A., Karunamoorthy, S. & Cao, W.-M. (1995), 'Finite state induced flow models. I – Two-dimensional thin airfoil', *Journal of Aircraft* **32**(2), 313–322.

Prasad, J. & Diaz, A. R. (2005), 'Synthesis of bistable periodic structures using topology optimization and a genetic algorithm', *Journal of Mechanical Design* **128**(6), 1298–1306.

Prusinkiewicz, P. (1986), 'Graphical applications of L systems', in *Proceedings of Graphics Interface '86 – Vision Interface '86*, pp. 247–253.

Prusinkiewicz, P. (1987), 'Applications of L systems to computer imagery', in *Graph Grammars and Their Applications to Computer Science*, Lecture Notes in Computer Sciences, Springer-Verlag, New York, pp. 534–548.

Prusinkiewicz, P. & Hanan, J. (1990), 'Visualization of botanical structures and processes using parametric L-systems', in D. Thalmann, ed., *Scientific Visualization and Graphics Simulation*, John Wiley & Sons, New York, pp. 183–201.

Prusinkiewicz, P. & Lindenmayer, A. (2004), *The Algorithmic Beauty of Plants*, Springer, New York.

Querin, O. M., Victoria, M., Alonso, C., Loyola, R. A. & Montrull, P. M. (2017), *Topology Design Methods for Structural Optimization*, Academic Press, London.

Querin, O. M., Young, V., Steven, G. P. & Xie, Y. M. (2000), 'Computational efficiency and validation of bi-directional evolutionary structural optimization', *Computer Methods in Applied Mechanics and Engineering* **189**(2), 559–573.

Querin, O., Steven, G. & Xie, Y. (1998), 'Evolutionary structural optimisation (ESO) using a bidirectional algorithm', *Engineering Computations* **15**(8), 1031–1048.

Reddy, J. N. (2019), *Introduction to the Finite Element Method*, McGraw-Hill Education, New York.

Roth-Nebelsick, A., Uhl, D., Mosbrugger, V. & Kerp, H. (2001), 'Evolution and function of leaf venation architecture: A review', *Annals of Botany* **87**, 553–566.

Rozenberg, G. & Salomaa, A. (1997), *Handbook of Formal Languages*, Springer Verlag, Berlin.

Rozvany, G. I. N. (1994), *Topology Optimization in Structural Mechanics*, Springer Verlag, Berlin.

Rozvany, G. I. N. (2001), 'Aims, scope, methods, history and unified terminology of computer-aided topology optimization in structural mechanics', *Structural and Multidisciplinary Optimization* **21**(2), 90–108.

Rozvany, G. I. N. (2009), 'A critical review of established methods of structural topology optimization', *Structural and Multidisciplinary Optimization* **37**(3), 217–237.

Rozvany, G. I. N., Bendsøe, M. P. & Kirsch, U. (1995), 'Layout optimization of structures', *Applied Mechanics Reviews* **48**(2), 41–119.

Schittkowski, K. (1983), 'On the convergence of a sequential quadratic programming method with an augmented Lagrangian line search function', *Mathematische Operationsforschung und Statistik. Series Optimization* **14**, 197–216.

Schmit, L. A. (1960), 'Structural design by systematic synthesis', in *Proceedings of the Second ASCE Conference on Electronic Computation*, New York, pp. 105–122.

Schoenauer, M. (1996), 'Shape representations and evolutionary schemes', in *Proceedings of the Fifth Annual Conference on Evolutionary Programming*, San Diego, CA.

Schramm, U. & Zhou, M. (2006), 'Recent developments in the commercial implementation of topology optimization', *IUTAM Symposium on Topological Design Optimization of Structures, Machines and Materials* **137**, 239–247.

Schutze, O., Esquivel, X., Lara, A. & Coello, C. A. C. (2012), 'Using the averaged hausdorff distance as a performance measure in evolutionary multiobjective optimization', *IEEE Transactions on Evolutionary Computation* **16**(4), 504–522.

Sethian, J. A. & Wiegmann, A. (2000), 'Structural boundary design via level set and immersed interface methods', *Journal of Computational Physics* **163**(2), 489–528.

Sharma, D., Deb, K. & Kishore, N. N. (2008), 'Towards generating diverse topologies of path tracing compliant mechanisms using a local search based multi-objective genetic algorithm procedure', in *2008 IEEE Congress on Evolutionary Computation (IEEE World Congress on Computational Intelligence)*, IEEE, pp. 2004–2011.

Shin, H., Todoroki, A. & Hirano, Y. (2013), 'Elite-initial population for efficient topology optimization using multi-objective genetic algorithms', *International Journal of Aeronautical and Space Sciences* **14**(4), 324–333.

Sigmund, O. (1997), 'On the design of compliant mechanisms using topology optimization', *Mechanics of Structures and Machines* **25**(4), 493–524.

Sigmund, O. (2007), 'Morphology-based black and white filters for topology optimization', *Structural and Multidisciplinary Optimization* **33**(4–5), 401–424.

Sigmund, O. (2011), 'On the usefulness of non-gradient approaches in topology optimization', *Structural and Multidisciplinary Optimization* **43**(5), 589–596.

Sigmund, O. & Petersson, J. (1998), 'Numerical instabilities in topology optimization: A survey on procedures dealing with checkerboards, mesh-dependencies and local minima', *Structural Optimization* **16**, 68–75.

Singh, B. & Chopra, I. (2008), 'Insect-based hover-capable flapping wings for micro air vehicles: Experiments and analysis', *AIAA Journal* **46**(9), 2115–2135.

Smith, H. & Norato, J. A. (2020), 'A MATLAB code for topology optimization using the geometry projection method', *Structural and Multidisciplinary Optimization* **62**, 1579–1594.

Sokolnikoff, I. S. (1951), *Tensor Analysis: Theory and Applications*, Wiley, New York.

Stanford, B., Beran, P. & Kobayashi, M. (2012), 'Aeroelastic optimization of flapping wing venation: A cellular division approach', *AIAA Journal* **50**(4), 938–951.

Szilard, A. L. & Quinton, R. E. (1979), 'An introduction for DOL systems by computer graphics', *The Science Terrapin* **4**, 8–13.

Tai, K. & Akhtar, S. (2005), 'Structural topology optimization using a genetic algorithm with a morphological geometric representation scheme', *Structural and Multidisciplinary Optimization* **30**(2), 113.

Tai, K. & Chee, T. H. (2000), 'Design of structures and compliant mechanisms by evolutionary optimization of morphological representations of topology', *Journal of Mechanical Design* **122**(4), 560–566.

Tai, K., Cui, G. Y. & Ray, T. (2002), 'Design synthesis of path generating compliant mechanisms by evolutionary optimization of topology and shape', *Journal of Mechanical Design* **124**(3), 492–500.

Tai, K. & Prasad, J. (2007), 'Target-matching test problem for multiobjective topology optimization using genetic algorithms', *Structural and Multidisciplinary Optimization* **34**(4), 333–345.

Tamaki, H., Kita, H. & Kobayashi, S. (1996), 'Multi-objective optimization by genetic algorithms: A review', in *Proceedings of IEEE International Conference on Evolutionary Computation*, pp. 517–522.

Tcherniak, D. & Sigmund, O. (2001), 'A web-based topology optimization program', *Structural and Multidisciplinary Optimization* **22**(3), 179–187.

Timoshenko, S. P. & Gere, J. M. (2009), *Theory of Elastic Stability*, McGraw-Hill, New York.

Turevsky, I. & Suresh, K. (2011), 'Efficient generation of Pareto-optimal topologies for compliance optimization', *International Journal for Numerical Methods in Engineering* **87**(12), 1207–1228.

van Dijk, N. P., Langelaar, M. & van Keulen, F. (2012), 'Explicit level-set-based topology optimization using an exact Heaviside function and consistent sensitivity analysis', *International Journal for Numerical Methods in Engineering* **91**(1), 67–97.

van Dijk, N. P., Maute, K., Langelaar, M. & van Keulen, F. (2013), 'Level-set methods for structural topology optimization: A review', *Structural and Multidisciplinary Optimization* **48**, 437–472.

Vanderplaats, G. N. & Moses, F. (1973), 'Structural optimization by methods of feasible directions', *Computers & Structures* **3**(4), 739–755.

Venturini, G. (2015), 'Ahkab 0.18 documentation', http://ahkab.readthedocs.io.

Vichnevetsky, R. & Bowles, J. B. (1982), *Fourier Analysis of Numerical Approximations of Hyperbolic Equations*, SIAM Studies in Applied Mathematics, Society for Industrial and Applied Mathematics, Philadelphia, PA.

Wallin, M., Ristinmaa, M. & Askfelt, H. (2012), 'Optimal topologies derived from a phase-field method', *Structural and Multidisciplinary Optimization* **45**, 171–183.

Wang, C., Zhao, Z., Zhou, M., Sigmund, O. & Zhang, X. S. (2021), 'A comprehensive review of educational articles on structural and multidisciplinary optimization', *Structural and Multidisciplinary Optimization* **64**, 2827–2880.

Wang, F., Lazarov, B. S. & Sigmund, O. (2011), 'On projection methods, convergence and robust formulations in topology optimization', *Structural and Multidisciplinary Optimization* **43**(6), 767–784.

Wang, M. Y., Wang, X. & Guo, D. (2003), 'A level set method for structural topology optimization', *Computer Methods in Applied Mechanics and Engineering* **192**(1–2), 227–246.

Wang, M. Y. & Zhou, S. (2004), 'Phase field: A variational method for structural topology optimization', *Computer Modeling in Engineering and Sciences* **6**, 547–566.

Wang, N. & Tai, K. (2008), 'Design of grip-and-move manipulators using symmetric path generating compliant mechanisms', *Journal of Mechanical Design* **130**(11), 103622.

Wang, S. & Wang, M. Y. (2006), 'Radial basis functions and level set method for structural topology optimization', *International Journal for Numerical Methods in Engineering* **65**(12), 2060–2090.

Wang, S. Y. & Tai, K. (2004), 'Graph representation for structural topology optimization using genetic algorithms', *Computers and Structures* **82**(20–21), 1609–1622.

Wang, S. Y. & Tai, K. (2005*a*), 'Bar-system representation for topology optimization using genetic algorithms', *Engineering Computations* **22**(1–2), 206–231.

Wang, S. Y. & Tai, K. (2005*b*), 'Structural topology design optimization using genetic algorithms with a bit-array representation', *Computer Methods in Applied Mechanics and Engineering* **194**(36–38), 3749–3770.

Wang, S. Y., Tai, K. & Wang, M. Y. (2006), 'An enhanced genetic algorithm for structural topology optimization', *International Journal for Numerical Methods in Engineering* **65**(1), 18–44.

Ward, R. M., Bielefeldt, B. R. & Hartl, D. J. (2020), 'Design of tailorable stiffness structures using L-system topology optimization', in *Behavior and Mechanics of Multifunctional Materials IX*, vol. 11377, International Society for Optics and Photonics, p. 113770R.

Wasserman, R. H. (2009), *Tensors and Manifolds: With Applications to Physics*, 2nd ed., Oxford University Press, Oxford.

Wu, J., Sigmund, O. & Groen, J. P. (2021), 'Topology optimization of multi-scale structures: A review', *Structural and Multidisciplinary Optimization* **63**, 1455–1480.

Wu, P., Stanford, B., Sällström, E., Ukeiley, L. & Ifju, P. (2011), 'Structural dynamics and aerodynamics measurements of biologically inspired flexible flapping wings', *Bioinspiration & Biomimetics* **6**(1), 016009.

Xie, Y. M. & Steven, G. P. (1997*a*), 'Basic evolutionary structural optimization', in '*Evolutionary Structural Optimization*', Springer, London, pp. 12–29.

Xie, Y. M. & Steven, G. P. (1997*b*), *Evolutionary Structural Optimization*, Springer, New York.

Xie, Y. M. & Steven, G. P. (1993), 'A simple evolutionary procedure for structural optimization', *Computers and Structures* **49**(5), 885–896.

Xing, X., Wei, P. & Wang, M. Y. (2010), 'A finite element-based level set method for structural optimization', *International Journal for Numerical Methods in Engineering* **82**(7), 805–842.

Yago, D., Cante, J., Lloberas-Valls, O. & Oliver, J. (2021), 'Topology optimization using the unsmooth variational topology optimization (UNVARTOP) method: An educational implementation in Matlab', *Structural and Multidisciplinary Optimization* **63**, 955–981.

Yoshimura, M., Shimoyama, K., Misaka, T. & Obayashi, S. (2016), 'Topology optimization using a Kriging-assisted genetic algorithm with a novel level set representation approach', in *7th European Congress on Computational Methods in Applied Sciences and Engineering, ECCOMAS Congress 2016*, National Technical University of Athens, pp. 3361–3376.

Zavala, V. M. & Flores-Tlacuahuac, A. (2012), 'Stability of multiobjective predictive control: A utopia-tracking approach', *Automatica* **48**(10), 2627–2632.

Zhou, M. & Rozvany, G. (2001), 'On the validity of ESO type methods in topology optimization', *Structural and Multidisciplinary Optimization* **21**, 80–83.

Zhu, B., Zhang, X. & Fatikow, S. (2014), 'A multi-objective method of hinge-free compliant mechanism optimization', *Structural and Multidisciplinary Optimization* **49**(3), 431–440.

Zhu, B., Zhang, X. & Fatikow, S. (2015), 'Filter the shape sensitivity in level set-based topology optimization methods', *Structural and Multidisciplinary Optimization* **51**(5), 1035–1049.

Zuo, Z. H. & Xie, Y. M. (2015), 'A simple and compact python code for complex 3D topology optimization', *Advances in Engineering Software* **85**, 1–11.

Zuo, Z. H., Xie, Y. M. & Huang, X. (2009), 'Combining genetic algorithms with BESO for topology optimization', *Structural and Multidisciplinary Optimization* **38**(5), 511–523.

Index

Printed by Integrated Books International,
United States of America